Periodic Table of the Elements with the Gmelin System Numbers

Each cell is shown as: atomic number · symbol · (Gmelin System Number)

1	2	3	4	5	6	7	8	9	10	11	12	13	14	15	16	17	18
1 H (2)																	2 He (1)
3 Li (20)	4 Be (26)											5 B (13)	6 C (14)	7 N (4)	8 O (3)	9 F (5)	10 Ne (1)
11 Na (21)	12 Mg (27)											13 Al (35)	14 Si (15)	15 P (16)	16 S (9)	17 Cl (6)	18 Ar (1)
19* K (22)	20 Ca (28)	21 Sc (39)	22 Ti (41)	23 V (48)	24 Cr (52)	25 Mn (56)	26 Fe (59)	27 Co (58)	28 Ni (57)	29 Cu (60)	30 Zn (32)	31 Ga (36)	32 Ge (45)	33 As (17)	34 Se (10)	35 Br (7)	36 Kr (1)
37 Rb (24)	38 Sr (29)	39 Y (39)	40 Zr (42)	41 Nb (49)	42 Mo (53)	43 Tc (69)	44 Ru (63)	45 Rh (64)	46 Pd (65)	47 Ag (61)	48 Cd (33)	49 In (37)	50 Sn (46)	51 Sb (18)	52 Te (11)	53 I (8)	54 Xe (1)
55 Cs (25)	56 Ba (30)	57** La (39)	72 Hf (43)	73 Ta (50)	74 W (54)	75 Re (70)	76 Os (66)	77 Ir (67)	78 Pt (68)	79 Au (62)	80 Hg (34)	81 Tl (38)	82 Pb (47)	83 Bi (19)	84 Po (12)	85 At (8a)	86 Rn (1)
87 Fr (25a)	88 Ra (31)	89*** Ac (40)	104 (71)	105 (71)													

*	NH$_4$ (23)

Lanthanides (39)

57** La	58 Ce	59 Pr	60 Nd	61 Pm	62 Sm	63 Eu	64 Gd	65 Tb	66 Dy	67 Ho	68 Er	69 Tm	70 Yb	71 Lu

Actinides

89*** Ac	90 Th (44)	91 Pa (51)	92 U (55)	93 Np (71)	94 Pu (71)	95 Am (71)	96 Cm (71)	97 Bk (71)	98 Cf (71)	99 Es (71)	100 Fm (71)	101 Md (71)	102 No (71)	103 Lr (71)

A Key to the Gmelin System is given on the Inside Back Cover

Gmelin Handbook of Inorganic Chemistry

8th Edition

Gmelin Handbook
of Inorganic Chemistry

8th Edition

Gmelin Handbuch der Anorganischen Chemie

Achte, völlig neu bearbeitete Auflage

Prepared and issued by	Gmelin-Institut für Anorganische Chemie der Max-Planck-Gesellschaft zur Förderung der Wissenschaften
	Director: Ekkehard Fluck

Founded by	Leopold Gmelin
8th Edition	8th Edition begun under the auspices of the Deutsche Chemische Gesellschaft by R. J. Meyer
Continued by	E. H. E. Pietsch and A. Kotowski, and by Margot Becke-Goehring

Springer-Verlag Berlin Heidelberg GmbH 1987

Gmelin-Institut für Anorganische Chemie
der Max-Planck-Gesellschaft zur Förderung der Wissenschaften

Organometallic Compounds in the Gmelin Handbook

The following listing indicates in which volumes these compounds are discussed or are referred to:

Ag   Silber B5 (1975)

Au   Organogold Compounds (1980)

Be   Organoberyllium Compounds 1 (1987)

Bi   Bismut-Organische Verbindungen (1977)

Co   Kobalt-Organische Verbindungen 1 (1973), 2 (1973), Kobalt Erg.-Bd. A (1961), B1 (1963), B2 (1964)

Cr   Chrom-Organische Verbindungen (1971)

Cu   Organocopper Compounds 1 (1985), 2 (1983), 3 (1986), 4 (1987)

Fe   Eisen-Organische Verbindungen A1 (1974), A2 (1977), A3 (1978), A4 (1980), A5 (1981), A6 (1977), A7 (1980), A8 (1985), B1 (partly in English; 1976), Organoiron Compounds B2 (1978), Eisen-Organische Verbindungen B3 (partly in English; 1979), B4 (1978), B5 (1978), Organoiron Compounds B6 (1981), B7 (1981), B8 to B10 (1985), B11 (1983), B12 (1984), Eisen-Organische Verbindungen C1 (1979), C2 (1979), Organoiron Compounds C3 (1980), C4 (1981), C5 (1981), C7 (1985), and Eisen B (1929 – 1932)

Ga   Organogallium Compounds 1 (1986)

Hf   Organohafnium Compounds (1973)

Nb   Niob B4 (1973)

Ni   Nickel-Organische Verbindungen 1 (1975), 2 (1974), Register (1975), Nickel B3 (1966) and C1 (1968), C2 (1969)

Np, Pu   Transurane C (partly in English; 1972)

Pb   Organolead Compounds 1 (1987) **present volume**

Pt   Platin C (1939), D (1957)

Ru   Ruthenium Erg.-Bd. (1970)

Sb   Organoantimony Compounds 1 (1981), 2 (1981), 3 (1982), 4 (1986)

Sc, Y, La to Lu   D6 (1983)

Sn   Zinn-Organische Verbindungen 1 (1975), 2 (1975), 3 (1976), 4 (1976), 5 (1978), 6 (1979), Organotin Compounds 7 (1980), 8 (1981), 9 (1982), 10 (1983), 11 (1984), 12 (1985), 13 (1986), 14 (1987)

Ta   Tantal B2 (1971)

Ti   Titan-Organische Verbindungen 1 (1977), 2 (1980), Organotitanium Compounds 3 (1984), 4 and Register (1984)

U   Uranium Suppl. Vol. E2 (1980)

V   Vanadium-Organische Verbindungen (1971), Vanadium B (1967)

Zr   Organozirconium Compounds (1973)

Gmelin Handbook of Inorganic Chemistry

8th Edition

Pb

Organolead Compounds

Part 1
Tetramethyllead

With 4 illustrations

AUTHOR   Friedo Huber, Universität Dortmund

EDITOR   Wolfgang Petz, Gmelin-Institut, Frankfurt am Main

System Number 47

Springer-Verlag Berlin Heidelberg GmbH 1987

LITERATURE CLOSING DATE: 1986
IN SOME CASES MORE RECENT DATA HAVE BEEN CONSIDERED

Library of Congress Catalog Card Number: Agr 25-1383

ISBN 978-3-662-10296-1   ISBN 978-3-662-10294-7 (eBook)
DOI 10.1007/978-3-662-10294-7

© by Springer-Verlag Berlin Heidelberg 1987
Originally published by Springer-Verlag, Berlin · Heidelberg · New York · Tokyo in 1987
Softcover reprint of the hardcover 8th edition 1987

Preface

Following the first synthesis of an organolead compound by Loewig in 1852, progress slowed markedly in organolead chemistry. Then, the discovery of the outstanding performance of organolead compounds as antiknock additives to automobile gasoline turned organolead chemistry into one of the main areas of organometallic chemistry. Large-scale industrial procedures for synthesizing tetraethyllead and other tetraalkyllead compounds and entirely new techniques, such as handling large amounts of sodium-lead alloys or electrolyzing Grignard solutions, were developed, and many other industrial projects as well as toxicological and environmental studies brought a tremendous increase of knowledge. Therefore, a compilation of the available data in organolead chemistry seemed justified.

In the present series only compounds containing at least one lead-to-carbon bond are considered; the simple inorganic cyanides, carbides, etc., are excluded.

The material is organized as follows:

1. Mononuclear compounds (compounds containing only one lead atom)
2. Dinuclear compounds
3. Oligo- and polynuclear compounds

Within the group of mononuclear compounds the material is arranged in a similar way as in the Gmelin volumes of the sister element tin, that is:

1.1· Compounds containing four Pb-C bonds
1.2 Compounds containing Pb-H bonds
1.3 Compounds containing bonds between lead and group 17 elements
1.4 Compounds containing bonds between lead and group 16 elements
1.5 Compounds containing bonds between lead and group 15 elements
1.6 Compounds containing bonds between lead and Si, Ge, or Sn
1.7 Compounds containing bonds between lead and group 13 elements
1.8 Compounds containing bonds between lead and group 2 elements
1.9 Compounds containing bonds between lead and group 1 elements
1.10 Compounds containing bonds between lead and transition elements
1.11 Other species

The arrangement in the groups of dinuclear and of oligo- and polynuclear compounds is analogous. Organolead(II) compounds are discussed in a separate final section.

Coordination compounds are considered in the sections on the appropriate uncomplexed compound. Those compounds that are associated by intermolecular coordination are dealt with in the form of their smallest known molecular entity.

The literature coverage was attempted to be as complete as possible but not all of the voluminous patent data have been regarded and certainly a series of publications that are not or not clearly abstracted by Chemical Abstracts (C.A.) might have not been found.

In the first volume of the series, which treats tetramethyllead, the literature is covered through 1986; C.A. has been evaluated including volume 106 of 1987.

I thank Professor Dr. Ekkehard Fluck, the director of the Gmelin Institute and all coworkers for their excellent cooperation. I am particularly indebted to the late Dr. Hubert Bitterer for his stimulating suggestions and patient guidance in the initial phase of writing this series. I thank Dr. Ulrich Krüerke and Dr. Wolfgang Petz for their commendable editing.

Dortmund, October 1987   Friedo Huber

Explanations, Abbreviations, and Units

Abbreviations are used in the text and in the tables; units are omitted in some tables for the sake of conciseness. This necessitates the following clarification:

Temperatures are given in °C, otherwise K stands for Kelvin. Abbreviations used with temperatures are m.p. for melting point, b.p. for boiling point, dec. for decomposition, and subl. for sublimation. Terms like 80°/0.1 mean the boiling or sublimation point at a pressure of 0.1 Torr. **Densities** d are given in g/cm^3. d_c and d_m distinguish calculated and measured values, respectively.

NMR represents **nuclear magnetic resonance**. Chemical shifts are given as δ values in ppm and positive to low field from the following reference substances: $Si(CH_3)_4$ for ^1H and ^{13}C, $BF_3 \cdot O(C_2H_5)_2$ for ^{11}B, $CFCl_3$ for ^{19}F, H_3PO_4 for ^{31}P, $Sn(CH_3)_4$ for ^{119}Sn, and $Pb(CH_3)_4$ for ^{207}Pb. Multiplicities of the signals are abbreviated as s, d, t, q (singlet to quartet), quint, sext, sept (quintet to septet), and m (multiplet); terms like dd (double doublet) and t's (triplets) are also used. Assignments referring to labelled structural formulas are given in the form C-4, H-3.5. Coupling constants J in Hz appear usually in parentheses behind the δ value, along with the multiplicity and the assignment, and refer to the respective nucleus. If a more precise designation is necessary, they are given as, e.g., $^nJ(C,H)$ or $J(1,3)$ referring to labelled formulas.

Optical spectra are labelled as IR (infrared), R (Raman), and UV (electronic spectrum including the visible region). IR bands and Raman lines are given in cm^{-1}; the assigned bands are usually labelled with the symbols ν for stretching vibration and δ for deformation vibration. Intensities occur in parentheses either in the common qualitative terms (s, m, w, vs, etc.) or as numerical relative intensities. The UV absorption maxima, λ_{max}, are given in nm followed by the extinction coefficient ε ($L \cdot cm^{-1} \cdot mol^{-1}$) or $\log \varepsilon$ in parentheses; sh means shoulder.

Photoelectron spectra are abbreviated PE, e.g., PE/He(I), with the ionization energies in eV.

Solvents or the **physical state** of the sample and the temperature (in °C or K) are given in parentheses immediately after the spectral symbol, e.g., R (solid), ^{13}C NMR (C_6D_6, 50°C), or at the end of the data if spectra for various media are reported. Common solvents are given by their formula (C_6H_{12} = cyclohexane) except THF, DMF, and HMPT, which represent tetrahydrofuran, dimethylformamide, and hexamethylphosphoric triamide, respectively.

The data of **mass spectra**, abbreviated MS, are given as m/e, relative intensity in parentheses, and fragment ions in brackets; $[M]^+$ is the molecular ion and m* represents a metastable peak.

Electron spin resonance is abbreviated as ESR. Radicals, e.g., $Pb(CH_3)_3^{\cdot}$, are characterized by their g-factors; hyperfine splittings a are given in G values.

References, quoted in the last column, are occasionally also placed in the first and second column if statements from different sources must be distinguished.

Figures give only selected parameters. Barred bond lengths (in Å) or angles are mean values for parameters of the same type.

Table of Contents

XII

Organolead Compounds

General References

Tetraorganolead compounds were the first organoelement compounds to gain large-scale applied importance. This was a consequence of their high efficiency as antiknock agents in gasoline. This property subsequently induced an appreciable part of the rapid development in automotive techniques in the last fifty to sixty years; however, as a consequence of the extensive consumption of leaded gasoline the lead burden in the environment was also increased. Public concern and legislative measures were therefore enacted to reduce or prohibit the use of tetraalkyllead compounds as antiknock additives in gasoline.

In the course of this development, and recalling that organolead compounds belong to the very first organoelement compounds to be studied, an extensive knowledge of the synthesis, the manufacture of antiknock additives with a very high technical and security standard, the health and environmental implications, and analysis of organolead compounds has been gathered. These facts are of general importance to other areas of organoelement chemistry, considering the position of lead as the heaviest element in the central main group of the periodic system. Consequently, reference to organolead compounds is made in numerous studies and treatises. It is therefore necessary to list general and relevant publications for those interested in organolead chemistry to gain a general survey, but also to search for correlations and relationships to comparable compounds.

The following compilation of references provides a survey of the literature on organolead chemistry. It is headed by a list of general literature on organoelement Group 14 compounds. The next part contains publications which refer mainly to organolead compounds. Both parts are divided into monographs, reviews and reports, and other general publications. Following this general survey, more specific lists with monographs, review articles, and general publications which concentrate on the physical properties, analysis, toxicology, uses, and, finally, on environmental aspects of organolead compounds are included; all parts are arranged chronologically.

Annual surveys concerning organolead compounds written by various authors will not be specified further as they have already appeared in:

Ann. Surv. Organometal. Chem. **1** [1964/65] 148/53, **2** [1965/66] 183/8, **3** [1966/67] 258/67.

Organometal. Chem. Rev. B **4** [1968] 394/404, B **5** [1969] 736/44, B **6** [1970] 541/55, B **9** [1972] 339/57.

J. Organometal. Chem. **48** [1972] 182/94, **62** [1973] 285/97, **95** [1975] 291/300, **109** [1976] 363/82, **274** [1984] 211/39.

J. Organometal. Chem. Libr. **4** [1977] 495/519, **6** [1978] 495/524, **8** [1979] 565/77, **10** [1980] 569/84, **13** [1982] 643/70, **14** [1984] 481/505.

Also valuable are collections of spectroscopic data such as Specialist Periodical Reports, Spectrosc. Prop. Inorg. Organometal. Compounds **1** [1968] to **19** [1986].

Organometallic Compounds of the Group 14 Elements

Monographs

Pascal, P., Nouveau Traité de Chimie Minerale, Vol. VIII, Germanium, Etain, Plomb, Masson, Paris 1963.

Lukevics, E. Y., Voronkov, M. G., Organic Insertion Reactions of Group IV Elements, Consultants Bureau, Plenum, New York 1966; corrected and updated English edition of: Lukevics, E., Voronkov, M. G., Gidrosililirovanie, Gidrogermilirovanie i Gidrostanni-lirovanie [(Addition Products from Organic and Inorganic) Hydrides of Silicon, Germanium, Tin, and Lead], Izd. Akad. Nauk Latv. SSR, Riga 1964, pp. 1/371.

Kochkin, D. A., Azerbaev, I. N., Tin and Lead Organic Monomers and Polymers, Nauka, Alma-Ata 1968.

MacDiarmid, A. G., Organometallic Compounds of the Group IV Elements, Vol. 1, The Bond to Carbon, Part II, Dekker, New York 1968.

Kocheshkov, K. A., Zemlyanskii, N. N., Sheverdina, N. I., Panov, E. M., Germanium, Tin, and Lead, in: Nesmeyanov, A. N., Kocheshkov, K. A., Methods of Elementoorganic Chemistry, 2nd Ed., Nauka, Moscow 1968.

Rochow, E. G., Abel, E. W., The Chemistry of Germanium, Tin, and Lead, Pergamon, Oxford 1975, pp. 1/146.

Danilov, S. N., Chemistry of Organometallic Compounds (Elements of Groups III to V), Nauka, Leningr. Otd., Leningrad 1976, pp. 1/243.

Aylett, B. J., Organometallic Compounds, 4th Ed., Vol. 1, The Main Group Elements, Part 2, Groups IV and V, Chapman & Hall, London 1979.

Harrison, P. G., Organometallic Compounds of Germanium, Tin and Lead, Chapman and Hall Chemistry Source Books, Chapman & Hall, London 1985, pp. 1/192.

Hartley, F. R., Patai, S., The Chemistry of the Metal-Carbon Bond, Vol. 1, The Structure, Preparation, Thermochemistry, and Characterization of Organometallic Compounds, Wiley, Chichester 1982; Vol. 2, The Nature and Cleavage of Metal-Carbon Bonds, Wiley, Chichester 1985.

Reviews and Reports

Kocheshkov, K. A., Organic Compounds of Germanium, Tin, and Lead, Usp. Khim. **3** [1934] 83/115.

Kocheshkov, K. A., Metalloorganic Compounds of Elements of the Fourth Group, Sint. Metody Obl. Metalloorg. Soedin. Inst. Org. Khim. Akad. Nauk SSSR **1947** No. 5, pp. 1/131; C.A. **1952** 11102.

Noltes, J. G., van der Kerk, G. J. M., Synthesis of Linear IVth Group Organometallic Polymers by Polyaddition, Chimia (Switz.) **16** [1962] 122/7.

Bradley, D. C., Polymeric Metal Alkoxides, Organometalloxanes, and Organome-talloxanosiloxanes, in: Stone, F. G. A., Graham W. A. G., Inorganic Polymers, Academic, New York 1962, pp. 410/46.

Ingham, R. K., Gilman, H., Organopolymers of Silicon, Germanium, Tin, and Lead, in: Stone, F. G. A., Graham, W. A. G., Inorganic Polymers, Academic, New York 1962, pp. 321/409.

Grosjean, M., Gielen, M., Nasielski, J., Chimie organométallique. Les réactions de substitution nucléophile sur les atomes du quatrième groupe, Ind. Chim. Belge **28** [1963] 721/30.

Azerbaev, I. N., Kochkin, D. A., Advances and Prospects of the Development of the Organotin and Organolead Chemistry, Vestn. Akad. Nauk Kaz. SSR **19** No. 10 [1963] 18/26.

Beattie, I. R., The Acceptor Properties of Quadripositive Silicon, Germanium, Tin, and Lead, Quart. Rev. [London] **17** [1963] 382/405.

Friswell, N. J., Gowenlock, B. G., Inorganic Hydrogen- and Alkyl-Containing Free Radicals, Groups II, III and IV, Advan. Free-Radical Chem. **1** [1965] 39/75.

Luijten, J. G. A., Rijkens, F., van der Kerk, G. J. M., Organometallic Nitrogen Compounds of Germanium, Tin, and Lead, Advan. Organometal. Chem. **3** [1965] 397/446.

Gilman, H., Atwell, W. H., Cartledge, F. K., Catenated Organic Compounds of Silicon, Germanium, Tin, and Lead, Advan. Organometal. Chem. **4** [1966] 1/94.

Gielen, M., Sprecher, N., Coordination au niveau des atomes de métal du groupe IVB. Intervention des orbitales d dans la réactivité des composés organométalliques, Organometal. Chem. Rev. **1** [1966] 455/89.

Lappert, M. F., Pyszora, H., Pseudohalides of Group IIIB and IVB Elements, Advan. Inorg. Chem. Radiochem. **9** [1966] 133/84.

Abel, E. W., Armitage, D. A., Organosulfur Derivatives of Silicon, Germanium, Tin, and Lead, Advan. Organometal. Chem. **5** [1967] 1/92.

Luneva, L. K., Ethynyl Derivatives of Silicon, Germanium, Tin, and Lead, Usp. Khim. **36** [1967] 1140/57; Russ. Chem. Rev. **36** [1967] 467/76.

Brook, A. G., Keto Derivatives of Group IV Organometalloids, Advan. Organometal. Chem. **7** [1968/69] 95/155.

Young, J. F., Transition Metal Complexes with Group IVB Elements, Advan. Inorg. Chem. Radiochem. **11** [1968] 91/152; C.A. **70** [1969] No. 53433.

Stone, F. G. A., Transition Metal Derivatives of Silicon, Germanium, Tin, and Lead, in: Ebsworth, E. A. V., Maddock, A. G., Sharpe, A. G., New Pathways in Inorganic Chemistry, Cambridge Univ. Press, Cambridge 1968, pp. 283/302.

Ebsworth, E. A. V., Physical Basis of the Chemistry of the Group IV Elements, in: MacDiarmid, A. G., Organometallic Compounds of the Group IV Elements, Vol. 1, Dekker, New York 1968, pp. 1/104.

Urry, G., Some Organometallic Radicals and Ion Radicals of the Group IV Elements, in: Kaiser, E. T., Kevan, L., Radical Ions, Interscience, New York 1968, pp. 275/99.

Mackay, K. M., Watt, R., Chain Compounds of Silicon, Germanium, Tin, and Lead, Organometal. Chem. Rev. A **4** [1969] 137/223.

Harrison, P. G., Metallostannoxanes and Related Compounds, Organometal. Chem. Rev. A **4** [1969] 379/478.

Neumann, W. P., Substituent Exchange Equilibria on Germanium, Tin, and Lead, Ann. N. Y. Acad. Sci. **159** [1969] 56/72.

Cadiot, P., Chodkiewicz, W., Acetylenic Derivatives of Groups IIIb, IVb, and Vb, in: Viehe, H. G., Chemistry of Acetylenes, Dekker, New York 1969, pp. 913/73.

4

Schumann, H., Organogermyl-, Organostannyl- und Organoplumbylphosphine, -arsine, -stibine und -bismutine, Angew. Chem. **81** [1969] 970/83; Angew. Chem. Intern. Ed. Engl. **8** [1969] 937/50.

Davis, D. D., Gray, C. E., Alkali Metal and Magnesium Derivatives of Organo-Silicon, -Germanium, -Tin, -Lead, -Phosphorus, -Arsenic, -Antimony and -Bismuth Compounds, Organometal. Chem. Rev. A **6** [1970] 283/318.

Seyferth, D., Divalent Carbon Insertions into Group IV Hydrides and Halides, Pure Appl. Chem. **23** [1970] 391/412.

Alexandrov, Yu. A., Oxidation of Organic Derivatives of Non-Transition Elements of Group IV (Other than Carbon) by Ozone, Organometal. Chem. Rev. A **6** [1970] 209/26.

Jackson, R. A., Silicon, Germanium, Tin, and Lead Radicals, Essays on Free-radical Chemistry, Chem. Soc. [London] Spec. Publ. No. 24 [1970] 295/321.

Baukov, Yu. I., Lutsenko, I. F., Organo-Element (Si, Ge, Sn, Pb) Derivatives of Ketoenols, Organometal. Chem. Rev. A **6** [1970] 355/445.

Sakai, S., Itoh, K., Ishii, Y., Addition Reactions of the Group IV Organometallic Compounds and Their Synthetic Applications, Yuki Gosei Kagaku Kyokaishi **28** [1970] 1109/26.

Abel, E. W., Illingworth, S. M., Phosphines, Arsines, Stibines, and Bismuthines Containing Silicon, Germanium, Tin or Lead, Organometal. Chem. Rev. A **5** [1970] 143/82.

Brooks, E. H., Cross, R. J., Group IVB Metal Derivatives of the Transition Elements, Organometal. Chem. Rev. A **6** [1970] 227/82.

Egorochkin, A. N., Vyazankin, N. S., Khorshev, S. Ya., Effect of $d_\pi - p_\pi$ Interaction in Organic Compounds of Group IVB Elements, Usp. Khim. **41** [1972] 828/51; Russ. Chem. Rev. **41** [1972] 425/38.

Carraher, C. E., Jr., Synthesis of Group IV Polymers by the Interfacial Technique, Inorg. Macromol. Rev. **1** [1972] 271/86.

Bloodworth, A. J., Tin and Lead, MTP [Med. Tech. Publ. Co.] Intern. Rev. Sci. Inorg. Chem. Ser. One **4** [1972] 275/354.

Glockling, F., Stobart, S. R., Organometallic Complexes Containing Group III (B to Tl) and Group IV (Si to Pb) Ligands, MTP [Med. Tech. Publ. Co.] Intern. Rev. Sci. Inorg. Chem. Ser. One **6** [1972] 63/120.

Orlov, V. Yu., Mass Spectra of Organometallic Compounds of Group IVb, Usp. Khim. **42** [1973] 1184/98; Russ. Chem. Rev. **42** [1973] 529/37.

Spalding, T. R., The Main Group IV Elements, in: Litzow, M. R., Spalding, T. R., Mass Spectrometry of Inorganic and Organometallic Compounds, Elsevier, Amsterdam 1973, pp. 207/319.

Abel, E. W., Dunster, M. O., Waters, A., Cyclopentadienyl Compounds of Silicon, Germanium, Tin, and Lead, J. Organometal. Chem. **49** [1973] 287/321.

Beletskaya, I. P., Butin, K. P., Ryabtsev, A. N., Reutov, O. A., Stability of Organo-Mercury, -Thallium, -Tin, and -Lead Complexes with Anionic and Neutral Ligands, J. Organometal. Chem. **59** [1973] 1/44.

Pant, B. C., Cycloalkanes Containing Heterocyclic Germanium, Tin, and Lead, J. Organometal. Chem. **66** [1974] 321/403.

Larrabee, R. B., Fluxional Main Group IV Organometallic Compounds, The Implications for Orbital Symmetry Rules, J. Organometal. Chem. **74** [1974] 313/64.

Kuivila, H. G., Alkali Metal Derivatives of Group IV Organometallics, Mechanistic Aspects of Reactions with Halides, Ann. N.Y. Acad. Sci. **239** [1974] 315/21.

Brandes, D., Blaschette, A., Organoelementperoxide von Elementen der 4. Hauptgruppe, J. Organometal. Chem. **78** [1974] 1/48.

Greninger, D., Kollonitsch, V., Kline, C. H., Willemsens, L. C., Cole, J. F., Lead Chemicals, International Lead Zinc Research Organization, Inc., New York 1975.

Harrison, P. G., Organo-Derivatives of Tin and Lead, MTP (Med. Tech. Publ. Co.) Intern. Rev. Sci. Inorg. Chem. Ser. Two **4** [1975] 81/118.

Majee, B., Interpretation of the Properties of Organo Derivatives of Silicon, Germanium, Tin, and Lead by the Del Re Method, Rev. Silicon Germanium Tin Lead Compounds **2** [1975/77] 5/80.

Devaud, M., Electrochemical Behavior of Organometallic Compounds of Group IV in Various Media, Rev. Silicon Germanium Tin Lead Compounds **2** [1975/77] 87/113.

Aleksandrov, Yu. A., Tarunin, B. I., Oxidation by Ozone of Heteroorganic Compounds of the Silicon Subgroup, Usp. Khim. **46** [1977] 1721/38; Russ. Chem. Rev. **46** [1977] 905/14.

Grimes, R. N., Group IV Carboranes, Rev. Silicon Germanium Tin Lead Compounds **2** [1975/77] 223/49.

Henderson, H. E., Drake, J. E., Synthesis and Reactivity of Organometallic Compounds Containing the Group IV to Group V Bond, Rev. Silicon Germanium Tin Lead Compounds **3** [1977/78] 145/234.

Fomina, N. V., Sheverdina, N. I., Kocheshkov, K. A., Radiation Effects in the Chemistry of Group IVB Elements (Silicon, Germanium, Tin, and Lead), Usp. Khim. **47** [1978] 428/43; Russ. Chem. Rev. **47** [1978] 238/46.

Bonny, A., Group IVB Derivatives of the Iron Triad Carbonyls, Coord. Chem. Rev. **25** [1978] 229/73.

Erchak, N. P., Furan Derivatives of Group IVA Elements, Usp. Khim. Furana **1978** 198/230, 281/91.

Mangravite, J. A., Allyl Derivatives of the Group IVA Metals and Mercury, J. Organometal. Chem. Libr. **7** [1979] 45/228.

Poller, R. C., Organic Compounds of Group IV Metals, in: Comprehensive Organic Chemistry, Vol. 3, Jones, D. N., Sulphur, Selenium, Silicon, Boron, Organometallic Compound, Pergamon, Oxford 1979, pp. 1061/109.

Ioffe, A. I., Nefedov, O. M., Structure and Reactivity of Neighbouring Analogs of Carbenes, Zh. Vses. Khim. Obshchestva **24** [1979] 475/84; C.A. **92** [1980] No. 57650.

Stedman, G., Compounds of Main Group Elements (Group III and Higher), Inorg. React. Mech. **7** [1979/81] 251/74.

Connolly, J. W., Hoff, C., Organic Compounds of Divalent Tin and Lead, Advan. Organometal. Chem. **19** [1981] 123/53.

Seyferth, D., Davies, A. G., Fischer, E. O., Normant, J. F., Reutov, O. A., Plenary Lectures from the Third International Conference on the Organometallic and Coordination Chemistry of Germanium, Tin, and Lead, Dortmund 1980, J. Organometal. Chem. Libr. **12** [1981] 181/376.

Armitage, D. A., Heterocyclic Rings Containing Silicon, Germanium, Tin or Lead, in: Katritzky, A. R., Rees, C. W., Meth-Cohn, O., Comprehensive Heterocyclic Chemistry, Vol. 1, Pt. 1, Pergamon, Oxford 1984, pp. 573/627.

Egorochkin, A. N., Spectroscopy of Organic Compounds of Silicon Subgroup Elements and Hyperconjugation, Usp. Khim. **53** [1984] 772/801; Russ. Chem. Rev. **53** [1984] 445/62.

Ustynyuk, Yu. A., Carbon-Carbon Metallotropic Tautomerism, in: Reutov, O. A., Advances in Organometallic Chemistry, Moscow 1984, pp. 30/72.

Gordetsov, A. S., Dergunov, Yu. I., The Synthesis and Properties of Silicon-, Germanium-, Tin- or Lead-containing s-Triazines, Usp. Khim. **54** [1985] 2076/106; Russ. Chem. Rev. **54** [1985] 1227/45.

Atassi, G., Antitumor and Toxic Effects of Silicon, Germanium, Tin and Lead Compounds, Rev. Silicon Germanium Tin Lead Compounds **8** [1985] 219/35.

Egorochkin, A. N., Spectroscopic Study of the Steric Effects in Organic Compounds of Silicon Subgroup Elements, Usp. Khim. **54** [1985] 1335/61; Russ. Chem. Rev. **54** [1985] 786/801.

Ng, S.-W., Zuckerman, J. J., Where are the Lone-Pair Electrons in Subvalent Fourth-Group Compounds?, Advan. Inorg. Chem. Radiochem. **29** [1985] 297/325.

Tomilov, A. P., Kargin, Yu. M., Chernykh, I. N., Electrochemistry of Organometallic Compounds (Groups IV, V, and VI Elements), Nauka, Moscow 1986, pp. 1/295; C.A. **106** [1987] No. 57762.

Mehrotra, R. C., Organogermanium, -tin, and -lead Chemistry with Sulfur Ligands, Silicon, Germanium, Tin Lead Comp. **9** [1986] 185/210.

Lappert, M. F., Heavy Atom Main Group IVA Analogs of Carbenes, Radicals, and Alkenes. The Use of Bulky Trimethylsilyl-Substituted Ligands, Silicon, Germanium, Tin Lead Comp. **9** [1986] 129/54.

Jutzi, P., π Bonding to Main-Group Elements, Advan. Organometal. Chem. **26** [1986] 217/95.

Petz, W., Transition-Metal Complexes with Derivatives of Divalent Silicon, Germanium, Tin, and Lead as Ligands, Chem. Rev. **86** [1986] 1019/47.

Dräger, M., Organometallic Compounds with Homonuclear and Heteronuclear Group IVB-Group IVB (Silicon Group) Bonds, Comments Inorg. Chem. **5** [1986] 201/14.

Other General Publications

Grohn, H., Paudert, R., Mechanochemische Reaktionen von Elementen der IV. Hauptgruppe mit einigen organischen Verbindungen, Z. Chem. [Leipzig] **3** [1963] 89/97.

Leusink, A. J., Noltes, J. G., Budding, H. A., van der Kerk, G. J. M., Synthesis of Group IV Organometallic Polymers and Related Compounds, AFML-TR-65-192, AD-629554 [1965] 1/98; C.A. **68** [1968] No. 13465.

Henry, M. C., Davidson, W. E., Organometallic Polymers, Ann. N.Y. Acad. Sci. **125** [1965] 172/82.

van der Kerk, G. J. M., Noltes, J. G., Hydride Additions, Ann. N.Y. Acad. Sci. **125** [1965] 25/42.

Atwell, W. H., Gilman, H., Catenated Organic Compounds of Silicon, Germanium, Tin, and Lead, Decompos. Organometal. Compounds Refract. Ceram. Met. Met. Alloys Proc. Intern. Symp., Dayton, Ohio, 1967, pp. 1/28.

Butin, K. P., Shishkin, V. N., Beletskaya, I. P., Reutov, O. A., Equilibriums and Rates of Disproportionation Reactions between Organometallic Compounds, V sb., Plenarnye Dokl. 11th Mendeleevsk. S'ezd Obshch. Prikl. Khim., Alma-Ata 1975 [1977], p. 193; C.A. **84** [1976] No. 163902.

Neumann, W. P., Die organischen Verbindungen und Komplexe von Germanium, Zinn und Blei, Naturwissenschaften **68** [1981] 354/59.

Cowley, A. H., Stable Compounds with Double Bonding between the Heavier Main-Group Elements, Accounts Chem. Res. **17** [1984] 386/92.

Organolead Compounds

Monographs

Willemsens, L. C., Organolead Chemistry. A Concise Review with Special Reference to the Literature Covering the Period January 1953 to July 1963, International Lead and Zinc Research Organization, Inc., New York 1964.

Willemsens, L. C., van der Kerk, G. J. M., Investigations in the Field of Organolead Chemistry, International Lead and Zinc Research Organization, Inc., Utrecht 1965.

Shapiro, H., Frey, F. W., The Organic Compounds of Lead, Interscience, New York 1968.

Kuhn, A. T., The Electrochemistry of Lead, Academic, London 1979.

Grandjean, P., Grandjean, E. C., Biological Effects of Organolead Compounds, CRC, Boca Raton, Fl., 1984.

Reviews and Reports

Calingaert, G., The Organic Compounds of Lead, Chem. Rev. **2** [1926] 43/83.

Zeller, Z., Organic Lead Compounds, Continental Met. Chem. Eng. **1** [1926] 17/9.

Schmidt, J., Organo-Blei-Verbindungen, in: Schmidt, J., Chemie in Einzeldarstellungen, Vol. 17, Schmidt, J., Organometallverbindungen, Pt. 2, Wiss. Verlagsges., Stuttgart 1934, pp. 274/305.

Krause, E., von Grosse, A., Blei, in: Die Chemie der metall-organischen Verbindungen, Bornträger, Berlin 1937, pp. 372/429.

Leeper, R. W., Summers, L., Gilman, H., Organolead Compounds, Chem. Rev. **54** [1954] 101/67.

Rochow, E. G., Hurd, D. T., Lewis, R. N., Lead, in: The Chemistry of Organometallic Compounds, Wiley, New York 1957, pp. 190/7.

Klema, F., Die Chemie der Organobleiverbindungen, Mitt. Chem. Forschungsinst. Wirtsch. Österr. **20** [1966] 1/6, 43/6.

Willemsens, L. C., van der Kerk, G. J. M., Synthesis and Properties of the Lead-Carbon Bond, in: MacDiarmid, A. G., Organometallic Compounds of the Group IV Elements, Vol. 1, Part 2, Dekker, New York 1968, pp. 191/229.

8

Neumann, W. P., Kühlein, K., Recent Developments in the Organic Chemistry of Lead, Preparations and Reactions of Compounds with Pb-C, Pb-H, Pb-N, and Pb-O Bonds, Advan. Organometal. Chem. **7** [1968/69] 241/312.

Piękoś, R., Organolead Compounds. Survey of Their Properties and Applications, Przemysl. Chem. **48** [1969] 255/60.

Frey, F. W., Shapiro, H., Commercial Organolead Compounds, Fortschr. Chem. Forsch. **16** [1970/71] 243/97.

Cook, S. E., Frey, F. W., Shapiro, H., Synthesis and Properties of the Lead-Halogen and Lead-Halogenoid Bond, in: MacDiarmid, A. G., Organometallic Compounds of the Group IV Elements, Vol. 2, Part 2, Dekker, New York 1972, pp. 149/97.

Abel, E. W., Lead, in: Bailar, J. C., Jr., Emeléus, H. J., Nyholm, R., Trotman-Dickenson, A. F., Comprehensive Inorganic Chemistry, Vol. 2, Pergamon, Oxford 1973, pp. 105/46.

Willemsens, L. C., Organolead Compounds, in: Greninger, D., Kollonitsch, V., Kline, C. H., Willemsens, L. C., Cole, J. F., Lead Chemicals, International Lead Zinc Research Organization, New York 1975, pp. 19/45.

Willemsens, L. C., Blei, in: Korte, F., Methodicum Chimicum, Vol. 7, Thieme, Stuttgart 1976, Zimmer, H., Niedenzu, K., Hauptgruppenelemente und deren Verbindungen, Academic, New York 1976, pp. 423/47.

Aylett, B. J., Lead, in: Aylett, B. J., Organometallic Compounds, 4th Ed., Vol. 1, Pt. 2, Groups IV and V, Chapman & Hall, London 1979, pp. 277/384.

de Vos, D., Wolters, J., Synthesis, Reactions and Physico-chemical Properties of Mono-organolead(IV) Compounds, Rev. Silicon Germanium Tin Lead Compounds **4** [1980] 209/43.

Hewitt, C. N., Organolead Compounds in the Environment, in: Craig, P. J., Organometallic Compounds in the Environment. Principles and Reactions, Longman, Burnt Mill, Harlow, Essex, 1986, pp. 160/97.

Other General Publications

Krause, E., Schmitz, M., Gemischte Blei- und Zinn-aryle und -aryl-alkyle und ihre Verwendung zur Darstellung von Silber-organoverbindungen, zugleich Beispiele für den Einfluß des Symmetriegrades auf die Eigenschaften chemischer Verbindungen, Ber. Deut. Chem. Ges. **52** [1919] 2150/64.

Brady, O. L., Organo-metallic Compounds. Sci. Progr. **16** [1922] 542/3; C.A. **1922** 2840.

McCorkle, M. R., Gilman, H., Ethynylmetallic and Substituted Ethynylmetallic Reactions in Various Solvents, Proc. Iowa Acad. Sci. **45** [1938] 133.

Finholt, A. E., Bond, A. C., Jr., Wilzbach, K. E., Schlesinger, H. I., The Preparation and Some Properties of Hydrides of Elements of the Fourth Group of the Periodic System and of Their Organic Derivatives, J. Am. Chem. Soc. **69** [1947] 2692/6.

Kay, J., Rowland, F. S., Formation of Volatile Compounds by ^{212}Pb Recoiling from α-Decay, J. Am. Chem. Soc. **80** [1958] 3165.

American Zinc Institute — Lead Industries Association, New York, Expanded Research Program, Quart. Rept. No. 4 [1960].

American Zinc Institute — Lead Industries Association, New York, Expanded Research Program, Research Digest No. 9 [1962].

American Zinc Institute — Lead Industries Association, New York, Expanded Research Program, Research Digest No. 10 [1962].

Henry, M. C., Noltes, J. G., Davidson, W., Krebs, A., Organometallic Compounds, AD-286653 [1962]; C.A. **60** [1964] 6859.

American Zinc Institute — Lead Industries Association, New York, Expanded Research Program, Research Digest No. 11 [1963].

Intern. Lead Zinc Research Organization, New York, Research Digest No. 12 [1963].

Intern. Lead Zinc Research Organization, New York, Research Digest No. 13 [1964].

Intern. Lead Zinc Research Organization, New York, Lead Chemistry, Research Digest No. 14 [1964].

Willemsens, L. C., van der Kerk, G. J. M., A Fundamental Study of Organolead Chemistry, Project No. LC-18, Rept. No. 20 [1964/65] to Intern. Lead Zinc Research Organization, New York.

van der Kerk, G. J. M., New Developments in Organolead Chemistry, Proc. 2nd Intern. Conf. Lead, Arnhem 1965 [1967], pp. 325/31.

Sekine, T., Yamura, A., Sugino, K., Mechanism of Hydrocarbon Formation in the Electrolytic Reduction of Acetone in Aqueous Sulfuric Acid, J. Electrochem. Soc. **112** [1965] 439/43.

Intern. Lead Zinc Research Organization, New York, Lead Chemistry, Research Digest No. 15 [1965].

Intern. Lead Zinc Research Organization, New York, Lead Chemistry, Research Digest No. 16 [1965].

Willemsens, L. C., van der Kerk, G. J. M., A Fundamental Study of Organolead Chemistry, Project No. LC-18, Rept. No. 24 [1965] to Intern. Lead Zinc Research Organization, New York.

Tomilov, A. P., Klyuev, B. L., Electroreduction of Acetone on Cathodes of a Lead-Tin Alloy, Elektrokhimiya **2** [1966] 1405/13; Soviet Electrochem. **2** [1966] 1284/91.

Intern. Lead Zinc Research Organization, Inc., New York, Lead Chemistry, Research Digest No. 17 [1966].

Intern. Lead Zinc Research Organization, Inc., New York, Lead Chemistry, Research Digest No. 18 [1966].

Willemsens, L. C., van der Kerk, G. J. M., A Fundamental Study of Organolead Chemistry, Project No. LC-18, Rept. No. 28 [1966] to Intern. Lead Zinc Research Organization Inc., New York.

Klema, F., Organobleiverbindungen, Chemiker-Ztg. **90** [1966] 106/7.

Huber, F., Chemistry and Stability of Lead(IV) Compounds, Angew. Chem. **79** [1967] 585; Angew. Chem. Intern. Ed. Engl. **6** [1967] 572.

Intern. Lead Zinc Research Organization, Inc., New York, Lead Chemistry, Research Digest No. 19 [1967].

Intern. Lead Zinc Research Organization, Inc., New York, Lead Chemistry, Research Digest No. 20 [1967].

Willemsens, L. C., van der Kerk, G. J. M., A Fundamental Study of Organolead Chemistry, Project No. LC-18, Rept. No. 32 [1967] to Intern. Lead Zinc Research Organization, Inc., New York.

Intern. Lead Zinc Research Organization, Inc., New York, Research Digest No. 21 [1968].

Intern. Lead Zinc Research Organization, Inc., New York, Research Digest No. 22 [1968].

van der Kerk, G. J. M., Recent Trends in Organolead Chemistry, Proc. 3rd Intern. Conf. Lead, Venice 1968, pp. 409/20.

Willemsens, L. C., van der Kerk, G. J. M., A Fundamental Study of Organolead Chemistry, Project No. LC-18, Rept. No. 36 [1968] to Intern. Lead Zinc Research Organization Inc., New York.

Intern. Lead Zinc Research Organization, Inc., New York, Lead Chemistry, Research Digest No. 23 [1969].

Intern. Lead Zinc Research Organization, Inc., New York, Lead Chemistry, Research Digest No. 24 [1969].

Willemsens, L. C., Noltes, J. G., van der Kerk, G. J. M., A Fundamental Study of Organolead Chemistry, Project No. LC-18, Rept. No. 40 [1969] to Intern. Lead Zinc Research Organization, Inc., New York.

Intern. Lead Zinc Research Organization, Inc., New York, Lead Chemistry, Research Digest No. 25 [1970].

Intern. Lead Zinc Research Organization, Inc., New York, Lead Chemistry, Research Digest No. 26 [1970].

Willemsens, L. C., Noltes, J. G., van der Kerk, G. J. M., A Fundamental Study of Organolead Chemistry, Project No. LC-18, Rept. No. 44 [1970] to Intern. Lead Zinc Research Organization, Inc., New York.

Intern. Lead Zinc Research Organization, Inc., New York, Lead Chemistry, Research Digest No. 27 [1971].

Intern. Lead Zinc Research Organization, Inc., New York, Lead Research Digest No. 28 [1971].

Intern. Lead Zinc Research Organization, Inc., New York, Lead Research Digest No. 29 [1972].

Buckle, J., Harrison, P. G., Ylid Complexes of Organotin and -lead Halides, J. Organometal. Chem. 49 [1973] C17/C18.

Somayajulu, G. R., Zwolinski, B. J., Generalized Treatment of Alkanes, J. Chem. Soc. Faraday Trans. II 1974 973/93.

Intern. Lead Zinc Research Organization, Inc., New York, Lead Research Digest No. 33 [1975].

Skell, P. S., AFOSR-75-0200 [1975], in: Blackborow, J. R., Young, D., Metal Vapor Synthesis in Organometallic Chemistry, Springer, Berlin 1979.

Taylor, R. T., Comparative Methylation Chemistry of Platinum, Palladium, Lead, and Manganese, PB-251553 [1976] 1/35; C.A. 87 [1977] No. 89872.

Rotenberg, Z. A., Rufman, N. M., Nature of Photocurrents in the Electroreduction of Acetone on a Lead Electrode in Acid Solutions, Elektrokhimiya **15** [1979] 349/53; Soviet Electrochem. **15** [1979] 293/8.

Karr, C., Jr., McCaskill, K. B., Deactivation of Hydrotreating Catalysts by Coal-Derived Organometallics, METC-RI-79-1 [1979] 1/34; C.A. **91** [1979] No. 177831.

Rufman, N. M., Rotenberg, Z. A., Features of the Kinetics of the Photodecomposition of Organolead Compounds on the Surface of a Lead Electrode, Elektrokhimiya **16** [1980] 364/9; Soviet Electrochem. **16** [1980] 309/14.

Jarvie, A. W. P., Whitmore, A. P., Methylation of Elemental Lead and Lead(II) Salts in Aqueous Solution, Environ. Technol. Letters **2** [1981] 197/204.

Kochkin, D. A., Dependence of Antifouling and Physicochemical Properties of Organotin (-lead) Monomers and Polymers on Chemical Structure (Including cis- and trans-Isomerism), Obrastanie Biokorroz. Vodn. Srede **1981** 164/9; C.A. **96** [1982] No. 99242.

Dewar, M. J. S., Holloway, M. K., Grady, G. L., Stewart, J. J. P., MNDO Calculations for Compounds Containing Lead, Organometallics **4** [1985] 1973/80.

Van Cleuvenbergen, R. J. A., Chakraborti, D., Adams, F. C., Occurrence of Monoalkyllead Species During the Speciation of Organolead, Anal. Chim. Acta **182** [1986] 239/44.

Physical Properties

Structure and Bonding

Vogel, A. I., Cresswell, W. T., Leicester, J., Bond Refractions for Tin, Silicon, Lead, Germanium and Mercury Compounds, J. Phys. Chem. **58** [1954] 174/7.

Long, L. H., Dissociation Energies of Metal-Carbon Bonds and the Excitation Energies of Metal Atoms in Combination, Pure Appl. Chem. **2** [1961] 61/9.

Skinner, H. A., The Strength of Metal-to-Carbon Bonds, Advan. Organometal. Chem. **2** [1964] 49/114.

Tel'noi, V. I., Rabinovich, I. B., Thermochemistry of Organic Compounds of Silicon, Germanium, and Tin, Zh. Fiz. Khim. **40** [1966] 1556/63; Russ. J. Phys. Chem. **40** [1966] 842/7.

Nagy, J., Reffy, J., Kuzman-Borbely, A., Palossy-Becker, K., Bond Structure in the Aryl Derivatives of the Elements of Group IV, J. Organometal. Chem. **7** [1967] 393/404.

Bokii, N. G., Struchkov, Yu. T., Structural Chemistry of Organic Compounds of the Nontransition Elements of Group IV [Si, Ge, Sn, Pb], Zh. Strukt. Khim. **9** [1968] 722/65; J. Struct. Chem. [USSR] **9** [1968] 633/72.

Aylett, B. J., The Stereochemistry of Main Group IV Elements, Progr. Stereochem. **4** [1969] 213/71.

Belloli, R., Resolution and Stereochemistry of Asymmetric Silicon, Germanium, Tin, and Lead Compounds, J. Chem. Educ. **46** [1969] 640/4.

Attridge, C. J., π-Bonding in Group IVB, Organometal. Chem. Rev. A **5** [1970] 323/53.

Shaw III, C. F., Allred, A. L., Nonbonded Interactions in Organometallic Compounds of Group IVB, Organometal. Chem. Rev. A **5** [1970] 95/142.

12

Bokii, N. G., Shklover, V. E., Struchkov, Yu. I., Structural Chemistry of Organic Derivatives of Nontransition Elements. Structural Chemistry of Silicon, Germanium, Tin, and Lead Organic Compounds, Itogi Nauki Tekh. Kristallokhim. **10** [1974] 94/148.

Tel'noi, V. I., Bond Strength in Organic Compounds of Nontransition Elements According to Thermochemical Data, Tr. Khim. Khim. Tekhnol. **1974** No. 1, pp. 28/39; C.A. **82** [1975] No. 72243.

Ratcliffe, C. I., Waddington, T. C., Internal Torsional Modes in Mixed Methyl Halogenocompounds of Group IV Elements Studied by Inelastic Neutron Scattering, J. Chem. Soc., Faraday Trans. II **72** [1976] 1840/50.

Brill, T. B., Nuclear Quadrupole Resonance as a Probe of Structure and Bonding in Organometallic Compounds, Advan. Nucl. Quadrupole Reson. **3** [1978] 131/83.

Skinner, H. A., The Thermochemistry of Organometallic Compounds, J. Chem. Thermodyn. **10** [1978] 309/20.

Glidewell, C., Intramolecular Non-Bonded Atomic Radii: New Data and Revised Radii for Heavy p-Elements, Inorg. Chim. Acta **36** [1979] 135/8.

Furmanova, N. G., Kuzmina, L. G., Struchkov, Yu. T., Structural Evidence of Coordination Interactions in Organic Derivatives of Mercury, Tin, and Lead, J. Organometal. Chem. Libr. **9** [1980] 153/87.

Tel'noi, V. I., Rabinovich, I. B., Thermochemistry of Organic Derivatives of Non-transition Elements, Usp. Khim. **49** [1980] 1137/73; Russ. Chem. Rev. **49** [1980] 603/22.

Mironov, V. F., Gar, T. K., Fedotov, N. S., Evert, G. E., Adamantane Structures in the Chemistry of Silicon, Germanium, and Tin, Usp. Khim. **50** [1981] 485/521; Russ. Chem. Rev. **50** [1981] 262/79.

Armstrong, D. R., Perkins, P. G., Calculations of the Electronic Structures of Organometallic Compounds and Homogeneous Catalytic Processes. Part I. Main Group Organometallic Compounds, Coord. Chem. Rev. **38** [1981] 139/275.

Haaland, A., The Determination of Molecular Structures of Organometallic Compounds by Gas Phase Electron Diffraction, in: Tsutsui, M., Ishii, Y., Yaozeng, H., Fundam. Res. Organomet. Chem. Proc. 1st China-Japan-U.S. Trilateral Semin. Organomet. Chem., Beijing, Peop. Rep. China, 1980 [1982].

Bruce, M. I., Index of Structures Determined by Diffraction Methods, in: Wilkinson, G., Stone, F. G. A., Abel, E. W., Comprehensive Organometallic Chemistry, Vol. 9, Pergamon, Oxford 1982, pp. 1209/520.

Wells, A. F., Structural Inorganic Chemistry, 5th Ed., Clarendon Press, Oxford 1984.

Dewar, M. J. S., Holloway, M. K., Grady, G. L., Stewart, J. J. P., MNDO Calculations for Compounds Containing Lead, Organometallics **4** [1985] 1973/80.

Imyanitov, N. S., Cone Angles of Ligands-Group IV and V Compounds, Koord. Khim. **11** [1985] 1171/8; Soviet J. Coord. Chem. **11** [1985] 663/70.

UV, IR, and Raman Spectra

Sheline, R. K., The Effective Methyl Mass and Its Use in Determining the Force Constants and Character of Metallo-Organic Bonds, J. Chem. Phys. **18** [1950] 602/6.

Noltes, J. G., Henry, M. C., Janssen, M. J., An Infrared Absorption Band Characteristic for Aromatic Compounds of Fourth Main Group Elements, Chem. Ind. [London] **1959** 298/9.

Rao, C. N. R., Ramachandran, J., Iah, M. S. C., Somasekhara, S., Rajakumar, T. V., Infrared and the Near Ultra-violet Absorption Spectra of Polyphenyl Derivatives of the Elements of Groups IVb and Vb, Nature **183** [1959] 1475/6.

Okawara, R., Sakiyama, M., Infrared Spectra of Organic Silicon, Germanium, Tin, and Lead Compounds, Kagaku Ryoiki Zokan No. 45 [1961] 127/64.

Rao, C. N. R., Ramachandran, J., Balasubramanian, A., The Infrared and the Near-Ultraviolet Absorption Spectra of Polyphenyl Derivatives of the Elements of Groups IVb and Vb, Can. J. Chem. **39** [1961] 171/9.

Vyshinskii, N. N., Rudnevskii, N. K., Vibrational Spectra of Certain Organometallic Compounds of the Group IV Elements, Opt. Spektroskopiya **10** [1961] 797/9; Opt. Spectrosc. [USSR] **10** [1961] 421/2.

La Paglia, S. R., The Electronic Spectra of the Group IV Tetraphenyls, J. Mol. Spectrosc. **7** [1961] 427/34.

Harrah, L. A., Ryan, M. T., Tamborski, C., Infrared Spectra of Phenyl Derivatives of Group IVb, Vb and VIIb Elements, Spectrochim. Acta **18** [1962] 21/37.

Kolninov, O. V., Zvonkova, Z. V., Dependence of the Electronic Absorption Spectra of Phenyl Derivatives of Elements in Groups IV and V on the Nuclear Potentials of the Elements, Zh. Fiz. Khim. **36** [1962] 2228/30; Russ. J. Phys. Chem. **36** [1962] 1199/201.

Chumaevskii, N. A., Vibrational Spectra of Compounds Containing Elements of the Carbon Subgroup, Usp. Khim. **32** [1963] 1152/75; Russ. Chem. Rev. **32** [1963] 509/22.

Obreimov, I. V., Chumaevskii, N. A., Valence Vibration Frequencies, E-C, in Organic Compounds Containing Elements of the IVth Group (E = Si, Ge, Sn), Zh. Strukt. Khim. **5** [1964] 59/63; J. Struct. Chem. **5** [1964] 51/4.

Gastilovich, E. A., Shigorin, D. N., Komarov, N. V., An Infrared Study of Acetylene Derivatives Containing Group IV Elements, Opt. Spektroskopiya **16** [1964] 46/51; Opt. Spectrosc. [USSR] **16** [1964] 24/6.

Gastilovich, E. A., Shigorin, D. N., Komarov, N. V., Valence Vibration Frequencies of $C \equiv C$ and CH in the Compounds $Me_3XC \equiv CH$, where X = C, Si, Sn, and Pb, Tr. Komis. Spektrosk. Akad. Nauk SSSR No. 3 [1964] 70/5; C.A. **65** [1966] 593.

Bogomolov, S. G., Veselkova, I. A., Lodochnikova, V. I., Carbon-Lead Band in Infrared Spectra, Tr. Komis. Spektrosk. Akad. Nauk SSSR No. 1 [1964] 475/82; C.A. **63** [1965] No. 13029.

Chumaevskii, N. A., Principles in the Vibrational Spectra of Organic Compounds of Elements of the IVB and VB Groups, Tr. Komis. Spektrosk. Akad. Nauk SSSR No. 3 [1964] 84/91; C.A. **64** [1966] 10592.

Shorygin, P. P., Petukhov, V. A., Nefedov, O. M., Kolesnikov, S. P., Shiryaev, V. I., Absorption Spectra of Molecules Consisting of Groups of the Same Type, Teor. Eksperim. Khim. **2** [1966] 190/5; Theor. Exptl. Chem. [USSR] **2** [1966] 146/9.

Bodiot, D., Spectres d'absorption infra-rouge des composés organométalliques de l'étain ou du plomb, Rev. Chim. Minerale **4** [1967] 957/75.

Tanaka, T., Vibrational Spectra of Organotin and Organolead Compounds, Organometal. Chem. Rev. A **5** [1970] 1/51.

14

Chumaevskii, N. A., Vibrational Spectra of Group IVB and VB Heteroorganic Compounds, Nauka, Moscow 1971.

Licht, K., Reich, P., Literature Data for IR, Raman, NMR Spectroscopy of Si, Ge, Sn and Pb Organic Compounds, VEB Deutscher Verlag der Wissenschaften, Berlin 1971.

Egorochkin, A. N., Vyazankin, N. S., Khorshev, S. Ya., Effect of $d\pi$-$p\pi$ Interaction in Organic Compounds of Group IVB-Elements, Usp. Khim. **41** [1972] 828/51; Russ. Chem. Rev. **41** [1972] 425/38.

Greenwood, N. N., Ross, E. J. F., Straughan, B. P., Index of Vibrational Spectra of Inorganic and Organometallic Compounds, Vol. 1 (1935 – 1960), Butterworths, London 1972.

Rodionov, A. N., Vibration Spectra and Structures of the Simplest Aromatic Derivatives of Group I – VI Elements, Usp. Khim. **42** [1973] 2152/75; Russ. Chem. Rev. **42** [1973] 998/1010.

Crompton, T. R., Chemical Analysis of Organometallic Compounds, Vol. 3, Elements of Group IVB, Lead, Academic Press, London 1974, pp. 99/189.

Rodionov, A. N., Syutkina, O. P., Smirnov, S. G., Panov, E. M., Kocheshkov, K. A., Vibrational Spectra and Structure of Certain Aromatic Derivatives of Mercury, Tin, and Lead, Izv. Akad. Nauk SSSR Ser. Khim. **1974** 1299/302; Bull. Acad. Sci. [USSR] Div. Chem. Sci. **1974** 1222/5.

Minaeva, N. A., Chumaevskii, N. A., The Use of Polyethylene Cells in the Longwave Region of the Infrared Spectrum, Zavodsk. Lab. **40** [1974] 966; Ind. Lab. [USSR] **40** [1974] 1157.

Razuvaev, G. A., Egorochkin, A. N., Koroshev, S. Ya., Kuznetsov, V. A., Spectroscopic Study of Intramolecular Interactions in Group IVA Heteroorganic Compounds, V sb., Ref. Dokl. Soobshch. 11th Mendeleevsk. S'ezd Obshch. Prikl. Khim. Alma-Ata 1975, Vol. 3, p. 208; C.A. **84** [1976] No. 187035.

Höfler, F., Modellberechnungen zu den Schwingungsspektren von Phenylsilanen und einigen verwandten Verbindungen, Monatsh. Chem. **107** [1976] 705/19.

Maslowsky, E., Jr., Vibrational Spectra of Organometallic Compounds, Wiley-Interscience, New York 1977.

Bordeau, M., Clement, C., Structural Study on Organosilicon Compounds by Depolarized Rayleigh Scattering. Determination of the Sign of Optical Anisotropies of Bonds and Groups with or without Silicon in the Aliphatic and Benzenic Series, J. Organometal. Chem. **202** [1980] 137/48.

Egorochkin, A. N., Khorshev, S. Ya., Spectroscopic Study of π-Acceptor Effects in Compounds of Silicon Subgroup Elements, Usp. Khim. **49** [1980] 1687/710; Russ. Chem. Rev. **49** [1980] 820/32.

Maslowsky, E., Jr., The Synthesis, Structure, and Vibrational Spectra of Organomethyl Compounds, Chem. Soc. Rev. **9** [1980] 25/40.

Sennikov, P. G., Egorochkin, A. N., Electronic Spectroscopy of Charge-transfer Complexes as a Method for the Investigation of Intramolecular Interactions in Organic and Organoelemental Compounds, Usp. Khim. **51** [1982] 561/85; Russ. Chem. Rev. **51** [1982] 317/31.

Crompton, T. R., Analysis of Organometallic Compounds: Spectroscopic Methods, in: Hartley, F. R., Patai, S., The Chemistry of the Metal-Carbon Bond, Vol. 1, Wiley, Chichester 1982, pp. 679/708.

Taylor, M. J., Infrared and Raman Spectroscopy of Organometallic Compounds, in: Hartley, F. R., Patai, S., The Chemistry of the Metal-Carbon Bond, Vol. 1, Wiley, Chichester 1982, pp. 775/812.

Eland, J. H. D., Photoelectron Spectroscopy, Butterworths, London 1984.

Magnetic Resonance

Brügel, W., Ankel, T., Krückeberg, F., Das Kernresonanzspektrum der Vinylgruppe, Z. Elektrochem. **64** [1960] 1121/55.

Dreeskamp, H., Kernmagnetische Spin-Spin-Kopplung in Tetramethylverbindungen, Z. Physik. Chem. [Frankfurt] **38** [1963] 121/4.

McCoy, C. R., Allred, A. L., Chemical Shifts of the Methyl Derivatives of the Representative Elements, J. Inorg. Nucl. Chem. **25** [1963] 1219/23.

Drago, R. S., Matwiyoff, N. A., Use of a Relationship between $J(^{13}C\text{-}H)$ and τ to Evaluate Anisotropic Contributions to τ-Anisotropies in the τ-Values of the Group IV Tetramethyl Derivatives, J. Organometal. Chem. **3** [1965] 62/9.

Fritz, H. P., Kreiter, C. G., Spektroskopische Untersuchungen an Organometallischen Verbindungen. ^{1}H-NMR-Spektren von σ-Cyclopentadienylen von Metallen der IV. Gruppe, J. Organometal. Chem. **4** [1965] 313/19.

Gallais, F., Labarre, J.-F., de Loth, P., Évolution de la rotation magnétique (Effet Faraday) de la liaison (C-E) et de quelques liaisons (E-E) en fonction de la position de l'élément E dans la classification périodique: les facteurs de la rotation de liaison, J. Chim. Phys. **64** [1967] 247/52.

Ham, N. S., Mole, T., The Application of NMR to Organometallic Exchange Reactions, Progr. Nucl. Magn. Resonance Spectry. **4** [1969] 91/192.

Jameson, C. J., Gutowsky, H. S., Systematic Trends in the Coupling Constants of Directly Bonded Nuclei, J. Chem. Phys. **51** [1969] 2790/803.

McFarlane, W., Nuclear Spin-Spin Coupling between Directly Bound Elements, Quart. Rev. [London] **23** [1969] 187/203.

McFarlane, W., Nuclear Magnetic Resonance, in: George, W. O., Spectroscopic Methods in Organometallic Chemistry, Butterworths, London 1970, pp. 61/94.

Licht, K., Reich, P., Literature Data for IR, Raman, NMR Spectroscopy of Si, Ge, Sn, Pb Organic Compounds, VEB Deutscher Verlag der Wissenschaften, Berlin 1971.

McFarlane, W., Nuclear Magnetic Double Resonance Spectroscopy, in: Nachod, F. C., Zuckerman, J. J., Determination of Organic Structures by Physical Methods, Vol. 4, Academic, New York 1971, pp. 139/93.

Pregosin, P. S., Randall, E. W., ^{13}C Nuclear Magnetic Resonance, in: Nachod, F. C., Zuckerman, J. J., Determination of Organic Structures by Physical Methods, Vol. 4, Academic, New York 1971, pp. 263/322.

Wells, P. R., NMR Spectra of the Heavier Elements, in: Nachod, F. C., Zuckerman, J. J., Determination of Organic Structures by Physical Methods, Vol. 4, Academic, New York 1971, pp. 233/62.

Sergeyev, N. M., Nuclear Magnetic Resonance Spectroscopy of Cyclopentadienyl Compounds, Progr. Nucl. Magn. Resonance Spectry. **9** [1973] 71/144.

Zimmer, H., Lankin, D. C., Kernmagnetische Resonanz von anderen Kernen als ^{1}H, in: Method. Chim. **1** Pt. 1 [1973] 359/89.

Mann, B. E., ^{13}C NMR Chemical Shift and Coupling Constants of Organometallic Compounds, Advan. Organometal. Chem. **12** [1974] 135/213.

Maciel, G. E., Pulse Fourier Transform NMR with Metal Nuclei, in: Axenrod, T., Webb, G. A., Nuclear Magnetic Resonance Spectroscopy of Nuclei Other than Protons, Wiley, New York 1974, pp. 347/75.

Voronov, V. K., Domnina, E. S., Glukhikh, V. I., Ivlev, Yu. N., Skortsova, G. G., Voronkov, M. G., Carbon-13 NMR of 1-Vinyl- and 1-Ethylimidazole Complexes with Triethyl-Group IV B Element Halides, Fiz. Mat. Metody Koord. Khim. Tezisy Dokl. 5th Vses. Soveshch., Kishinev, U.S.S.R., 1974 [1974], p. 95.

Barbieri, G., Benassi, R., Taddei, F., Long-Range Proton-Metal and Carbon-Metal Coupling Constants in Organometallic Compounds. An Empirical Interpretation, Gazz. Chim. Ital. **105** [1975] 807/26.

Gansow, O. A., Vernon, W. D., Carbon-13 Nuclear Magnetic Resonance Studies of Organometallic and Transition Metal Complex Compounds, Top. Carbon-13 NMR Spectrosc. **2** [1976] 269/341.

Chisholm, M. H., Godleski, S., Applications of Carbon-13 NMR in Inorganic Chemistry, Progr. Inorg. Chem. **20** [1976] 299/436.

Inamoto, N., Masuda, S., Tokumaru, K., Tori, K., Yoshida, M., Yoshimura, Y., Substituent Effects on C-13 Chemical Shifts of Substituted Benzenes. Periodical Dependency of Substituent Electronegativity Effects upon Ring-Carbon Chemical Shifts, Tetrahedron Letters **1976** 3711/4.

Shapiro, B. L., Mohrmann, L. E., NMR Spectral Data: A Compilation of Aromatic Proton Chemical Shifts in Mono- and Di-Substituted Benzenes, J. Phys. Chem. Ref. Data **6** [1977] 919/91.

Harris, R. K., Kennedy, J. D., McFarlane, W., Group IV — Silicon, Germanium, Tin, and Lead, in: Harris, R. K., Mann, B. E., NMR and the Periodic Table, Academic, London 1978, pp. 309/77.

Brill, T. B., Nuclear Quadrupole Resonance as a Probe of Structure and Bonding in Organometallic Compounds, Advan. Nucl. Quadrupole Reson. **3** [1978] 131/83.

Ewing, D. F., ^{13}C Substituent Effects in Monosubstituted Benzenes, Org. Magn. Resonance **12** [1979] 499/524.

Hawkes, G. E., Applications of Nuclear Shielding, Nucl. Magn. Resonance **10** [1981] 18/55.

Pachler, K. G. R., Theoretical Aspects of Spin-Spin Couplings, Nucl. Magn. Resonance **10** [1981] 56/73.

Radeglia, R., Steinborn, D., Taube, R., One-Bond Spin-Spin Coupling Constants and the s-Character of Hybrid Orbitals, Z. Chem. [Leipzig] **21** [1981] 365/6.

Knorr, R., Inductive Substituent Constants of σ_I^j from Olefinic Geminal Proton-Proton NMR Coupling Constants, Tetrahedron **37** [1981] 929/37.

Davies, J. A., Multinuclear Magnetic Resonance Methods in the Study of Organometallic Compounds, in: Hartley, F. R., Patai, S., The Chemistry of the Metal-Carbon Bond, Vol. 1, Wiley, Chichester 1982, pp. 813/918.

Fedin, É. I., Fedorov, L. A., Nuclear Spin-Spin Metal-Carbon Interaction and Polarization of the s-Electron Core of Nontransition Metal Atoms in Organometallic Compounds, Dokl. Akad. Nauk SSSR **267** [1982] 1159/62; Dokl. Phys. Chem. Proc. Acad. Sci. USSR **262/267** [1982] 1007/10.

Dechter, J. J., NMR of Metal Nuclides. The Main-Group Metals, Progr. Inorg. Chem. **29** [1982] 285/385.

Wrackmeyer, B., Carbon Carbon and Metal Carbon Indirect Nuclear Spin-Spin Coupling Constants $^1J(^{13}C^{13}C)$ and $^1J(M^{13}C)$ in Organometallic Compounds, Spectrosc. Intern. J. **1** [1982] 201/8.

Granger, P., NMR of Less Common Nuclei, in: Laszlo, P., NMR of Newly Accessible Nuclei, Vol. 2, Academic, New York 1983, pp. 386/417.

Harris, R. K., Solution-State NMR Studies of Group IV Elements (Other than Carbon), NATO ASI Ser. C **103** [1983] 343/59; C.A. **99** [1983] No. 132397.

Mitchell, T. N., On the Relation between Element NMR Chemical Shifts in the Fourth Main Group, J. Organometal. Chem. **255** [1983] 279/85.

Hawkes, G. E., Applications of Nuclear Shielding, Nucl. Magn. Resonance **13** [1984] 21/63.

Fedorov, L. A., NMR Spectroscopy of Organometallic Compounds: Testing the Use of NMR in the Chemistry of Organic Compounds of Heavy Nontransition Metals, Nauka, Moscow 1984, pp. 1/248.

Fedorov, L. A., Effect of Substituents in R_nM Organometallic Compounds and Changes in Direct Constants of the $^1J_{M^{-13}C}$ Spin-Spin Interaction, Zh. Strukt. Khim. **25** [1984] 35/42; J. Struct. Chem. [USSR] **25** [1984] 538/44.

Fedorov, L. A., Direct $^1J_{\cdot M^{-13}C}$ Spin-Spin Interaction Constants and Structure of Symmetric Organic Derivatives of Heavy Nontransition Metals, Zh. Strukt. Khim. **25** [1984] 43/8; J. Struct. Chem. [USSR] **25** [1984] 379/84.

Harris, R. K., Reams, P., Packer, K. J., Aspects of High-Resolution Multinuclear Magnetic Resonance of Solid Organometallic Compounds, J. Mol. Struct. **141** [1986] 13/25.

Other Physical Properties

Costa, G., Considerations on the Polarographic Behavior of Organometallic Compounds of the Type $R_{n-1}Me^n$-X, Ann. Chim. [Rome] **41** [1951] 207/20.

Cresswell, W. T., Leicester, J., Vogel, A. I., Bond Refractions for Compounds of Tin, Silicon, Lead, Mercury, and Germanium, Chem. Ind. [London] **1953** 19.

Vogel, A. I., Cresswell, W. T., Leicester, J., Bond Refractions for Tin, Silicon, Lead, Germanium, and Mercury Compounds, J. Phys. Chem. **58** [1954] 174/7.

Dibeler, V. H., Mass Spectra of the Tetramethyl Compounds of Carbon, Silicon, Germanium, Tin, and Lead, J. Res. Natl. Bur. Std. **49** [1952] 235/9.

18

Lutskii, A. E., Molecular Constants and the Physical Properties of Liquids. The Boiling Points, Zh. Fiz. Khim. **30** [1956] 396/406.

Shimizu, K., Intramolecular Potential of M(CH$_3$)$_4$-Type Molecules, Nippon Kagaku Zasshi **77** [1956] 1284/7.

Bothorel, P., Molecular Anisotropy and Orientation of the Phenyl Rings in Polyphenyl Compounds, Ann. Chim. [Paris] [13] **4** [1959] 669/712.

Good, W. D., Scott, D. W., Combustion in a Bomb of Organometallic Compounds, in: Skinner, H. A., Experimental Thermochemistry, Vol. 2, Interscience, New York 1962, pp. 57/76.

Lautsch, W. F., Tröber, A., Zimmer, W., Mehner, L., Linck, W., Lehmann, H.-M., Brandenburger, H., Körner, H., Metzschker, H.-J., Wagner, K., Kaden, R., Energetische Daten metallorganischer Verbindungen, Pt. 1: Verbrennungs- und Bildungsenthalpien, Z. Chem. [Leipzig] **3** [1963] 415/21.

Pascal, P., Gallais, F., Labarre, J. F., Sur une systématique de susceptibilités diamagnétiques de liaisons en série aliphatique, Compt. Rend. **256** [1963] 335/9.

Lautsch, W. F., Tröber, A., Körner, H., Wagner, K., Kaden, R., Blase, S., Energetische Daten metallorganischer Verbindungen, Pt. 2: Verdampfungs- und Bindungsenthalpien, Z. Chem. [Leipzig] **4** [1964] 441/54.

Lautsch, W. F., Tröber, A., Körner, H., Wagner, K., Kaden, R., Blase, S., Energetische Daten metallorganischer Verbindungen. Pt. 3: Eigene Meßmethodik und Fehlerbetrachtungen, Z. Chem. [Leipzig] **6** [1966] 171/81.

Chambers, D. B., Glockling, F., Light, J. R. C., Mass Spectra of Organometallic Compounds, Quart. Rev. [London] **22** [1968] 317/37.

Lappert, M. F., Simpson, J., Spalting, T. R., A Mass Spectrometric Study of Some Group IVB Organometallic Compounds Containing Metal-Metal Bonds, in: George, W. O., Spectroscopic Methods in Organometallic Chemistry, Butterworths, London 1970, pp. 197/8.

Boyer-Donzelot, M., Boyer, P., Incréments de polarisation diélectrique pour quelques liaisons simples d'atomes avec le carbone, Bull. Soc. Chim. France **1971** 1172/4.

Quac Dang Cheu, Calculation of the Heats of Combustion of Organometallic Compounds Containing Group IV Elements, Vestn. Mosk. Univ. Khim. **26** No. 3 [1971] 307/9; Moscow Univ. Chem. Bull. **26** No. 3 [1971] 31/2.

Quac Dang Cheu, Baburina, I. I., Molecular Polarisability of Alkylplumbines and Alkylgermanes, Zh. Fiz. Khim. **46** [1972] 2156; Russ. J. Phys. Chem. **46** [1972] 1235/6.

Bursey, J. T., Bursey, M. M., Kingston, D. G. I., Intramolecular Hydrogen Transfer in Mass Spectra. Rearrangements in Aliphatic Hydrocarbons and Aromatic Compounds, Chem. Rev. **73** [1973] 191/234.

Orlov, V. Yu., Mass Spectra of Organometallic Compounds of Group IVb, Usp. Khim. **42** [1973] 1184/9; Russ. Chem. Rev. **42** [1973] 529/37.

Maslova, V. A., Rabinovich, I. B., Heat Capacity, Phase Transitions, and Thermodynamic Functions of Alkyl Compounds of Nontransition Elements, Tr. Khim. Khim. Tekhnol. **1974** No. 1, pp. 40/63.

Meites, L., Zuman, P., Scott, W. J., Campbell, B. H., Kardos, A. M., Electrochemical Data, Pt. 1, Organic, Organometallic, and Biochemical Substances, Vol. A, Wiley, New York 1974.

Glockling, F., Main-Group Organometallics and Metal-Metal Bonded Compounds, in: Charalambous, J., Mass Spectrometry of Metal Complexes, Butterworths, London 1975, pp. 87/103.

Mairanovskii, S. G., Polarography of Organoelementary Compounds of Nontransition Elements, Usp. Khim. **45** [1976] 604/39; Russ. Chem. Rev. **45** [1976] 298/317.

Pilcher, G., Thermochemistry of Organometallic Compounds Containing Metal-Carbon Linkages, MTP [Med. Tech. Publ. Co.] Intern. Rev. Sci. Phys. Chem. Ser. Two **10** [1975] 45/80.

Majee, B., Interpretation of the Properties of Organo Derivatives of Silicon, Germanium, Tin, and Lead by the Del Re Method, Rev. Silicon Germanium Tin Lead Compounds **2** [1975/77] 5/80.

Vogel, L., Kehlen, H., Excess Enthalpies of Binary Liquid Systems of Symmetrical Tetraalkyls of Silicon, Tin, and Lead, 1st Czech. Conf. Calorim. [Lect. Short Commun.], Liblice, Czech., 1977, pp. C4-1/C4-4.

Ribeiro da Silva, M. A. V., Reis, A. M. M. V., Thermochemistry of the Metal-Carbon Bond. Critical Review of Bond Energy Data of Organometallic Compounds and of Transition Metal Carbonyls, Rev. Port. Quim. **20** [1978] 47/62.

Steele, W. V., The Standard Enthalpy of Formation of Tetraphenylsilane and the Ph-M Mean Bond-Dissociation Energies of the Group IV Elements, J. Chem. Thermodyn. **10** [1978] 445/52.

Pedley, J. B., Rylance, J., Sussex-N.P.L. Computer Analyzed Thermochemical Data: Organic and Organometallic Compounds, University of Sussex, Brighton 1979, pp. 1/201.

Daolio, S., Foffani, M. T., Mengoli, G., Mass Spectrometric Identification of Unusual Lead and Tin Alkyls Produced by the Electrochemical Reduction of Methyl-Ethyl Ketone at Lead/Tin Alloys, Intern. J. Mass Spectrom. Ion Phys. **48** [1983] 67/70.

Zorin, A. D., Feshchenko, I. A., Tsinovoi, Yu. N., Karataev, E. N., Thermodynamic Substantiation of the Feasibility of Obtaining Metals from Organometallic Compounds, Poluch. Anal. Chist. Veshchestv **1983** 3/6; C.A. **101** [1984] No. 32234.

Ducros, M., Sannier, H., Determination of Vaporization Enthalpies of Liquid Organic Compounds. Application to Organometallic Compounds, Thermochim. Acta **75** [1984] 329/40.

Burmakov, V. M., Kuznetsova, T. V., Feshchenko, I. A., Tsinovoi, Yu. N., Density, Viscosity, and Surface Tension of Lower Alkyl Derivatives of Germanium, Tin, and Lead, Poluch. Anal. Chist. Veshchestv **1984** 78/80; C.A. **104** [1986] No. 213577.

Analysis

Cholak, J., Analytical Methods for Determination of Lead, Arch. Environ. Health **8** [1964] 222/31.

Linch, A. L., Davis, R. B., Stalzer, R. F., Anzilotti, W. F., Studies of Analytical Methods for Lead-in-Air Determination and Use with an Improved Self-Powered Portable Sampler, Am. Ind. Hyg. Ass. J. **25** [1964] 81/93.

Gunn, E. L., Determination of Lead in Gasoline by X-Ray Fluorescence by Using an Internal Intensity Reference, Appl. Spectrosc. **19** [1965] 96/9.

Siniramed, C., Renzanigo, F., Comparative Study of Several Methods for Determination of Lead in Gasoline, Riv. Combust. **19** [1965] 351/62.

Hasegawa, K., Kajikawa, M., Okamoto, N., A Rapid Method for Determination of Lead and Bromine in Gasoline by X-Ray Fluorescence, Bunseki Kagaku **14** [1965] 717/20; C.A. **63** [1965] 11207.

Trent, D. J., The Determination of Lead in Gasoline by Atomic Absorption Spectroscopy, At. Absorpt. Newsl. **4** [1965] 348/50.

Kreshkov, A. P., Kuchkarev, E. A., Spectrographic Determination of Germanium, Tin, and Lead in Metalloorganic Compounds, Zavodsk. Lab. **32** [1966] 558/9; C.A. **65** [1966] No. 6292.

Bolanowska, W., A Method of Determination of Triethyllead in Blood and Urine, Chem. Anal. [Warsaw] **12** [1967] 121/9.

Campbell, K., Moss, R., The Determination of Trace Amounts of Lead in Crude Oil and Petroleum Products, J. Inst. Petrol. **53** No. 521 [1967] 194/200.

Green, L. E., Characterization of Lead Alkyls in Petroleum Products by Using High-Temperature Electron-Capture Gas Chromatography, Facts Methods **8** No. 4 [1967] 4/7; C.A. **68** [1968] No. 88740.

Prohaska, G., Determination of the Total Lead Content of Gasoline, Nafta [Zagreb] **19** [1968] 461/4; C.A. **70** [1969] No. 30576.

Kashiki, M., Yamazoe, S., Oshima, S., Determination of Lead in Gasoline by Atomic Absorption Spectroscopy, Anal. Chim. Acta **53** [1971] 95/100.

Galtieri, A., Gas Chromatography. Use in Fuels and Their Derivatives, Riv. Combust. **25** [1971] 417/27.

Marti, M., Polarographic Determination of Microamounts of Lead in Gasoline, Riv. Combust. **25** [1971] 252/55; C.A. **75** [1971] No. 99649.

Tanaka, K., Fukaya, K., Yoshitani, K., Fukui, S., Kanno, S., Hygienic Chemical Studies on Pollution by Harmful Substances. Atomic Absorption Photometric Determination of Lead in Gasoline, Eisei Kagaku **17** [1971] 393/7; C.A. **76** [1972] No. 156412.

Garcia Escolar, L., Paz Castro, M., Determination of Lead in Gasolines, Ion [Madrid] **32** [1972] 100/1.

Weiss, M. K., Lead, in: Snell, F. D., Ettre, L. S., Encyclopedia of Industrial Chemical Analysis, Vol. 15, Interscience, New York 1972, pp. 161/201.

Campbell, K., Palmer, J. M., The Determination of Trace Amounts of Lead in Petroleum Products by Iodine Monochloride Extraction and Atomic Absorption Analysis, J. Inst. Petrol. **58** [1972] 193/200.

Rüssel, H., Tölg, G., Anwendung der Gaschromatographie zur Trennung und Bestimmung anorganischer Stoffe, Fortschr. Chem. Forsch. **33** [1972] 1/74.

Mizuno, K., Shiio, H., X-Ray Fluorescent Analysis of Alkyl Lead in Air with Active Carbon, Bunseki Kagaku **21** [1972] 271/3.

Suzuki, T., Isogai, Y., Fukami, T., Ito, H., Determination of Lead in Gasoline, Nagoyashi Kogyo Kenkyusho Kenkyu Hokoku No. 45 [1971] 24/6; C.A. **75** [1971] No. 89767.

Chernoplekova, V. A., Sakharov, V. M., Sakodynskii, K. I., Gas Chromatography of Organometallic Compounds of Group I – IV of the Periodic System, Usp. Khim. **42** [1973] 2274/98; Russ. Chem. Rev. **42** [1973] 1063/77.

Anonymous, Standard Method of Test for Lead in Gasoline, Polarographic Method, D 1269-61 (Reapproved 1968), Ann. Book ASTM Stand., Vol. 17, Am. Soc. Test. Mater., Philadelphia 1973, pp. 446/50.

Carboni, C., De Lindemann, L., Determination of Organic Lead in the Air, Ann. Ist. Super. Sanita **9** (Spec. Number 1) [1973] 534/9; C.A. **83** [1975] No. 120258.

Leisey, F. A., Continuous Analyzer for Lead in Gasoline, ISA Trans. **12** [1973] 78/81.

Hodkova, M., Holle, B., Determination of Trace Amounts of Lead in Gasolines, Ropa Uhlie **15** [1973] 207/13; C.A. **79** [1973] No. 68349.

Forino, G., Detection of Lead in Gasolines, Rass. Chim. **25** [1973] 344/6; C.A. **80** [1974] No. 110578.

Takeuchi, S., Determination of Lead in Gasoline by the Oxygen Bomb and A. C. Polarographic Method, Sekiyu Gakkaishi **16** [1973] 913/15; C.A. **80** [1974] No. 61801.

Purdue, L. J., Enrione, R. E., Thompson, R. J., Bonfield, B. A., Determination of Organic and Total Lead in the Atmosphere by Atomic Absorption Spectrometry, Anal. Chem. **45** [1973] 527/30.

Crompton, T. R., Chemical Analysis of Organometallic Compounds, Vols. 2 and 3: Elements of Group IVB, Academic, London 1974.

Larson, J. A., Short, M. A., Bonfiglio, S., Allie, W., X-Ray Fluorescence Technique for the Analysis of Lead in Gasoline, X-Ray Spectrom. **3** [1974] 125/9.

Bowen, B. C., Foote, H., Total Lead in Gasolines by Atomic Absorption Spectrophotometry, Inst. Pet. [Tech. Pap.] IP 74-010 [1974]; C.A. **86** [1977] No. 192206.

Heistand, R. N., Shaner Jr., W. C., Automated Atomic Absorption Determination of Low Levels of Lead in Gasoline, At. Absorpt. Newsl. **13** [1974] 65/7; C.A. **81** [1974] No. 172648.

Price, B. J., Field, K. M., Direct Analysis of Sulfur and Lead in Fuels by Nondispersive X-Ray Fluorescence, Am. Lab. **6** [1974] 62/4, 66; C.A. **82** [1975] No. 75301.

Harrison, R. M., Perry, R., Slater, D. H., An Adsorption Technique for the Determination of Organic Lead in Street Air, Atmos. Environ. **8** [1974] 1187/94.

Blears, D. G., Coventry, R. J., Development of Continuous Organic Lead-in-Air Monitors, Proc. 5th Symp. Chem. Process Hazards Spec. Ref. Plant Design, London 1974; Inst. Chem. Eng. Symp. Ser. A **39** [1974] 322/37; C.A. **81** [1974] No. 140424, **86** [1977] No. 95106.

Lukasiewicz, R. J., Berens, P. H., Buell, B. E., Rapid Determination of Lead in Gasoline by Atomic Absorption Spectrometry in the Nitrous Oxide – Hydrogen Flame, Anal. Chem. **47** [1975] 1045/9.

McCorriston, L. L., Ritchie, R. K., Determination of Lead in Gasoline by Atomic Absorption Spectrometry Using a Total Consumption Burner, Anal. Chem. **47** [1975] 1137/9.

Katou, T., Nakagawa, R., Determination of Alkylleads by Combined System of GC/AAS, Yokohama Kokuritsu Daigaku Kankyo Kagaku Kenkyu Senta Kiyo **1** [1975] 19/24.

Anonymous, Standard Method of Test for Lead in Gasoline, Gravimetric Method, D 526-70, Ann. Book ASTM Stand., Vol. 17, Am. Soc. Test. Mater., Philadelphia 1975, pp. 275/8.

Robinson, J. W., Rhodes, L., Wolcott, D. K., The Determination and Identification of Molecular Lead Pollutants in the Atmosphere, Anal. Chim. Acta **78** [1975] 474/8.

Sawicki, C. R., Seminar Summary: Sampling and Analysis of the Various Forms of Atmospheric Lead, PB-240620 [1975] 1/14.

Madec, M., La Villa, F., Determination of Lead Traces in Naphthas, Rev. Inst. Fr. Petrol. **31** [1976] 687/701.

Russell, T. J., Campbell, K., Rapid Determination of Lead in Petroleum Products by Atomic Absorption, Compend. Deut. Ges. Mineraloelwiss. Kohlechem. **76/77** [1976] 986/94; C.A. **88** [1978] No. 9358.

Kaegler, S. H., Tillmanns, V., Hebild, F., First Experience with "Roentgenboy" for Lead Determination in Gasolines, Compend. Deut. Ges. Mineraloelwiss. Kohlechem. **76/77** [1976] 995/1008; C.A. **88** [1978] No. 9359.

Teller, S., A Radioisotope Immersion Probe for Continuous or Discrete Measurements of Sulphur in Crude Oils and Lead in Refinery Products, Proc. ERDA Symp. X-Gamma-Ray Sources Appl., Ann Arbor, Mich., 1976, pp. 194/7.

Tausch, H., Die Grundlagen der Gaschromatographie und ihre Anwendung zur Spurenanalyse flüchtiger organischer Bleiverbindungen in der Atmosphäre, Ber. Oesterr. Studienges. Atomenerg. SGAE-2636 [1976] 1/11.

Chau, Y. K., Wong, P. T. S., An Element- and Speciation-specific Technique for the Determination of Organometallic Compounds, Environ. Anal. Papers 3rd Ann. Meet. Fed. Anal. Chem. Spectrosc. Soc., Philadelphia 1976 [1977], pp. 215/25.

Reamer, D. C., O'Haver, T. C., Zoller, W. H., The Use of a Gas Chromatograph — Microwave Plasma Detector for the Detection of Alkyl Lead and Selenium Compounds in the Atmosphere, NBS-SP-464 [1977] 609/12; C.A. **89** [1978] No. 48156.

MacCrehan, W. A., Durst, R. A., Bellama, J. M., HPLC with Electrochemical Detection for the Measurement of Trace Organometallic Species in Environmental Water Samples, 4th Joint Conf. Sens. Environ. Pollut. Conf. Proc., New Orleans 1977 [1978], pp. 635/6.

Schmidt, U., Huber, F., Spektralphotometrische Bestimmung von Blei(II)-, sowie Dialkylblei- und Trialkylbleiverbindungen in geringen Konzentrationen, Anal. Chim. Acta **98** [1978] 147/9.

Beccaria, A. M., Mor, E. D., Poggi, G., A Method for the Analysis of Traces of Inorganic and Organic Lead Compounds in Marine Sediments, Ann. Chim. [Rome] **68** [1978] 607/17.

Reamer, D. C., Zoller, W. H., O'Haver, T. C., Gas Chromatograph — Microwave Plasma Detector for the Determination of Tetraalkyllead Species in the Atmosphere, Anal. Chem. **50** [1978] 1449/53.

Coker, D. T., A Simple, Sensitive Technique for Personal and Environmental Sampling and Analysis of Lead Alkyl Vapours in Air, Ann. Occup. Hyg. **21** [1978] 33/8.

Simon, N., Welebir, A. J., Aldridge, M. H., Trace Organometallics in Water, AD-A 058566 [1978] 1/54; C.A. **90** [1979] No. 156857.

Fouzder, N. B., Fleet, B., The Electrochemical Behavior and Analysis of Organometallic Compounds of Mercury, Tin, Lead, and Germanium, in: Smyth, W. F., Polarography of Molecules of Biological Significance, Academic, London 1979, pp. 261/93.

MacCrehan, W. A., Durst, R. A., Bellama, J. M., Study of Organometal Speciation in Water Samples Using Liquid Chromatography with Electrochemical Detection, NBS-SP-519 [1979] 57/63.

Reamer, D. V., Methods for the Determination of Atmospheric Tetraalkyllead and Alkyl Selenide Species Using a GC-Microwave Plasma Detector, Diss. Univ. Maryland 1978; Diss. Abstr. Intern. B **40** [1979] 710/11.

Rohbock, E., Müller, J., Methode zur Messung gasförmiger Bleialkylverbindungen in Außenluft, Mikrochim. Acta **1979** I 423/34.

Chau, Y. K., Wong, P. T. S., Bengert, G. A., Kramar, O., Determination of Tetraalkyllead Compounds in Water, Sediment, and Fish Samples, Anal. Chem. **51** [1979] 186/88.

Radziuk, B., Thomassen, Y., van Loon, J. C., Chau, Y. K., Determination of Alkyl Lead Compounds in Air by Gas Chromatography and Atomic Absorption Spectrometry, Anal. Chim. Acta **105** [1979] 255/62.

Berenguer, V., Guinon, J. L., de la Guardia, M., Rapid Determination of Lead in Petrol by Atomic Absorption Spectrometry of Emulsified Samples, Z. Anal. Chem. **294** [1979] 416.

Mitchell, W. J., Midgett, M. R., Measuring Inorganic and Alkyl Lead Emissions from Stationary Sources, J. Air Pollution Control Assoc. **29** [1979] 959/62.

Brinckman, F. E., Blair, W. R., Speciation of Metals in Used Oils: Recent Progress and Environmental Implications of Molecular Lead Compounds in Used Crankase Oils, NBS-SP-556 [1979] 25/38.

Campiglio, A., Potentiometric Microdetermination of Lead(II) with an Ion-Selective Lead Electrode and its Application to the Analysis of Organic Lead Compounds, Mikrochim. Acta **1979** I 267/78.

Campiglio, A., Über die Mineralisierung nichtflüchtiger organischer Bleiverbindungen, Mikrochim. Acta **1979** I 395/98.

Robinson, J. W., Rhodes, L. J., Sources of Inorganic Molecular Lead in the Ambient Atmosphere, Spectrosc. Letters **12** [1979] 781/807.

Anonymous, Standard Test Method for Lead in Gasoline, Volumetric Chromate Method, ANSI/ASTM D 2547-70 (Reapproved 1977), Ann. Book ASTM Stand., Vol. 24, Am. Soc. Test. Mater., Philadelphia 1979, pp. 453/6.

Anonymous, Standard Test Method for Lead in Gasoline by X-Ray Spectrometry, ANSI/ASTM D 2599-71 (Reapproved 1976), Ann. Book ASTM Stand., Vol. 24, Am. Soc. Test. Mater., Philadelphia 1979, pp. 521/4.

Anonymous, Standard Test Method for Trace Amounts of Lead in Gasoline, ANSI/ASTM D 3116-72 (Reapproved 1977), Ann. Book ASTM Stand., Vol. 25, Am. Soc. Test. Mater., Philadelphia 1979, pp. 31/5.

Anonymous, Standard Test Method for Lead in Gasoline by Atomic Absorption Spectrometry, ANSI/ASTM D 3237-79, Ann. Book ASTM Stand., Vol. 25, Am. Soc. Test. Mater., Philadelphia 1979, pp. 112/4.

Anonymous, Standard Test Method for Lead in Gasoline — Iodine Monochloride Method, ANSI/ASTM D 3341-79, Ann. Book ASTM Stand., Vol. 25, Am. Soc. Test. Mater., Philadelphia 1979, pp. 236/9.

Radziuk, B., Thomassen, Y., Butler, L. R. P., Van Loon, J. C., Chau, Y. K., A Study of Atomic Absorption and Atomic Fluorescence Atomization Systems as Detectors in the Gas Chromatographic Determination of Lead, Anal. Chim. Acta **108** [1979] 31/8.

Noden, F. G., The Determination of Tetraalkyllead Compounds and Their Degradation Products in Natural Water, Lead Mar. Environ. Proc. Intern. Experts Discuss., Rovinj, Yugoslav., 1977 [1980], pp. 83/91.

Hewitt, A. G., The Control of Toxic Atmospheres in Selected Batch and Continuous Chemical Processes, Symp. Pap. Inst. Chem. Eng. North West. Branch, Manchester 1980, Vol. 2, pp. 3-1/3-7.

Cruz, R. B., Lorouso, C., George, S., Thomassen, Y., Kinrade, J. D., Butler, L. R. P., Lye, J., Van Loon, J. C., Determination of Total, Organic Solvent Extractable, Volatile and Tetraalkyllead in Fish, Vegetation, Sediment and Water Samples, Spectrochim. Acta B **35** [1980] 775/83.

Birch, J., Harrison, R. M., Laxen, D. P. H., A Specific Method for 24—48 Hour Analysis of Tetraalkyl Lead in Air, Sci. Total Environ. **14** [1980] 31/42.

de Jonghe, W., Chakraborti, D., Adams, F., Graphite-Furnace Atomic Absorption Spectrometry as a Metal-Specific Detection System for Tetraalkyllead Compounds Separated by Gas-Liquid Chromatography, Anal. Chim. Acta **115** [1980] 89/101.

Bet-Wei Zeng, Determination of Tetraalkyllead in Environmental Samples, Huanjing Kexue **1** [1980] 55/7; C.A. **94** [1981] No. 149736.

Frank, H. A., Lead Alkyl Components as Discriminating Factors in the Comparison of Gasolines, J. Forensic Sci. Soc. **20** [1980] 285/92.

Zabairova, R. A., Bortnikov, G. N., Gorina, F. A., Yashin, Ya. I., Chromatographic Determination of Organolead Compounds, Khim. Elementoorg. Soedin. [Gor'kii] No. 8 [1980] 15/26; C.A. **96** [1982] No. 96770.

United States Environmental Protection Agency (Washington, D.C.), Lead-in-Gasoline Test Procedure, Fed. Regist. **46** No. 66 [1981] 20698/703; C.A. **94** [1981] No. 194621.

Chan, L., The Determination of Tetraalkyl Lead Compounds in Petrol Using Combined Gas Chromatography—Atomic Absorption Spectrometry, Forensic Sci. Intern. **18** [1981] 57/62.

Jiang, S. G., Chakraborti, D., De Jonghe, W., Adams, F., Atomic-Absorption Spectrophotometric Determination of Volatile Organolead Compounds in the Atmosphere, Z. Anal. Chem. **305** [1981] 177/80.

Chakraborti, D., Jiang, S. G., Surkijn, P., De Jonghe, W., Adams, F., Determination of Tetraalkyllead Compounds in Environmental Samples by Gas Chromatography—Graphite Furnace Atomic-Absorption Spectrometry, Anal. Proc. London **18** [1981] 347/50.

Zabairova, R. A., Bortnikov, G. N., Bochkarev, V. N., Gorina, F. A., Samarin, K. M., Gas Chromatographic Analysis in the Manufacture of Organolead Compounds, Usp. Gazov. Khromatogr. **1981** No. 6, pp. 278/83; C.A. **98** [1983] No. 154695.

Aldridge, W. N., Street, B. W., Spectrophotometric and Fluorimetric Determination of Tri- and Di-organotin and -organolead Compounds Using Dithizone and 3-Hydroxyflavone, Analyst [London] **106** [1981] 60/8.

Yamauchi, H., Arai, F., Yamamura, Y., Determination of Triethyllead, Diethyllead and Inorganic Lead in Urine by Hydride Generation-Flameless Atomic Absorption Spectrometry, Ind. Health **19** [1981] 115/24.

Hodges, D. J., Skelding, D., Determination of Lead in Urine by Atomic-Absorption Spectroscopy with Electrothermal Atomisation, Analyst [London] **106** [1981] 299/304.

Zabairova, R. A., Bortnikov, G. N., Gorina, F. A., Bochkarev, V. N., Samarin, K. M., Gas Chromatographic Methods for the Analytical Monitoring of Organolead Compound Manufacture, Khim. Elementoorg. Soedin. [Gor'kii] **1981** 86/98; C.A. **97** [1982] No. 137936.

Crompton, T. R., Gas Chromatography of Organometallic Compounds, Plenum, New York 1982.

Ebdon, L., Ward, R. W., Leathard, D. A., Approaches to Trace-Metal Speciation in Environmental Samples, Anal. Proc. [London] **19** [1982] 110/4.

De Jonghe, W. R. A., Adams, F. C., Measurements of Organic Lead in Air — A Review, Talanta **29** [1982] 1057/67.

de Jonghe, W., Adams, F., Organolead Compounds in the Atmosphere, COST 61a bis, 2nd Discuss. Meeting Working Party 1: Detection, Identification and Analysis of Air Pollutants, Vienna 1982, pp. 17/23.

Andersson, K., Levin, J. O., et al., Sampling and Analysis of Organic Substances in the List of Limit Values IX: Tetramethyl Lead and Tetraethyl Lead, HSE Trans — 10311 [1982] 1/16; Brit. Rept. Trans. Thesis. [1983].

West, N. G., Purnell, C. J., Brown, R. H., Withers, E., The Measurement of Low Concentrations of Organic and Inorganic Gaseous Contaminants in Occupational Environments by X-Ray Spectrometry (XRS), Advan. X-Ray Anal. **25** [1982] 181/7.

Neagoe, S., Antonescu, L., Determination of Lead in Gasolines by Atomic Absorption Spectrophotometry, Rev. Chim. [Bucharest] **33** [1982] 850/3; C.A. **98** [1983] No. 6095.

Institute of Petroleum, The Determination of Lead in Gasoline Atomic Absorption Spectrophotometry, IP-82-008 [1982]; C.A. **98** [1983] No. 74992.

United States Environmental Protection Agency (Washington, D.C.), Lead-in-Gasoline Test Procedure, Fed. Regist. **47** No. 4 [1982] 764/8; C.A. **96** [1982] No. 54933.

Lowry, J. H., Meszaros, T. J., Conlon, L., Automated Atomic Absorption Determination of Lead in Gasoline, J. Autom. Chem. **4** [1982] 112/5.

Jin, J., Li, F., Determination of Trace Lead in Gasoline by Tantalum Ribbon Flameless Atomic Absorption Spectrometry, Fenxi Huaxue **10** [1982] 487/8, 512; C.A. **98** [1983] No. 19256.

Campbell, W. C., Applications of Atomic Absorption Spectrometry in the Petroleum Industry, in: Cantle, J. E., Techniques and Instrumentation in Analytical Chemistry, Vol. 5, Atomic Absorption Spectrometry, Elsevier, Amsterdam 1982, pp. 285/306.

Frigerio, I. J., McCormick, M. J., Symons, R. K., The Determination of Lead in Petrol by Atomic Absorption Spectrometry, Anal. Chim. Acta **143** [1982] 261/4.

Epstein, M. S., Determination of Ultratrace Levels of Lead in Reference Fuels by Graphite Furnace Atomic Absorption, At. Spectrosc. **4** [1983] 62/3; C.A. **98** [1983] No. 200930.

Chau, Y. K., Wong, P. T. S., Kramar, O., The Determination of Dialkyllead, Trialkyllead, Tetraalkyllead and Lead(II) Ions in Water by Chelation/Extraction and Gas Chromatography/Atomic Absorption Spectrometry, Anal. Chim. Acta **146** [1983] 211/7.

Torsi, G., Palmisano, F., Electrostatic Capture of Gaseous Tetraalkyllead Compounds and Their Determination by Electrothermal Atomic-Absorption Spectrometry, Analyst [London] **108** [1983] 1318/22.

Crippen, R. C., GC/LC, Instruments, Derivatives in Identifying Pollutants and Unknowns, Pergamon, New York 1983.

Rajkovic, M.B., Jovanovic, M. S., Modification of the Standard Method for Determination of Lead in Gasolines, Glas. Hem. Drus. Beograd **48** [1983] 789/90; C.A. **100** [1984] No. 212645.

Scott, D. R., Holboke, L. E., Hadeishi, T., Determination of Lead in Gasoline by Zeeman Atomic Absorption Spectrometry, Anal. Chem. **55** [1983] 2006/7.

De Jonghe, W. R. A., Van Mol, W. E., Adams, F. C., Determination of Trialkyllead Compounds in Water by Extraction and Graphite Furnace Atomic Absorption Spectrometry, Anal. Chem. **55** [1983] 1050/4.

Sneddon, J., Collection and Atomic Spectroscopic Measurement of Metal Compounds in the Atmosphere: A Review, Talanta **30** [1983] 631/48.

Braman, R. S., Chemical Speciation, Chem. Anal. (N.Y.) **64** [1983] 1/59.

Bond, A. M., Bradbury, J. R., Howell, G. N., Hudson, H. A., Hanna, P. J., Strother, S., Electrochemical Reduction of the Trimethyllead(IV) Cation in Seawater, J. Electroanal. Chem. Interfacial Electrochem. **154** [1983] 217/28.

Dmitriev, M. T., Bykhovskii, M. Ya., Brande, A., Chromatoatomic Absorption Measurement of Toxic Organometallic Compounds in Hygienic Research Studies, Gig. Sanit. **1983** No. 12, pp. 43/6.

Colombini, M. P., Fuoco, R., Papoff, P., Electrochemical Speciation and Determination of Organometallic Species in Natural Waters, Sci. Total Environ. **37** [1984] 61/70.

Harrison, R. M., Recent Advances in Air Pollution Analysis, CRC Crit. Rev. Anal. Chem. **15** [1984] 1/61.

Robinson, J. W., Boothe, E. D., Speciation of Organo Lead Compounds by T.L.C. and High Performance Liquid Chromatography-Atomic Absorption Spectroscopy. Decomposition of TEL in Sea Water, Spectrosc. Letters **17** [1984] 689/712.

Grandjean, P., Olsen, N. B., Lead, in: Vercruysse, A., Hazardous Metals in Human Toxicology, Techniques and Instrumentation in Analytical Chemistry, Vol. 4, Elsevier, Amsterdam 1984, pp. 153/69.

Scott, D. R., Application of SIMCA (Soft Independent Modeling of Class Analogy) Pattern Recognition to Air Pollutant Analytical Data, EPA/600/D-84/271, PB-85-124683-GAR [1984/85] 1/41; C.A. **102** [1985] No. 214218.

Zabairova, R. A., Bortnikov, G. N., Bochkarev, V. N., Gorina, F. A., Samarin, K. M., Gas Chromatographic Analysis of Antiknock Liquids, Fiz. Khim. Metody Anal. Gor'kiy **1984** 54/6; C.A. **104** [1986] No. 53195.

Berg, S., Jonsson, A., Analysis of Airborne Organic Lead, in: Grandjean, P., Biological Effects of Organolead Compounds, CRC, Boca Raton, Fla., 1984, pp. 33/42.

Palmer, J. M., Measuring Lead in Gasoline: A Review of Methods, Pet. Rev. **38** [1984] Nr. 449, 41/4.

Holding, S. T., Palmer, J. M., Determination of Lead in Gasoline by Atomic-absorption Spectrophotometry — Evaluation of Standard Methods, Analyst [London] **109** [1984] 507/10.

Zorin, A. D., Zanozina, V. F., Feshchenko, I. A., Tumanova, A. N., Spectrographic Determination of Impurities in Alkyl Lead Compounds, Zavodsk. Lab. **50** [1984] 44.

Bühler, A. E., Determination of Lead in Premium Gasoline, Prax. Naturwiss. Chem. **33** [1984] 165/9.

Riggin, R. M., Technical Assistance Document: The Use of Portable Volatile Organic Compound Analyzer for Leak Detection, PB-84-179993 [1984]; C.A. **101** [1984] No. 194214.

Kyuregyan, S. K., Smirnov, B. V., Determination of Lead Content in Leaded Gasolines, Metody Anal. Issled. Ispyt. Neftei Nefteproduktov Nestandart. Metod. M **1984** 7/10; C.A. **102** [1985] No. 48362.

Kyuregyan, S. K., Determination of Lead in Straight-run Gasolines for Export, Metody Anal. Issled. Ispyt. Neftei Nefteproduktov Nestandart. Metod. M **1984** 10/3; C.A. **102** [1985] No. 48363.

Jennen, A., Delafortrie, A., Verdoodt, D., Jacobs, T., Dourte, P., Determination of Trace Amounts of Organic Lead in Fish, Rev. Agric. [Brussels] **37** [1984] 1025/7.

Hewitt, C. N., Harrison, R. M., A Sensitive, Specific Method for the Determination of Tetraalkyllead Compounds in Air by Gas Chromatography/Atomic Absorption Spectrometry, Anal. Chim. Acta **167** [1985] 277/87.

Aneva, Z., Iancheva, M., Simultaneous Extraction and Determination of Traces of Lead and Arsenic in Petrol by Electrothermal Atomic Absorption Spectrometry, Anal. Chim. Acta **167** [1985] 371/4.

Kazaryan, S. A., Radchenko, E. D., Silin, A. V., Comparison of the Precision of Methods for Determining the Lead Concentration in Petroleum Products, Khim. Tekhnol. Topl. Masel **1985** No. 3, pp. 44/7; C.A. **102** [1985] No. 169303.

Coleman, M. F. M., The Determination of Lead in Gasoline by Atomic Absorption Spectrometry, J. Chem. Educ. **62** [1985] 261/2.

Petrakakis, M., Boesch, P., Haerdi, W., Simultaneous Determination of Lead and Bromine in Gasoline by Energy Dispersive X-Ray Fluorescence. Use of a New Method of Stripping the Spectra, Analusis **13** [1985] 279/85.

Uden, P. C., Gas Chromatography of Inorganic Compounds, Organometallics, and Metal Complexes, Chem. Anal. [N.Y.] **78** [1985] 229/84.

MacDonald, J. C., High-Performance Liquid Chromatography of Inorganics and Organometallics, Chem. Anal. [N.Y.] **78** [1985] 285/99.

Harrison, R. M., Radojevic, M., Hewitt, C. N., Measurements of Alkyllead Compounds in the Gas and Aerosol Phase in Urban and Rural Atmospheres, Sci. Total Environ. **44** [1985] 235/44.

Harrison, R. M., Chemical Speciation and Reaction Pathways of Metals in the Atmosphere, Advan. Environ. Sci. Technol. **17** [1986] 319/33.

Cardarelli, E., Cifani, M., Mecozzi, M., Sechi, G., Analytical Application of Emulsions. Determination of Lead in Gasoline by Atomic-Absorption Spectrophotometry, Talanta **33** [1986] 279/80.

Banerjee, S., A New Method for the Atomic-Absorption Determination of Lead Blended as Lead Alkyls in Motor Spirit, Talanta **33** [1986] 358/9.

Taylor, C. G., Trevaskis, J. M., Determination of Lead in Gasoline by a Flow-Injection Technique with Atomic Absorption Spectrometric Detection, Anal. Chim. Acta **179** [1986] 491/6.

Røyset, O., Thomassen, Y., Activated Carbon as Adsorbent for Alkyllead in Air, Anal. Chim. Acta **188** [1986] 247/55.

Donard, O. F. X., Randall, L., Rapsomanikis, S., Weber, J. H., Developments in the Speciation and Determination of Alkylmetals (Sn, Pb) Using Volatilization Techniques and Chromatography-Atomic Absorption Spectroscopy, Intern. J. Environ. Anal. Chem. **27** [1986] 55/67.

Van Cleuvenbergen, R. J. A., Chakraborti, D., Adams, F. C., Occurrence of Monoalkyllead Species During the Speciation of Organolead, Anal. Chim. Acta **182** [1986] 239/44.

Patents

Shell Internationale Research Maatschappij N.V., Determination of Lead Alkyls in Liquid Hydrocarbons and Apparatus Therefore, Neth. Appl. 65-11077 [1965/67]; C.A. **68** [1968] No. 31816.

Snyder, L. J., Henderson, S. R., Ethyl Corp., Determination of Tetrahydrocarbon Lead Impurities in Gases, U.S. 3071446 [1962/63]; C.A. **58** [1963] No. 6200.

Leisey, F. A., Standard Oil Co. (Indiana), Determination of Alkylleads in Gasoline, U.S. 3462244 [1966/69]; C.A. **71** [1969] No. 126973.

Walker, A. O., Nalco Chemical Co., Apparatus and Method for Monitoring the Presence of Volatile Organic Lead Compounds, U.S. 3870469 [1973/75]; C.A. **83** [1975] No. 136502.

Snyder, L. J., Ethyl Corp., Determination of Antiknock Compounds in Gasoline, U.S. 3912454 [1974/75]; C.A. **84** [1976] No. 62223.

Zelaskowski, C. A., Mobil Oil Corp., Trace Lead Analysis, U.S. 3934976 [1974/76]; C.A. **84** [1976] No. 167189.

Olson, D. C., Shell Oil Co., Electrochemical Detector for Lead Alkyls, U.S. 3960690 [1974/76]; C.A. **85** [1976] No. 145574.

Olson, D. C., Shell Oil Co., Electrochemical Detection for Lead Alkyls, U.S. 4012290 [1974/77]; C.A. **86** [1977] No. 174115.

Buchholz, J. C., General Motors Corp., On-Board Detection of Antiknock Compounds in Automotive Gasoline, U.S. 4203807 [1979/80]; C.A. **93** [1980] No. 152946.

Zorin, A. D., Umilin, V. A., Vanchagova, V. K., Institute of Chemistry, Academy of Sciences, U.S.S.R., A Gas-chromatographic Analysis of Thermally Unstable Organometallic Compounds, U.S.S.R. 519628 [1972/76]; C.A. **85** [1976] No. 153565.

Klopov, B. N., Cherpak, A. G., Muratova, R. D., Kuznetsova, L. N., Korneeva, A. T., Determining the Lead Content in Aviation and Automobile Gasoline, U.S.S.R. 1065349 [1982/84]; C.A. **100** [1984] No. 141998.

Radchenko, E. D., Kyuregyan, S. K., Kazaryan, S. A., Atomic Absorption Determination of Lead in Gasolines, U.S.S.R. 1109603 [1983/84]; C.A. **101** [1984] No. 232982.

Toxicology and Biocidal Uses

Collier, W. A., Zur experimentellen Therapie der Tumoren. Die Wirksamkeit einiger metall-organischer Blei- und Zinn-Verbindungen, Z. Hyg. Infektionskrankh. **110** [1929] 169/74.

Wurtz, J. G., Lead Poisoning, Hahnemannian Monthly **67** [1932] 846/54; C.A. **1933** 781.

Hand, B. M., Chronic Lead Poisoning, Hahnemannian Monthly **74** [1939] 733/48; C.A. **1939** No. 8801.

Cantarow, A., Trumper, M., Lead Poisoning, Williams & Wilkins, Baltimore 1944.

McCombie, H., Saunders, B. C., Toxic Organo-Lead Compounds, Nature **159** [1947] 491/4.

Sroka, K. H., Zur Frage der Bleientgiftung, Arch. Metallk. **3** [1949] 429/32.

Cumings, J. N., Heavy Metals and the Brain, Blackwell, Oxford 1959.

Fleming, A. J., Industrial Hygiene and Medical Control Procedures, Arch. Environ. Health **8** [1964] 266/70.

Bolanowska, W., Toxicology of Tetraethyllead, Med. Pracy **16** No. 6 [1965] 476/83.

Sanders, L. W., Lead Excretion and Health of Antiknock Blenders, Arch. Environ. Health **10** [1965] 886/92.

Cremer, J. E., Toxicology and Biochemistry of Alkyllead Compounds, Occup. Health Rev. **17** [1965] 14/9.

Fowler, D. G., Facts about Lead and Industrial Hygiene, J. Occup. Med. **7** [1965] 324/9.

Tepper, L. P., Under what Circumstances is Direct Contact with Pb Dangerous?, U.S. Public Health Serv. Publ. No. 1440 [1965] 59/62; C.A. **65** [1966] No. 11224/5.

Berman, E., The Biochemistry of Lead: Review of the Body Distribution and Methods of Lead Determination, Clin. Pediat. [Philadelphia] **5** [1966] 287/91.

Ohmori, K., Studies on Absorption and Excretion of Organic Lead Compounds. The Absorption and Excretion of Tetramethyllead and Mixed Alkyl Lead, Nippon Eiseigaku Zasshi **22** [1967] 376/82.

Ohmori, K., et al., The Affinity of Alkyl Lead Compounds on Tissues, Yokohama Igaku **18** [1967] 535/9.

Bolanowska, W., Piotrowski, J., Garczyński, H., Triethyllead in the Biological Material in Cases of Acute Tetraethyllead Poisoning, Arch. Toxikol. **22** [1967] 278/82.

Bolanowska, W., Distribution and Excretion of Triethyllead in Rats, Brit. J. Ind. Med. **25** [1968] 203/8.

Bolanowska, W., Garczyński, H., Metabolism of Tetraethyllead in Rabbits, Med. Pracy **19** No. 3 [1968] 235/43.

Schwanecke, R., Unfall- und Gesundheitsgefahren durch Antiklopfmittel auf der Basis bleiorganischer Verbindungen, Zentralbl. Arbeitsmed. Arbeitsschutz **18** [1968] 69/78.

Barnes, J. M., Magos, L., The Toxicology of Organometallic Compounds, Organometal. Chem. Rev. A **3** [1968] 137/150.

Browning, E., Toxicity of Industrial Metals, 2nd Ed., Butterworths, London 1969.

Ohmori, K., Absorption and Excretion of Organic Lead Compounds. Distribution of Lead in Blood Constituents, Yokohama Igaku **20** [1969] 210/3.

Sijpesteijn, A. K., Luijten, J. G. A., Van der Kerk, G. J. M., Organometallic Fungicides, in: Torgeson, D. C., Fungicides, An Advanced Treatment, Vol. 2, Academic, New York 1969, pp. 331/66.

Kochkin, D. A., Fungicidal and Antibacterial Properties of Organotin and Organolead Monomers and Polymers, Vopr. Vodn. Toksikol. **1970** 121/4.

Thayer, J. S., Some Biological Aspects of Organometallic Compounds, J. Chem. Educ. **48** [1971] 806/8.

Casida, J. E., Kimmel, E. C., Holm, B., Widmark, G., Oxidative Dealkylation of Tetra-, Tri-, and Dialkyltins and Tetra- and Trialkylleads by Liver Microsomes, Acta Chem. Scand. **25** [1971] 1497/9.

Ahlberg, J., Ramel, C. Wachtmeister, C. A., Organolead Compounds Shown to be Genetically Active, Ambio **1** [1972] 29/31.

Hall, S. K., Lead Pollution and Poisoning, Environ. Sci. Technol. **6** [1972] 30/5.

Blokker, P. C., A Literature Survey on some Health Aspects of Lead Emissions from Gasoline Engines, Atmos. Environ. **6** [1972] 1/18.

Stöfen, D., Less Noted European Papers on Lead, Proc. Intern. Symp. Environ. Health Aspects Lead, Luxembourg 1972 [1973], pp. 441/8; C.A. **80** [1974] No. 500.

Mizoi, Y., Tatsuno, Y., Hishida, S., Morigaki, T., Nakanishi, K., Tetraalkyllead Poisoning, An Autopsy Report on the Cases at the "Boston Maru", Nippon Hoigaku Zasshi **27** [1973] 371/86.

Akatsuka, K., Tetraalkyl Lead Poisoning, Sangyo Igaku **15** [1973] 3/66.

Przybylowski, J., Chronic Toxic Action of Leaded Gasoline 78 Vapors. 1. Changes in the Liver and Central Nervous System, Biul. Sluzby Sanit. Epidemiol. Wojewodztwa Katowickiego **17** [1973] 27/46; C.A. **83** [1975] No. 127024.

Przybylowski, J., Zajusz, K., Chronic Toxic Action of Leaded Gasoline 78 Vapors. 2. Changes in the Lungs and Heart, Biul. Sluzby Sanit. Epidemiol. Wojewodztwa Katowickiego **17** [1973] 47/50; C.A. **81** [1974] No. 146499.

Barth, D., Berlin, A., Engel, R., Recht, P., Smeets, J., Proc. Intern. Symp. Environ. Health Aspects Lead, Luxembourg 1972 [1973].

Chow, T. J., Our Daily Lead, Chem. Brit. **9** [1973] 258/63.

Goyer, R. A., Rhyne, B. C., Pathological Effects of Lead, Intern. Rev. Exp. Pathol. **12** [1973] 1/77.

Anon., Task Group on Metal Accumulation, Accumulation of Toxic Metals with Special Reference to Their Absorption, Excretion and Biological Half-Times, Environ. Physiol. Biochem. 3 [1973] 65/107.

Przybylowski, J., Pierzchala, W., Panasiewicz, M., Chronic Toxic Actions of Leaded Gasoline 78 Vapors. 3. Blood and Blood Platelet Morphology, Leukocyte Alkaline Phosphatase Activity, and Serum Electrolyte Levels, Biul. Sluzby Sanit. Epidemiol. Wojewodztwa Katowickiego 17 [1973] 61/6; C.A. 81 [1974] No. 146500.

Przybylowski, J., Matuszewski, W., Kaminski, K., Chronic Toxic Action of Leaded Gasoline 78 Vapors. 4. Monoamine Oxidase (MAO) and Lactic Dehydrogenase (LDH) Activity in Blood and Tissues, Biul. Sluzby Sanit. Epidemiol. Wojewodztwa Katowickiego 17 [1973] 67/70; C.A. 81 [1974] No. 146501.

Przybylowski, J., Pathomechanisms of the Prolonged Poisoning with Gasoline and Antiknock Compounds, Pol. Tyg. Lek. 29 [1974] 1139/40.

Robinson, T. R., Delta-Aminolevulinic Acid and Lead in Urine of Lead Antiknock Workers, Arch. Environ. Health 28 [1974] 133/8.

Waldron, H. A., Stöfen, D., Sub-Clinical Lead Poisoning, Academic, London 1974.

Thayer, J. S., Organometallic Compounds and Living Organisms, J. Organometal. Chem. 76 [1974] 265/95.

Fishbein, L., Mutagens and Potential Mutagens in the Biosphere. Metals. Mercury, Lead, Cadmium, and Tin, Sci. Total Environ. 2 [1974] 341/71.

Kawai, T., Nagano, Y., Aoyagi, K., Koyano, M., Chemical Substances and Toxicological Tests, Rodo Eisei 15 [1974] 6/25; C.A. 83 [1975] No. 72802.

Linch, A. L., Biological Monitoring for Industrial Exposure to Tetraalkyl Lead, Am. Ind. Hyg. Assoc. J. 36 [1975] 214/9.

Cole, J. F., Safety in Handling Lead Compounds, in: Greninger, D., Kollonitsch, V., Kline, C. H., Willemsens, L. C., Cole, J. F., Lead Chemicals, International Lead Zinc Research Organization, Inc., New York 1975, pp. 315/22.

Moore, P. J., Pridmore, S. A., Gill, G. F., Total Blood Lead Levels in Petrol Vendors, Med. J. Aust. 1 [1976] 438/40.

Rotunno, R., Tarantino, M., Bonsignore, D., Environmental Lead Pollution. An Epidemiologic Study in Road Gasoline Station Workers, Lav. Um. 28 [1976] 65/72.

Urushibara, S., Experimental Studies on the Effect of Alkyl Lead in Leaded Gasoline on Cerebral Serotonin Metabolism, Tokyo Jikeikai Ika Daigaku Zasshi 91 [1976] 189/94.

Kehoe, R. A., Pharmacology and Toxicology of Heavy Metals: Lead, Pharmacol. Ther. A 1 [1976] 161/88.

Oxley, G. R., Protective Clothing in the Context of Health Protection against Toxic Chemicals, Ann. Occup. Hyg. 19 [1976] 163/7.

Przybylowski, J., Kowalski, W., Podolecki, A., Effect of Chronic Experimental Poisoning with Leaded and Straight Gasoline Vapors on the Circulatory System, Patol. Pol. 27 [1976] 149/56; C.A. 85 [1976] No. 88185.

Nordberg, G. F., Effects and Dose-Response Relationships of Toxic Metals, Elsevier, Amsterdam 1976.

32

Goyer, R. A., Mushak, P., Lead Toxicity Laboratory Aspects, Advan. Mod. Toxicol. **2** [1977] 41/77.

Tsuchiya, K., Toxicology of Metals, Lead, EPA-600-1-77-022 [1977]; Toxicol. Met. **2** [1977] 242/300; C.A. **93** [1980] No. 62625.

Grandjean, P., Nielsen, T., Organic Lead Compounds — an Overlooked Pollution Danger?, Nat. Verden No. 10 [1977] 363.

Poklis, A., Burkett, C. D., Gasoline Sniffing: A Review, Clin. Toxicol. **11** [1977] 35/41.

Grandjean, P., Nielsen, T., Organic Lead Compounds — Pollution and Toxicology, SNV PM 879, Swedish Environmental Protection Agency, Stockholm 1977.

Maddock, B. G., Taylor, D., The Acute Toxicity and Bioaccumulation of Some Lead Alkyl Compounds in Marine Animals, in: Branica, M., Konrad, Z., Lead Marine Environ. Proc. Intern. Experts Discuss., Rovinj, Yugosl., 1977 [1980], pp. 233/61.

Kochkin, D. A., Relationship of Biological Activity of Organotin (or -lead) Compounds from Chemical Structures (Including cis and trans-Isomerism), Compt. Rend. 4th Congr. Intern. Corros. Marine Salissures, Antibes/Juan-les-Pins 1976 [1977], pp. 281/4.

van der Kerk, G. J. M., Metallo-Organic Fungicides, ACS Symp. Ser. No. 37 [1977] 123/52.

de Silva, P. E., Donnan, M. B., Petrol Vendors, Capillary Blood Lead Levels and Contamination, Med. J. Australia **1** [1977] 344/7.

U. N. Environment Programme and World Health Organization, Environmental Health Criteria 3, Lead, World Health Organization, Geneva 1977.

Huber, F., Schmidt, U., Kirchmann, H., Aqueous Chemistry of Organolead and Organothallium Compounds in the Presence of Microorganisms, ACS Symp. Ser. No. 82 [1978] 65/81.

Posner, H. S., Damstra, T., Nriagu, J. O., Human Health Effects of Lead, in: Nriagu, J. O., The Biogeochemistry of Lead in the Environment, Pt. B, Elsevier Biomed./North-Holland, Amsterdam 1978, pp. 173/223.

Nielsen, T., Jensen, K. A., Grandjean, P., Organic Lead in Normal Human Brains, Nature **274** [1978] 602/3.

Malizia, E., Stacchini, A., Costantini, S., Baldini, M., Giordano, R., Organic Lead Levels in Fishes, Mussels, Water, and Sediments in the Area of Wreckage of M/B "Cavtat" Near Otranto, Riv. Tossicol. Sper. Clin. **8** [1978] 431/43.

Weiss, B., The Behavioral Toxicology of Metals, Fed. Proc. Fed. Am. Soc. Exptl. Biol. **37** [1978] 22/7.

Hammond, P. B., Metabolism and Metabolic Action of Lead and Other Heavy Metals, in: Oehme, F. W., Toxicity of Heavy Metals in the Environment, Pt. 1, Dekker, New York 1978, pp. 87/99.

Green, V. A., Wise, G. W., Callenbach, J. C., Lead Poisoning, in: Oehme, F. W., Toxicity of Heavy Metals in the Environment, Pt. 1, Dekker, New York 1978, pp. 123/41.

Aldridge, W. N., The Biological Properties of Organo-Germanium, -Tin, and -Lead Compounds, Organomet. Coord. Chem. Germanium Tin Lead Pleanary Lect. 2nd Intern. Conf., Nottingham 1977 [1978], pp. 9/30; C.A. **91** [1979] No. 84293.

Coodin, F. J., Boeckx, R., Lead Poisoning from Sniffing Gasoline, New Engl. J. Med. **298** [1978] 347.

Seshia, S. S., Rajani, K. R., Boeckx, R. L., Chow, P. N., The Neurological Manifestations of Chronic Leaded Gasoline Sniffing, Develop. Med. Child Neurol. **20** [1978] 323/34.

Valpey, R., Sumi, S. M., Copass, M. K., Goble, G. J., Acute and Chronic Progressive Encephalopathy Due to Gasoline Sniffing, Neurology **28** [1978] 507/10.

Botré, C., De Zorzi, C., Malizia, E., Melchiorri, P., Stacchini, E., Tiravanti, G., Study and Evaluation of Organic Lead Levels in Fishes and Phytoplankton Near Otranto, Arch. Toxicol. Suppl. No. 1 [1978] 157.

Swuste, P., Lood daar klopt iets niet: Toxicologische aspekten van loodverbindingen, Leiden Univ., Leiden, Neth., 1978.

Tsuchiya, K., Lead, in: Friberg, L., Nordberg, G. F., Vouk, V. B., Handbook on the Toxicology of Metals, Elsevier Biomed./North-Holland, Amsterdam 1979, pp. 451/84.

Clarkson, T. W., Effects — General Principles Underlying the Toxic Action of Metals, in: Friberg, L., Nordberg, G. F., Vouk, V. B., Handbook on the Toxicology of Metals, Elsevier Biomed./North-Holland, Amsterdam 1979, pp. 99/117.

Kazantzis, G., Lilly, L. J., Mutagenic and Carcinogenic Effects of Metals, in: Friberg, L., Nordberg, G. F., Vouk, V. B., Handbook on the Toxicology of Metals, Elsevier Biomed./ North-Holland, Amsterdam 1979, pp. 237/72.

Cope, R. F., Pancamo, B. P., Rinehart, W. E., ter Haar, G. L., Personnel Monitoring for Tetraalkyl Lead in the Workplace, Am. Ind. Hyg. Assoc. J. **40** [1979] 372/9.

Grandjean, P., Nielsen, T., Organolead Compounds: Environmental Health Aspects, Residue Rev. **72** [1979] 97/148.

Prockop, L., Neurotoxic Volatile Substances, Neurology **29** [1979] 862/5.

Hunter, A. G. W., Thompson, D., Evans, J. A., Is There a Fetal Gasoline Syndrome?, Teratology **20** [1979] 75/9.

Moeschlin, S., Organische Bleiverbindungen, in: Moeschlin, S., Klinik und Therapie der Vergiftungen, Thieme, Stuttgart 1980, pp. 97/9.

Sax, N. I., Dangerous Properties of Industrial Materials, 5th Ed., Van Nostrand Reinhold, New York 1979.

Nosov, V. N., Kolosova, L. V., Features of a Toxicological Curve as a Function of the Chemical Structure of Hetero-Organic Compounds, Biol. Nauki [Moscow] **1979** 79/101; Biol. Abstr. **69** [1980] No. 27133.

Kolosova, L. V., Nosov, V. N., Dobrovol'skii, I. P., Structure of Hetero-Organic Compounds and Their Toxic Effect on Daphnia, Samoochishchenie Bioindik. Zagryaz. Vod Tr. 3rd Vses. Soveshch. Sanit. Gidrobiol., Moscow 1977 [1980], pp. 184/93; C.A. **93** [1980] No. 180461; C.A. **94** [1981] No. 786.

Lauwerys, R., Buchet, J. P., Roels, H., Biological Methods for Surveillance of Workers Exposed to Various Industrial Toxic Substances, Cah. Med. Trav. **17** [1980] 91/7.

Singhal, R. L., Thomas, J. A., Lead Toxicity, Urban & Schwarzenberg, Baltimore-Munich 1980.

Blumer, W., Reich, T., Leaded Gasoline — A Cause of Cancer, Environ. Intern. **3** [1980] 465/71.

Nieboer, E., Richardson, D. H. S., The Replacement of the Nondescript Term 'Heavy Metals' by a Biologically and Chemically Significant Classification of Metal Ions, Environ. Pollut. B **1** [1980] 3/26.

Boudene, C., Recent Data on the Toxicity of Lead, Pollut. Atmos. No. 85 [1980] 62/70.

Gerber, G. B., Leonard, A., Jacquet, P., Toxicity, Mutagenicity and Teratogenicity of Lead, Mutation Res. **76** [1980] 115/41.

Goyer, R. A., Lead, Disorders of Mineral Metabolism, Vol. 1 of Trace Minerals, Bronner, F., Coburn, J., Academic, New York 1981, pp. 159/99.

Röderer, G., Fate and Toxicity of Tetraalkyl Lead and Its Derivatives in Aquatic Environments, Heavy Met. Environ. 3rd Intern. Conf., Amsterdam 1981, pp. 250/3.

Kochkin, D. A., Dependence of Antifouling and Physicochemical Properties of Organotin (-lead) Monomers and Polymers on Chemical Structure (Including cis- and trans-Isomerism), Obrastanie Biokorroz. Vodn. Srede **1981** 164/9; C.A. **96** [1982] No. 99242.

Hertz, M. M., Bolwig, T. G., Grandjean, P., Westergaard, E., Lead Poisoning and the Blood-Brain Barrier, Acta Neurol. Scand. **63** [1981] 286/96.

Harman, A. W., Frewin, D. B., Priestly, B. G., Induction of Microsomal Drug Metabolism in Man and in the Rat by Exposure to Petroleum, Brit. J. Ind. Med. **38** [1981] 91/7.

Grandjean, P., Andersen, O., Toxicity of Lead Additives, Lancet **1982** II 333/4.

Demayo, A., Taylor, M. C., Taylor, K. W., Hodson, P. V., Toxic Effects of Lead and Lead Compounds on Human Health, Aquatic Life, Wildlife Plants, and Lifestock, CRC Crit. Rev. Environ. Control **12** [1982] 257/305.

Bulavko, G. I., Effect of Various Lead Compounds on Soil Microflora, Izv. Sibirsk. Otd. Akad. Nauk SSSR Ser. Biol. Nauk **1982** No. 1, pp. 79/86; C.A. **97** [1982] No. 50872.

Johnson, M. S., Pluck, H., Hutton, M., Moore, G., Accumulation and Renal Effects of Lead in Urban Populations of Feral Pigeons, Columba livia, Arch. Environ. Contam. Toxicol. **11** [1982] 761/7.

Anonymous, Lead in Petrol and Elsewhere, Lancet **1982** I 1337/8.

Russell Jones, R., Lead in Petrol and Elsewhere, Lancet **1982** I 1464.

Barry, P. S. I., Lead in Petrol, Lancet **1982** II 94.

Winneke, G., Neurobehavioural and Neuropsychological Effects of Lead, Lancet **1982** II 550.

Anonymous, Lead in Petrol, Brit. Med. J. **1982** 529.

Anonymous, Lead in Petrol: Again, Brit. Med. J. **1982** 1506.

Grandjean, P., Health Significance of Organolead Compounds, in: Rutter, M., Russell Jones, R., Lead Versus Health, Wiley, Chichester 1983, pp. 179/89.

Corradetti, E., Mannozzi, A., Marinelli, G., Blood Lead and Urine Lead in the Evaluation of Lead Exposure in People Working at Gasoline Pumps, Boll. Chim. Unione Ital. Lab. Prov. Parte Sci. **34** [1983] 259/67.

Skorka, G., Bien, E., Neurotoxicological Effects of Lead and Mercury, Pharmazie **38** [1983] 709/16.

Kehoe, R., Lead Alkyl Compounds, in: Encyclopaedia of Occupational Health and Safety, 3rd Ed., International Labour Office, Geneva 1983, pp. 1197/99.

Winchester, R. V., A Review of Lead Hazards in the Motor Service and Repair Industry, Chem. New Zealand **47** [1983] 28/30.

Rabinowitz, M., Needleman, H. L., Petrol Sales and Umbilical Cord Blood Lead Levels in Boston, Massachusetts, Lancet **1983** I 63.

Rutter, M., Russel Jones, R., Lead Versus Health, Sources and Effects of Low Level Lead Exposure, Wiley, Chichester 1983.

Dmitriev, M. T., Bykhovskii, M. Ya., Brande, A., Chromatoatomic Absorption Measurement of Toxic Organometallic Compounds in Hygienic Research Studies, Gig. Sanit. **1983** No. 12, pp. 43/6.

Lehnert, G., Szadkowski, D., Die Bleibelastung des Menschen, Verlag Chemie, Weinheim 1983.

Kolosova, L. V., Nosov, V. N., Cumulative Toxicity and its Relationship with Adaptive Reactions, Reakts. Gidrobiontov Zagryaz. **1983** 128/34.

Sandhu, G. K., Sandhu, G. K., Toxicology of Organometallic Compounds of Silicon, Germanium, Tin and Lead, J. Chem. Sci. **9** [1983] 36/50.

Hodson, P. V., Whittle, D. M., Wong, P. T. S., Borgman, U., Thomas, R. L., Chau, Y. K., Nriagu, J. O., Hallett, D. J., Lead Contamination of the Great Lakes and Its Potential Effects on Aquatic Biota, Advan. Environ. Sci. Technol. **14** [1984] 335/69.

Thayer, J. S., Organometallic Compounds in Living Organisms, Academic, New York 1984.

Doeltz, M. K., Mackie, M., Rich, P. A., Lent, D., Sigman, C. C., Helmes, C. T., A Study of Organometallic Compounds for the Selection of Candidates for Carcinogen Bioassay, J. Environ. Sci. Health A **19** [1984] 27/65.

Zimmermann, H.-P., Röderer, G., Doenges, K. H., Influence of Triethyl Lead on the In Vitro and In Vivo Assembly and Disassembly of Microtubules from Mammalian Cells, J. Submicrosc. Cytol. **16** [1984] 203/5.

Caplun, E., Petit, D., Picciotto, E., Lead in Petrol, Endeavour **8** [1984] 135/44.

Rabinowitz, M., Needleman, H., Burley, M., Finch, H., Rees, J., Lead in Umbilical Blood, Indoor Air, Tap Water, and Gasoline in Boston, Arch. Environ. Health **39** [1984] 299/301.

Bress, W. C., In Vitro and In Vivo Percutaneous Absorption of Organoleads, Lead Salts, and Inorganic Lead in Guinea Pig and Human Autopsy Skin, Diss. St. John's Univ., Jamaica, N.Y., 1984; C.A. **101** [1984] No. 105414.

Swartzwelder, H. S., Altered Responsiveness to Alcohol after Exposure to Organic Lead, Alcohol **1** [1984] 181/3.

Kuna, R. A., Ulrich, C. E., Subchronic Inhalation Toxicity of Two Motor Fuels, J. Am. Coll. Toxicol. **3** [1984] 217/29.

Ewers, U., Merian, E., Schutzvorschriften und -richtlinien betreffend Metalle und Metallverbindungen, in: Merian, E., Metalle in der Umwelt, Verlag Chemie, Weinheim 1984, pp. 283/9.

Walsh, T. J., Tilson, H. A., Neurobehavioral Toxicology of the Organoleads, Neurotoxicology **5** [1984] 67/86.

Grandjean, P., Biological Effects of Organolead Compounds, CRC, Boca Raton, Fla., 1984.

Röderer, G., Doenges, K. H., Zimmermann, H.-P., Ultrastructural Aspects of the Neurotoxic Action of Alkyl Lead Compounds, Naunyn-Schmiedeberg's Arch. Pharmacol., Abstr. 102, R 26 [1984].

Röderer, G., Different Physicochemical and Toxic Properties of Inorganic and Organic (Alkylated) Compounds of Lead, Abstr. Intern. Symp. Bioavailability Environ. Chem., Schmallenberg-Grafschaft 1984, p. 9.

Haeffner, E. W., Zimmermann, H.-P., Hoffmann, C. J. K., Influence of Triethyllead on the Activity of Enzymes of the Ascites Tumor Cell Plasma Membrane and Its Microviscosity, Toxicol. Letters **23** [1984] 183/8.

Konat, G., Triethyllead and Cerebral Development: An Overview, Neurotoxicology **5** [1984] 87/96.

Reichlmayr-Lais, A. M., Kirchgessner, M., Lead, Biochem. Elem. **3** [1984] 367/87; C.A. **102** [1985] No. 126708.

Jendryczko, A., Effect of Garage Storage of Gasoline on Blood Lead Levels in Drivers, Bromatol. Chem. Toksykol. **17** [1984] 261/3; C.A. **102** [1985] No. 137093.

Winder, C., The Developmental Neurotoxicology of Lead, MTP, Lancaster, UK, pp. 1/150.

Woolley, D. E., Cranmer, J. M., Neurotoxicology of Lead: Reviews and Recent Advances, Neurotoxicology **5** [1984] 1/361.

Tera, O., Schwartzman, D. W., Watkins, T. R., Identification of Gasoline Lead in Children's Blood Using Isotopic Analysis, Arch. Environ. Health **40** [1985] 120/3.

Joumard, R., Chiron, M., Responsibility of the Automobile for the Level of Lead in Organisms, Med. Hyg. No. 1632 [1985] 3480/2.

Silbergeld, E. K., Neurotoxicology of Lead, Drug Chem. Toxicol. **1984**; Neurotoxicology **3** [1985] 299/322.

Röderer, G., Leaded Gasoline-Problem: Mechanism of Action of Triethyllead, Biol. Unserer Zeit **15** [1985] 129/33.

Atassi, G., Antitumor and Toxic Effects of Silicon, Germanium, Tin and Lead Compounds, Rev. Silicon Germanium Tin Lead Compounds **8** [1985] 219/35.

Osborn, D., Young, W., Bird Mortalities on the River Mersey: Laboratory Studies of the Toxicity and Effects on Essential Metals of Triethyl and Trimethyl Lead, Trace Elem. Man Anim. TEMA 5 Proc. 5th Intern. Symp., Aberdeen, U.K., 1984 [1985], pp. 870/1; C.A. **105** [1986] No. 185363.

Osborn, D., Mass Bird Mortalities on the Mersey Estuary UK — Incident, Investigation, Resolution and Follow-up, NATO Conf. Ser. I **11** [1985] 59/93; C.A. **105** [1986] No. 84725.

Wilson, K. W., Head, P. C., Jones, P. D., Mersey Estuary (U.K.) Bird Mortalities — Causes, Consequences and Correctives, Water Sci. Technol. **18** [1986] 171/80; C.A. **105** [1986] No. 55901.

Zarling, D., Occupational Exposure to Lead and Its Effect on the Oral System, Z. Gesamte Hyg. Grenzgeb. **32** [1986] 558/6.

Röderer, G., Toxicity and Mechanism of Action of Triethyllead, Naturwiss. Rundschau **39** [1986] 213/5.

Ibels, L. S., Pollock, C. A., Lead Intoxication, Med. Toxicol. **1** [1986] 387/410.

Uses

Charch, W. H., Mack, E., Jr., Boord, C. E., Antiknock Materials, Ind. Eng. Chem. **18** [1926] 334/40.

Grote, G., Antiklopfmittel, Petroleum Z. **22** [1926] 1344/7; C.A. **1928** 1034.

Schmidt, H., Arzneimittelsynthetische Studien über Blei, Med. Chem. Abhandl. Med. Chem. Forschungsstätten I.G. Farbenind. **3** [1936] 418/28.

Steiger, B., Über metallorganische Verbindungen — besonders Bleitetraäthyl — in der Mineralölindustrie, Petroleum Z. **33** No. 32 [1937] 1/7.

Wormser, F. E., Lead, a Basic Natural Resource. Lead Pigments, Science Counselor **7** [1941] 39/40, 58/9; C.A. **1941** 5331.

Beatty, H. A., Lovell, W. G., Antiknock Agents, Ind. Eng. Chem. **41** [1949] 886/8.

Boyd, T. A., Pathfinding in Fuels and Engines, SAE Quart. Trans. **4** [1950] 182/95.

Felt, A. E., Kerley, R. V., Sumner, H. C., Fuel Additives and Engine Durability, Prepr. Soc. Automot. Eng. Natl. West Coast Meet., Los Angeles 1954.

Smith, H. V., Stabilizers for Vinyl Polymers, Brit. Plast. **27** [1954] 213/7.

Harwood, J. H., Industrial Applications of the Metal Organic Compounds, Ind. Chemist **35** [1959] 348/50, 375/8, 433/7, 483/5, 580/2, **36** [1960] 30/4, 74/6, 114/8, 176/80, 223/8, 285/90, 325/30, 391/6, 437/42; C.A. **1961** 1967.

Little, A. D., Applications of Organic Lead Compounds 1960, Summary Report to Lead Industries Association 1961.

Schildwächter, H., Additive in Mineralölprodukten, Erdöl Kohle **13** [1960] 901/5.

Hesselberg, H. E., Howard, J. R., Antiknock Behavior of Alkyl Lead Compounds, SAE Trans. **69** [1961] 5/16.

Sturgis, B. M., Additives in Petroleum Fuels — A Decade of Progress, Prepr. Am. Chem. Soc. Div. Petrol. Chem. **6** [1961] A-51/A-66; C.A. **58** [1963] 12348.

Richardson, W. L., Barusch, M. R., Kautsky, G. J., Steinke, R. E., An Improved Gasoline Antidetonant, Ind. Eng. Chem. **53** [1961] 305.

Stormont, D. H., Which — TEL, MEL, TELMEL, or MLA — for Maximum Road Octane? Oil Gas J. **60** No. 13 [1962] 189, 192/3, 195.

Morris, W. E., Antiknock Performance of Lead Alkyl Mixtures in Today's Gasolines, SAE Prepr. 547C [1962] 1/8; C.A. **60** [1964] 11819.

Farnsworth, I. R., Boddy, J., Chemical Additives for Gasolines, Petroleum [London] **25** [1962] 156/9, 161.

Harwood, J. H., Lead, in: Harwood, J. H., Industrial Application of the Organometallic Compounds, Chapman & Hall, London 1963, pp. 124/39.

38

Goodacre, G. L., Foord, D., Blei-alkyle als Antiklopfmittel für Autobenzin, Acta Chim. Acad. Sci. Hung. **36** [1963] 235/53.

Radtke, S. F., Fortschritte der internationalen Bleiforschung, Metall **16** [1962] 758/64.

Gelius, R., Neuere Entwicklungen auf dem Gebiet metallorganischer, insbesondere blei-haltiger Antiklopfmittel, Freiberger Forschungsh. A No. 264 [1963] 19/32.

Richardson, W. L., Ryason, P. R., Kautsky, G. J., Barusch, M. R., 9th Symp. Intern. Combust., Ithaca, N. Y., 1962 [1963], pp. 1023/33.

Anonymous, Organo-Lead Job Hunt, Chem. Week **1964** 95, 97.

Normant, H., Aspects industriels de la chimie de quelques composés organo-métalliques, Bull. Assoc. Franc. Techn. Petrole No. 168 [1964] 673/89.

Ziegfeld, R. L., Importance and Uses of Lead, Arch. Environ. Health **8** [1964] 202/12.

Marshall, E. F., Wirth, R. A., Uses for Organometallics in Fuels and Lubricants, Ann. N.Y. Acad. Sci. **125** [1965] 198/217.

Podall, H. E., Mitchell Jr., M. M., The Use of Organometallic Compounds in Chemical Vapor Deposition, Ann. N.Y. Acad. Sci. **125** [1965] 218/28.

Barusch, M. R., Macpherson, J. H., Engine Fuel Additives, in: McKetta Jr., J. J., Advances in Petroleum Chemistry and Refining, Vol. 10, Interscience, New York 1965, pp. 457/546.

Korn, O., Woggon, H., Nachweis von Organo-Metallstabilisatoren in PVC-Materialien, Ernäh-rungsforschung **10** [1965] 57/65.

Anonymous, Organolead Compounds. Potential Raw Material for Surface Coatings, Paint Manuf. **35** [1965] 36/8.

Anonymous, Organobleiverbindungen auf neuen Wegen, Chem. Ind. [Düsseldorf] **17** [1965] 272.

van der Kerk, G. J. M., New Developments in Organolead Chemistry, Ind. Eng. Chem. **58** No. 10 [1966] 29/35.

Ruiz, J. L., New Applications of Lead. Organic Compounds, Rev. Met. [Madrid] **2** [1966] 370/2.

Braun, D., Sung Bong Chang, Thallmaier, M., Über die Wirksamkeit von bleiorganischen Verbindungen als Hitzestabilisatoren für Polyvinylchlorid, Gummi Asbest Kunstst. **19** [1966] 1353/6.

Klema, F., Organobleiverbindungen, Chemiker-Ztg. **90** [1966] 106/7.

Harwood, J. H., Developments in Organometallics, Chem. Process. Eng. [London] **47** [1966] 65/9.

Giesen, M., Metallorganische Verbindungen als Biocide für Anstrichstoffe, Congr. FATIPEC **8** [1966] 185/96.

Rabek, J. F., The Newest Achievements in the Research and Synthesis of Heatstable Polymers, Przemysl Chem. **46** [1967] 130/4.

Carr, D. S., Some Market Potentials for Organolead Compounds, Chem. Ind. [London] **1967** 1854/7.

Clark, M. C., Practical Evaluation of Road Performance of European Gasolines Containing Different Lead Alkyls, Rev. Ass. Fr. Tech. Petrole No. 183 [1967] 67/76.

Carr, D. S., Organolead Compounds, Paint Varn. Prod. **58** (2) [1968] 23/8.

Beatty, H. A., Organolead Compounds as Lubricant Additives, Chem. Ind. [London] **1968** 733/6.

Anonymous, Organolead Compounds: Big Market Tomorrow?, Metals Week **39** No. 43 [1968] 10/1.

Walden, C. C., Allen, I. V. F., Organolead Compounds as Wood Preservatives, Australasian Eng. **1969** 21/7.

Henry, M. C., Pant, B. C., Organolead Chemistry: Syntheses and Applications, U.S. Army Natick Laboratories, Natick, Mass., Techn. Rept. 69-77-CE [1969].

Klimsch, P., Kühnert, P., Synergetische Effekte bei der Stabilisierung von PVC mit Organozinnverbindungen, Plaste Kautschuk **16** [1969] 242/51.

Thinius, K., Stabilisierung und Stabilisatoren von Plastwerkstoffen, Verlag Chemie, Weinheim 1969.

Frey, F. W., Shapiro, H., Commercial Organolead Compounds, Fortschr. Chem. Forsch. **16** [1971] 243/97.

Kochkin, D. A., Churakov, V. P., Voronkov, N. A., Petri, V. N., Azerbaev, I. N., Antifungus Protection of Wood with Organotin and Organolead Compounds and Evaluation of Their Effectiveness, Khim. Atsetilena Tr. 3rd Vses. Konf., Dushanbe, USSR, 1968 [1972], pp. 366/70.

Razuvaev, G. A., Gribov, B. G., Domrachev, G. A., Salamantin, B. A., Organometallic Compounds in Electronics, Nauka, Moscow 1972.

Gurevich, E. S., Modern Chemical Marine Fouling-Prevention Methods, Probl. Biol. Povrezhdenii Obrastanii Mater. Izdelii Sooruzhenii **1972** 217/25.

Churakov, B. P., Petri, V. N., Wood Preservatives, Tr. Ural. Lesotekh. Inst. No. 26 [1972] 123/36; C.A. **79** [1973] No. 55073.

Phillip, A. T., Marine Science Aids the Development of Antifouling Coatings, Austral. OCCA Proc. News **1973** 17/22.

Porter, F. D., Lead in Petrol — What Price Its Removal, Chem. Brit. **10** [1974] 61/2.

Austin, G. T., Industrially Significant Organic Chemicals, Pt. 8, Chem. Eng. [New York] **81** No. 15 [1974] 107/16.

Skinner, C. E., Organolead Compounds as Fungicides and Algicides in Paints, Mod. Ytbehandling **10** [1975] 25/7.

Prilutskaya, N. V., Organometallic Polymers for Varnishes: Synopsis of Lectures, Khark. Politekh. Inst., Kharkov, USSR, **1975** 1/28; C.A. **85** [1976] No. 34717.

Allen, D. M., The Photochemistry of Organometallics with Applications to Imaging Systems, J. Photogr. Sci. **24** [1976] 61/7.

Dick, R. J., Nowacki, L. J., Sherrard, J. R., New Marine Coatings Technology Applied to the Protection of Buoys, C. R. 4th Congr. Intern. Corros. Marine Salissures, Antibes and Juan-les-Pins, Fr., 1976 [1977], pp. 145/54.

Shustova, O. A., Gladyshev, G. P., The Mechanisms of the Stabilisation of Thermostable Polymers, Usp. Khim. **45** [1976] 1695/724; Russ. Chem. Rev. **45** [1976] 865/82.

Carraher, C. E., Jr., Organometallic Condensation Polymers, Organometal. Polymer. Symp., New Orleans 1977 [1978], pp. 79/85.

Carraher, C. E., Jr., Reese, C. D., Lead(IV) Polyesters from a New Solution System, Organometal. Polymer. Symp., New Orleans 1977 [1978], pp. 101/6.

Borden, D. G., Photocrosslinkable Organometallic Polyesters, Organometal. Polymer. Symp., New Orleans 1977 [1978], pp. 115/27.

Facchetti, S., Isotope Study of Lead in Petrol, Manage. Control Heavy Met. Environ. Intern. Conf., London 1979, pp. 95/102.

Tsutsui, T., Marine Antifouling Agents and Paints, Toso to Toryo No. 309 [1979] 47/50; C.A. **91** [1979] No. 159067.

Christensen, G., Pedersen, C. M., Unexpected Chemical Reactions in Paints, Skand. Tidskr. Faerg. Lack **26** [1980] 5, 7, 9, 11.

Negishi, E., Organometallics in Organic Synthesis, Vol. 1, General Discussions and Organometallics of Main Group Metals in Organic Synthesis, Wiley, New York 1980, pp. 1/532.

Michalski, G. W., Unzelman, G. H., Effective Use of Antiknocks During the 1980s, Proc. Refin. Dept. Am. Pet. Inst. **58** [1979] 185/203.

Shuptrine, G. R., Lead in Petrol. Automotive Energy and Crude Oil Resources, Papers Congr. Australian New Zealand Assoc. Advan. Sci., **50** [1980] Paper No. 129.

Emelyanov, V. E., Levinson, G. I., Grebenshchikov, V. P., Golosova, V. F., Use of Alkyllead Antiknock Additives in the Production of Automobile Gasolines, Prisadki Toplivam M **1980** 92/6; C.A. **95** [1981] No. 153306.

Domrachev, G. A., Suvorova, O. N., The Formation of Inorganic Coatings in the Decomposition of Organometallic Compounds, Usp. Khim. **49** [1980] 1671/86; Russ. Chem. Rev. **49** [1980] 810/9.

Ryabova, L. A., Thin Films from Organometallic Compounds, Current Top. Mater. Sci. **7** [1981] 587/642.

Spencer, E. H., Brandberg, Å., Optimum Octane Number of Alcohol Gasoline Blends, Proc. 5th Intern. Symp. Alcohol Fuels Technol., Vol. 2, Auckland, New Zealand, 1982, pp. 159/66.

Hall, C. A., Willoughby, V. S., French, B. J., Effect of Lead Antiknock Regulations on Gasoline Aromatics and Aromatic Exhaust Emissions, NATO ASI Ser. C No. 112 [1983] 59/76.

Cole, J. F., Bleiadditive im Vergaserkraftstoff — Erhaltenswerter Marktsektor der Bleinachfrage, Metall **37** [1983] 927/9.

Joumard, R., Chiron, M., Delsey, J., Lambert, J., Lead File: Car Fuel Additives, N-84-27341 [1983] 1/76; C.A. **102** [1985] No. 100089, 190123.

Perry, R., McIntyre, A. E., Lester, J. N., Clark, A., Vehicle Emission Controls and Energy — The Role of Aromatics and Lead Compounds, NATO ASI Ser. C No. 112 [1983] 247/58.

Kahsnitz, R., The Rational Utilization of Fuels in Private Transport (RUFIT): Extrapolation to Unleaded Gasoline Case, Heavy Metals Environ. 4th Intern. Conf., Heidelberg 1983, pp. 124/7.

French, F. J., The Role of Lead in Energy Conservation, Heavy Metals Environ. 4th Intern. Conf., Heidelberg 1983, pp. 136/9.

Franco, G., Riganti, V., Lead in Gasolines: Some Considerations of Technical and Public Health Problems, Not. Ecol. **2** No. 8 [1984] 9/15; C.A. **102** [1985] No. 11481.

Unzelman, G. H., Michalski, G. W., Octane Improvement Update — Refinery Processing, Antiknocks, and Oxygenates, Natl. Pet. Refin. Assoc. Tech. Pap. **1984** 1/55; C.A. **101** [1984] No. 75468.

Salvi, G., Casalini, A., Lead and Aromatic Hydrocarbons in Gasoline and the Composition of Exhaust Gases, Riv. Combust. **38** [1984] 159/83.

U.S. Environmental Protection Agency, Regulation of Fuels and Fuel Additives; Lead Phase Down, Fed. Regist. **49** [1984] 31032/50; C.A. **101** [1984] No. 136152.

U.S. Environmental Protection Agency, Regulation of Fuels and Fuel Additives; Gasoline Lead Content, Fed. Regist. **50** [1985] 9386/99; C.A. **102** [1985] No. 208561.

Her Majesty's Stationery Office, House of Lords, Lead in Petrol and Vehicle Emissions, HMSO, London 1985, pp. 1/192.

Barnard, J. A., Bradley, J. N., Flame and Combustion, 2nd Ed., Chapman & Hall, London 1985.

Hancock, E. G., Technology of Gasoline, Blackwell, Oxford 1985.

Griffiths, R. J. M., Organometallics for Chemical Vapor Deposition, Chem. Ind. [London] **1985** 247/51.

Hawthorne, H. M., Yick, S., Some Effects of Retrofit Additives on Automotive Oil Performance in Sliding Concentrated Contact Lubrication Tests, Wear **103** [1985] 103/18.

Troitskii, V. B., Troitskaya, L. S., Thermal Decomposition and Stabilisation of Poly(vinyl Chloride), Usp. Khim. **54** [1985] 1287/311; Russ. Chem. Rev. **54** [1985] 755/69.

Razuvaev, G. A., Use of Organometallic Compounds for the Production of Inorganic Coatings and Materials, Nauka, Moscow 1986.

Environmental Aspects

Hirschler, D. A., Gilbert, L. F., Nature of Lead in Automobile Exhaust Gas, Arch. Environ. Health **8** [1964] 297/313.

MacPhee, R. D., Eye, M. G., Parkinson, E. E., A Method for Monitoring Organic Lead in the Atmosphere, Air Pollution Control District, County of Los Angeles, in: Cholak, J., Arch. Environ. Health **8** [1964] 222/31.

Cholak, J., Further Investigations of Atmospheric Concentration of Lead, Arch. Environ. Health **8** [1964] 314/24.

Anonymous, Survey of Lead in the Atmosphere of Three Urban Communities, U.S. Department of Health, Education, and Welfare, Public Health Service, Cincinnati, January 1965.

Snyder, L. J., Determination of Trace Amounts of Organic Lead in Air, Composite Sample Method, Anal. Chem. **39** [1967] 591/5.

Hepple, P., Lead in the Environment, Appl. Sci., Barking, Essex, and Inst. Pet., London 1971.

42

Pierrard, J. M., Crane, R. A., Effect of Some Gasoline Compositional Factors on Atmospheric Visibility and Soiling, SAE [Soc. Automot. Eng.] Tech. Papers No. 720253 [1972] 1/27; C.A. **82** [1975] No. 63776.

Laveskog, A., Organolead Compounds in Auto Exhaust and Street Air. Alkyllead Compounds. Tetramethyllead and Tetraethyllead, AB Atomenergic, TPM-BIL-64 [1972].

Clayton, P., Ellis, D. J., Palmer, P. L., Potter, C. J., Wallin, S. C., The Evaluation of a Filter for the Removal of Lead from Exhausts of Petrol Engines, LR 170 (AP), Warren Spring Laboratory, England 1972.

Pundir, B. P., Iyer, N. V., Goel, P. K., Lead Antiknocks and Air Pollution, Chem. Eng. World **7** [1972] 59/64.

Barry, P. S. I., Harrison, G. F., Motorenbenzin, Blei und Luftverschmutzung, Erdoel Erdgas Z. **88** [1972] 326/33.

Lee, J. A., Lead Pollution from a Factory Manufacturing Anti-Knock Compounds, Nature **238** [1972] 165/6.

National Research Council, Lead: Airborne Lead in Perspective, Committee on Biological Effects of Atmospheric Pollutants, Division of Medical Sciences, National Academy of Sciences, Washington, D.C., 1972.

Hall, S. K., Lead Pollution and Poisoning, Environ. Sci. Technol. **6** [1972] 30/5.

Bryce-Smith, D., Behavioural Effects of Lead and Other Heavy Metal Pollutants, Chem. Brit. **8** [1972] 240/3.

Colwill, D. M., Hickman, A. J., The Concentration of Volatile and Particulate Lead Compounds in the Atmosphere: Measurements at Four Road Sites, PB-221663/8 [1973] 1/8; C.A. **80** [1974] No. 18930.

Ewing, B. B., Pearson, J. E., Lead in the Environment, Advan. Environ. Sci. Technol. **3** [1974] 1/126; C.A. **82** [1975] No. 26623.

Robinson, J. W., Wolcott, D. K., Simultaneous Determination of Particulate and Molecular Lead in the Atmosphere, Environ. Letters **6** [1974] 321/33.

Magi, F., Facchetti, S., Garibaldi, P., Gasoline with Isotopically Differentiated Lead Added, Isotop. Ratios Pollut. Source Behav. Indic. Proc. Symp., Vienna 1974 [1975], pp. 109/19.

Wood, J. M., Biological Cycles for Toxic Elements in the Environment, Science **183** [1974] 1049/52.

Harrison, R. M., Perry, R., Slater, D. H., The Contribution of Organic Lead Compounds to Total Lead Levels in Urban Atmospheres, EUR-5360 [1974/75] 1783/8; C.A. **87** [1977] No. 156309.

Anonymous, Lead in the Environment and Its Significance to Man, Department of the Environment, Central Unit on Environmental Pollution, HMSO, London 1974.

Griffin, T. B., Knelson, J. H., Lead, in: Coulston, F., Korte, F., Suppl. Vol. II, Environmental Quality and Safety, Thieme, Stuttgart, and Academic, New York 1975.

Huntzicker, J. J., Friedlander, S. K., Davidson, C. I., Material Balance for Automobile-Emitted Lead in Los Angeles Basin, Environ. Sci. Technol. **9** [1975] 448/57.

Brinckman, F. E., Iverson, W. P., Chemical and Bacterial Cycling of Heavy Metals in the Estuarine System, ACS Symp. Ser. No. 18 [1975] 319/42.

Sittig, M., Lead, in: Sittig, M., Environmental Sources and Emissions Handbook, Noyes Data Corporation, Park Ridge 1975, pp. 77/81.

Edwards, H. W., Rosenvold, R. J., Wheat, H. G., Sorption of Organic Lead Vapor on Atmospheric Dust Particles, Trace Subst. Environ. Health **9** [1975] 197/205.

Harrison, R. M., Organic Lead in Street Dusts, J. Environ. Sci. Health A **11** [1976] 417/23.

Wood, J. M., The Biochemistry of Toxic Elements in Aqueous Systems, Biochem. Biophys. Perspect. Marine Biol. **3** [1976] 407/31.

Goetz, L. E., Springer, A., Transfer of Air-Borne Lead to Water, Chemistry Division, IRC, Ispra 1976.

Yamamura, Y., Yamauchi, H., Yoshida, M., Hirayama, F., Ohkura, E., Atmospheric Concentrations of Lead Alkyls in Shipping Containers Carrying Antiknock Compounds, Sangyo Igaku **18** [1976] 480/1.

Ridley, W. P., Dizikes, L. J., Wood, J. M., Biomethylation of Toxic Elements in the Environment, Science **197** [1977] 329/32.

Harrison, R. M., Perry, R., The Analysis of Tetraalkyl Lead Compounds and their Significance as Urban Air Pollutants, Atmos. Environ. **11** [1977] 847/52.

Harrison, R. M., Laxen, D. P. H., Organolead Compounds Adsorbed upon Atmospheric Particulates: A Minor Component of Urban Air, Atmos. Environ. **11** [1977] 201/3.

Robinson, J. W., Kiesel, E. L., Concentrations of Molecular and Organic Lead in the Atmosphere, J. Environ. Sci. Health A **12** [1977] 411/22.

Potter, H. R., Jarvie, A. W. P., Markall, R. N., Detection and Determination of Alkyl Lead Compounds in Natural Waters, Water Pollut. Contr. **76** [1977] 123/8.

Anonymous, Environmental Health Criteria 3, Lead, World Health Organization, Geneva 1977.

U.S. Environmental Protection Agency, Control Techniques for Lead Air Emissions, EPA-450-2-77-012 [1977].

Boggess, W. R., Wixson, B. G., Lead in the Environment, NSF/RA-770214 [1977] 1/265; C.A. **89** [1978] No. 203385.

Robinson, J. W., The Analysis of Tetra-Alkyl Lead Compounds and Their Significance as Urban Air Pollutants, Atmos. Environ. **12** [1978] 957; Harrison, R. M., Perry, R., Author's Reply, Atmos. Environ. **12** [1978] 957/8.

Noden, F. G., The Determination of Tetraalkyllead Compounds and Their Degradation Products in Natural Water, Lead Marine Environ. Proc. Intern. Experts Discuss., Rovinj, Yugosl., 1977 [1980], pp. 83/91.

Wood, J. M., Lead in the Marine Environment: Some Biochemical Considerations, Lead Marine Environ. Proc. Intern. Experts Discuss., Rovinj, Yugosl., 1977 [1980], pp. 299/303.

Harrison, G. F., The Cavtat Incident, Lead Marine Environ. Proc. Intern. Experts Discuss., Rovinj, Yugosl., 1977 [1980], pp. 305/17.

Cleaver, J. W., Dispersion of Lead Alkyls from Pools Located on the Sea-Bed, Lead Marine Environ. Proc. Intern. Experts Discuss., Rovinj, Yugosl., 1977 [1980], pp. 325/43.

Bernhard, M., The Relative Importance of Lead as a Marine Pollutant, Lead Marine Environ. Proc. Intern. Experts Discuss., Rovinj, Yugosl., 1977 [1980], pp. 345/52.

44

Saxena, J., Howard, P. H., Environmental Transformation of Alkylated and Inorganic Forms of Certain Metals, Advan. Appl. Microbiol. **21** [1977] 185/226.

Grandjean, P., Nielsen, T., Organic Lead Compounds, Pollution and Toxicology, SNV PM 879, Swedish Environmental Protection Agency, Stockholm 1977.

Allvin, B., Berg, S., Analysis of Tetraalkyl Lead in Street Air, SNV PM 907, Swedish Environmental Protection Agency, Stockholm 1977, pp. 1/16.

Skogerboe, R. K., Dick, D. L., Lamothe, P. J., Evaluation of Filter Inefficiencies for Particulate Collection under Low Loading Conditions, Atmos. Environ. **11** [1977] 243/9.

Bell, M. A., Ewing, R. A., Lutz, G. A., Holoman, V. L., Paris, B., Reviews of the Environmental Effects of Pollutants: VII. Lead, PB-80-121072 [1978]; Gov. Rept. Announce. Index [U.S.] **80** [1980] 763.

Brinckman, F. E., Bellama, J. M., Organometals and Organometalloids. Occurrence and Fate in the Environment, ACS Symp. Ser. No. 82 [1978] 1/447.

Cardin, D. J., Organometallic Compounds in Biological Chemistry, Organometal. Chem. **7** [1978] 400/13.

Mahaffey, K. R., Environmental Exposure to Lead, in: Nriagu, J. O., The Biogeochemistry of Lead in the Environment, Pt. B, Elsevier Biomed./North-Holland, Amsterdam 1978, pp. 1/36.

Jaworski, J. F., Effects of Lead in the Environment. Quantitative Aspects, National Research Council Canada, Ottawa, Ont., 1978, pp. 1/779; C.A. **91** [1979] No. 135353.

Malizia, E., Stacchini, A., Costantini, S., Baldini, M., Giordano, R., Organic Lead Levels in Fishes, Mussels, Water, and Sediments in the Area of Wreckage of M/B. "Cavtat" near Otranto, Riv. Tossicol. Sper. Clin. **8** [1978] 431/43.

Nriagu, J. O., The Biogeochemistry of Lead in the Environment, Vol. 1, Pt. A, Ecological Cycles, Vol. 1, Pt. B, Biological Effects, Elsevier Biomed./North-Holland, Amsterdam 1978.

Oehme, F. W., Toxicity of Heavy Metals in the Environment, Pt. 1, Dekker, New York 1978.

Harrison, R. M., Laxen, D. P. H., Sink Processes for Tetraalkyllead Compounds in the Atmosphere, Environ. Sci. Technol. **12** [1978] 1384/92.

Harrison, R. M., Laxen, D. P. H., Natural Source of Tetraalkyllead in Air, Nature **275** [1978] 738/40.

De Jonghe, W., Adams, F., The Determination of Organic and Inorganic Lead Compounds in Urban Air by Atomic-Absorption Spectrometry with Electrothermal Atomization, Anal. Chim. Acta **108** [1979] 21/30.

Chau, Y. K., Wong, P. T. S., Bengert, G. A., Kramar, O., Determination of Tetraalkyllead Compounds in Water, Sediment, and Fish Samples, Anal. Chem. **51** [1979] 186/88.

Botré, C., De Zorzi, C., Malizia, E., Melchiorri, P., Stacchini, E., Tiravanti, G., Study and Evaluation of Organic Lead Levels in Fishes and Phytoplankton Near Otranto, Toxicol. Aspects Food Safety, Arch. Toxicol. Suppl. **1** [1978] 157.

Grandjean, P., Nielsen, T., Organolead Compounds: Environmental Health Aspects, Residue Rev. **72** [1979] 97/148.

Beijer, K., Jernelöv, A., Sources, Transport, and Transformation of Metals in the Environment, in: Friberg, L., Nordberg, G. F., Vouk, V. B., Handbook on the Toxicology of Metals, Elsevier Biomed./North-Holland, Amsterdam 1979, pp. 47/63.

Butler, J. D., Air Pollution Chemistry, Academic, London 1979.

Turner, D., Leaded Gasoline and the Environment, Manage. Control Heavy Met. Environ. Intern. Conf., London 1979, pp. 109/12.

Harrison, R. M., Laxen, D. P. H., Birch, J., Tetraalkyllead in Air: Sources, Sinks, and Concentrations, Manage. Control Heavy Met. Environ. Intern. Conf., London 1979, pp. 257/61.

Rohbock, E., Measurements of the Gaseous Organic Lead Portion in Urban Air, Manage. Control Heavy Met. Environ. Intern. Conf., London 1979, pp. 386/9.

Tiravanti, G., Boari, G., Potential Pollution of a Marine Environment by Lead Alkyls: The Cavtat Incident, Environ. Sci. Technol. 13 [1979] 849/54.

Förstner, U., Metal Transfer between Solid and Aqueous Phases, in: Förstner, U., Wittmann, G. T. W., Metal Pollution in the Aquatic Environment, Springer, Berlin 1979, pp. 197/270.

Hodges, D. J., Noden, F. G., The Determination of Alkyl Lead Species in Natural Waters by Polarographic Techniques, Manage. Control Heavy Met. Environ. Intern. Conf., London 1979, pp. 408/11.

Fergusson, J. E., Tetraethyl Lead in Petrol, Soil Health 38 [1979] 38/9.

Rohbock, E., Georgii, H.-W., Müller, J., Measurements of Gaseous Lead Alkyls in Polluted Atmospheres, Atmos. Environ. 14 [1980] 89/98.

De Jonghe, W. R. A., Adams, F. C., Organic and Inorganic Lead Concentrations in Environmental Air in Antwerp, Belgium, Atmos. Environ. 14 [1980] 1177/80.

Robinson, J. W., Comments: Measurements of Gaseous Lead Alkyls in Polluted Atmospheres, Atmos. Environ. 14 [1980] 1207.

Rohbock, E., Georgii, H.-W., Müller, J., Reply to Comments: Measurements of Gaseous Lead Alkyls in Polluted Atmospheres, Atmos. Environ. 14 [1980] 1207/8.

Anonymous, Origin of Lead Pollution from Developments in Engineering Technology, in: National Research Council, Committee on Lead in the Human Environment, Lead in the Human Environment, National Academy of Sciences, Washington 1980.

Cruz, R. B., Lorouso, C., George, S., Thomassen, Y., Kinrade, J. D., Butler, L. R. P., Lye, J., Van Loon, J. C., Determination of Total, Organic Solvent Extractable, Volatile and Tetraalkyllead in Fish, Vegetation, Sediment and Water Samples, Spectrochim. Acta B 35 [1980] 775/83.

National Research Council, Washington, D.C. [U.S.A.], Committee on Lead in the Human Environment, Lead in the Human Environment, PB-82-117136 [1980]; Gov. Rept. Announce. Index [U.S.] 82 [1982] 269.

Charlou, J. L., Comportement geochimique des alkyls plomb dans l'eau de mer, CNEXO-COB [1980], Contrat CNEXO/ENSCR No. 79/5943, pp. 1/30.

Branica, M., Konrad, Z., Lead in the Marine Environment, Pergamon, Oxford 1980.

Glockling, F., Organometallic Compounds in Relation to Pollution, Anal. Proc. [London] 17 [1980] 417/22.

Cardin, D. J., Organometallic Compounds in Biological Chemistry, Organometal. Chem. **8** [1980] 434/45.

Geraci, S., Montanari, M., Di Cintio, R., Alkyllead Pollution along Eastern Salentina Coast (Adriatic Sea) after "Cavtat" Incident, Mem. Biol. Marina Oceanogr. **10** Suppl. [1980] 195/206.

Tiravanti, G., Rozzi, A., Dall'Aglio, M., Delaney, W., Dadone, A., The "Cavtat" Accident: Evaluation of Alkyl Lead Pollution by Simulation and Analytical Studies, Prog. Water Technol. **12** [1980] 49/65.

Chau, Y. K., Wong, P. T. S., Kramar, O., Bengert, G. A., Cruz, R. B., Kinrade, J. O., Lye, J., Van Loon, J. V., Occurrence of Tetraalkyllead Compounds in the Aquatic Environment, Bull. Environm. Contam. Toxicol. **24** [1980] 265/69.

Jernelöv, A., Wennergren, G., Studies of Concentrations of Methyl Mercury in Sediments from the St. Clair System and Rate of Biological Methylation in Incubated Samples of Sediments, Inst. Vatten Luftvårdsforsk. Publ. B IVL **531** [1980].

Craig, P. J., Metal Cycles and Biological Methylation, in: Hutzinger, O., The Handbook on Environmental Chemistry, Vol. A1, The Natural Environment and the Biogeochemical Cycles, Springer, Berlin 1980, pp. 169/227.

Harrison, R. M., Laxen, D. P. H., Lead Pollution, Causes and Control, Chapman & Hall, London 1981.

Craig, P., Biomethylation: Pollution Amplified, New Scientist **90** [1981] 694/7.

Jarvie, A. W. P., Whitmore, A. P., Methylation of Elemental Lead and Lead(II) Salts in Aqueous Solution, Environ. Technol. Letters **2** [1981] 197/204.

De Jonghe, W. R. A., Chakraborti, D., Adams, F. C., Measurements of Gaseous Lead Alkyls in Polluted Atmospheres. Comments, Atmos. Environ. **15** [1981] 421/2.

Rohbock, E., Georgii, H.-W., Müller, J., Measurements of Gaseous Lead Alkyls in Polluted Atmospheres. Reply to Comments, Atmos. Environ. **15** [1981] 422.

Harrison, R. M., Laxen, D. P. H., Measurements of Gaseous Lead Alkyls in Polluted Atmospheres. Comments, Atmos. Environ. **15** [1981] 422/3.

Rohbock, E., Georgii, H.-W., Müller, J., Measurements of Gaseous Lead Alkyls in Polluted Atmospheres. Reply to Comments, Atmos. Environ. **15** [1981] 423/4.

Rohbock, E., Determination of Gaseous Alkylleads in the Air of Large Cities, Atomspektrom. Spurenanal. Vortr. Kolloq. [1981/82] 267/74; C.A. **98** [1983] No. 131453.

Craig, P. J., Wood, J. M., The Biological Methylation of Lead: An Assessment of the Present Position, in: Lynam, D. R., Piantanida, L. G., Cole, J. F., Environmental Lead, Academic, New York 1981, pp. 333/49.

Chau, Y. K., Wong, P. T. S., Some Environmental Aspects of Organo-Arsenic, Lead, and Tin, NBS-SP-618 [1981] 65/80.

Birnie, S. E., Hodges, D. J., Determination of Ionic Alkyl Lead Species in Marine Fauna, Environ. Technol. Letters **2** [1981] 433/42.

Lynam, D. R., Piantanida, L. G., Cole, J. F., Environmental Lead, Academic, New York 1981.

Jarvie, A. W. P., Markall, R. N., Potter, H. R., Decomposition of Organolead Compounds in Aqueous Systems, Environ. Res. **25** [1981] 241/9.

Brondi, M., Dall'Aglio, M., Ghiara, E., Mignuzzi, C., Tiravanti, G., Environmental Studies on Lead Alkyl Release in Sea Water by the "Cavtat" Wreck, Sci. Total Environ. **19** [1981] 21/31.

De Jonghe, W., Chakraborti, D., Adams, F., Occurrence and Fate of Tetraalkyllead Compounds in the Atmosphere, Heavy Metals Environ. 3rd Intern. Conf., Amsterdam 1981, pp. 72/5.

Gibson, M. J., Farmer, J. G., Tetraalkyl Lead in the Urban Atmosphere of Glasgow, Environ. Technol. Letters **2** [1981] 521/30.

De Jonghe, W. R. A., Chakraborti, D., Adams, F. C., Identification and Determination of Individual Tetraalkyllead Species in Air, Environ. Sci. Technol. **15** [1981] 1217/22.

Jiang, S. G., Chakraborti, D., De Jonghe, W., Adams, F., Atomic-Absorption Spectrophotometric Determination of Volatile Organolead Compounds in the Atmosphere, Z. Anal. Chem. **305** [1981] 177/80.

Bryce-Smith, D., Environmental Lead and the Analyst, TrAC Trends Anal. Chem. **1** [1982] 199/203.

Johnson, M. S., Pluck, H., Hutton, M., Moore, G., Accumulation and Renal Effects of Lead in Urban Populations of Feral Pigeons, Columba livia, Arch. Environ. Contam. Toxicol. **11** [1982] 761/7.

Newland, L. W., Daum, K. A., Lead, in: Hutzinger, O., The Handbook on Environmental Chemistry, Vol. B3, Anthropogenic Compounds, Springer, Berlin 1982, pp. 1/16.

Brinckman, F. E., Olson, G. J., Iverson, W. P., The Production and Fate of Volatile Molecular Species in the Environment: Metals and Metalloids, in: Goldberg, E. D., Atmospheric Chemistry, Dahlem Konferenzen, Springer, Berlin 1982, pp. 231/49.

Thayer, J. S., Brinckman, F. E., The Biological Methylation of Metals and Metalloids, Advan. Organometal. Chem. **20** [1982] 313/56.

De Jonghe, W. R. A., Adams, F. C., Measurements of Organic Lead in Air, A Review, Talanta **29** [1982] 1057/67.

Craig, P. J., Environmental Aspects of Organometallic Chemistry, in: Wilkinson, G., Stone, F. G. A., Abel, E. W., Comprehensive Organometallic Chemistry, Vol. 2, Pergamon, Oxford 1982, pp. 979/1020.

De Jonghe, W., Adams, F., Organolead Compounds in the Atmosphere, COST 61a bis, 2nd Disc. Meeting Working Party 1. Detection, Identification and Analysis of Air Pollutants, Vienna 1982, pp. 17/23.

Craig, P. J., Organometallic Compounds in the Environment, in: Harrison, R. M., Pollution, Causes, Effects and Control, Spec. Publ. Roy. Soc. Chem. No. 44 [1983] 277/322.

Alzieu, C., Thibaud, Y., Marine Pollution by Organometallics: Derivatives of Mercury, Lead, and Tin, Bull. Acad. Natl. Med. [Paris] **167** [1983] 473/82.

Rabinowitz, M., Needleman, H. L., Petrol Sales and Umbilical Cord Blood Lead Levels in Boston, Massachusetts, Lancet **1983** I 63.

Jarvie, A. W. P., Whitmore, A. P., Markall, R. N., Potter, H. R., Lead Biomethylation, An Elusive Goal, Environ. Pollut. B **6** [1983] 81/94.

Anagnostopoulos, A., Air Lead and Dust in Thessaloniki, in: Rutter, M., Russell Jones, R., Lead Versus Health, Wiley, Chichester 1983, pp. 107/13.

Bull, K. R., Every, W. J., Freestone, P., Hall, J. R., Osborn, D., Cooke, A. S., Stowe, T., Alkyl Lead Pollution and Bird Mortalities on the Mersey Estuary, Environ. Pollut. A **31** [1983] 239/59.

Laxen, D. P. H., The Chemistry of Metal Pollutants in Water, in: Harrison, R. M., Pollution, Causes, Effects and Control, Spec. Publ. Roy. Soc. Chem. No. 44 [1983] 104/23.

U.S. Environmental Protection Agency, Air Quality Criteria for Lead: Vols. I through IV — Review Draft, PB 84-144591 [1983].

Vuorinen, A., Emission of Lead by Highway Traffic: Lead Fixation and Speciation, Heavy Met. Environ. 4th Intern. Conf., Heidelberg 1983, pp. 159/62.

De Jonghe, W. R. A., Van Mol, E. E., Adams, F. C., Occurrence and Fate of Organic Lead Compounds in the Hydrosphere, Heavy Met. Environ. 4th Intern. Conf., Heidelberg 1983, pp. 166/9.

Sneddon, J., Collection and Atomic Spectroscopic Measurement of Metal Compounds in the Atmosphere: A Review, Talanta **30** [1983] 631/48.

Berdicevsky, I., Shachar, M., Yannai, S., Conversion of Inorganic Lead into a Highly Toxic Organic Derivative by Marine Microorganisms, Arch. Toxicol. Suppl. No. 6 [1983] 285/91.

Brinckman, F. E., Environmental Inorganic Chemistry of Main Group Elements with Special Emphasis on Their Occurrence as Methyl Derivatives, Environ. Inorg. Chem. Papers US-Italy Joint Semin. Workshop, San Miniato, Italy, 1983 [1985], pp. 195/238; C.A. **103** [1985] No. 218393.

Jiang, S., Alkylation of Heavy Metals and the Environment, Huanjing Huaxue **2** [1983] 9/14; C.A. **99** [1983] No. 109895.

Craxford, S. R., Pollution from Lead in Petrol, Oil Petrochem. Pollut. **1** [1983] 285/90.

Olson, B. H., Bakterien und Pilze, Biologische Umwandlung von Metallverbindungen, in: Merian, E., Metalle in der Umwelt, Verlag Chemie, Weinheim 1984, pp. 141/51.

Hewitt, C. N., Harrison, R. M., De Mora, S. J., Comment on the Atmospheric Distribution of Lead Over a Number of Marine Regions, Marine Chem. **15** [1984] 189/90.

Chester, R., Sharples, E. J., Murphy, K., Saydam, A. C., Sanders, G. S., Reply to Comment on the Atmospheric Distribution of Lead Over a Number of Marine Regions, Marine Chem. **15** [1984] 191.

Grandjean, P., Biological Effects of Organolead Compounds, CRC, Boca Raton, Fla., 1984.

Riley, J. P., Towner, J. V., The Distribution of Alkyl Lead Species in the Mersey Estuary, Marine Pollut. Bull. **15** [1984] 153/8.

Caplun, E., Petit, D., Picciotto, E., Lead in Gasoline, Recherche **15** [1984] 270/80.

Shou-Gui Jiang, Ci-Guang Ma, Huai-Chuan Liu, Ji-Rong Ge, Min Li, Adams, F. C., Winchester, J. W., Absence of Tetraalkyl Lead Vapors in the Atmosphere of Beijing, China, Atmos. Environ. **18** [1984] 2553/6.

Nriagu, J. O., Simmons, M. S., Toxic Contaminants in the Great Lakes, Advan. Environ. Science Technol. **14** [1984].

Ewers, U., Schlipköter, H.-W., Blei, in: Merian, E., Metalle in der Umwelt, Verlag Chemie, Weinheim 1984, pp. 351/73.

Faulstich, H., Stournaras, C., Is Triethyllead Causing Damage in West German Forests? Naturw. Rundschau **37** [1984] 398/401; C.A. **102** [1985] No. 41131.

Turner, D. R., Relationships Between Biological Availability and Chemical Measurements, in: Sigel, H., Metal Ions in Biological Systems, Vol. 18, Circulation of Metals in the Environment, Dekker, New York 1984, pp. 137/64.

Diehl, K. H., Rosopulo, A., Kreuzer, W., Lead Contents in Tissues, Organs, and Eggs of Laying Hens in the Area of Organic Lead Compound Emissions, Arch. Lebensmittelhyg. **36** [1985] 113/6.

Chau, Y. K., Wong, P. T. S., Bengert, G. A., Dunn, J. L., Glen, B., Occurrence of Alkyllead Compounds in the Detroit and St. Clair Rivers, J. Great Lakes Res. **11** [1985] 313/9; C.A. **104** [1986] No. 95050.

Wood, J. M., Microbiological Strategies in Resistance to Metal Ion Toxicity, in: Sigel, H., Metal Ions in Biological Systems, Vol. 18, Circulation of Metals in the Environment, Dekker, New York 1984, pp. 333/51.

Harrison, R. M., Radojević, M., Determination of Tetraalkyl and Ionic Alkyllead Compounds in Environmental Samples by Butylation and Gas Chromatography-Atomic Absorption, Environ. Technol. Letters **6** [1985] 129/36.

Craig, P. J., Rapsomanikis, S., Methylation of Tin and Lead in the Environment: Oxidative Methyl Transfer as a Model for Environmental Reactions, Environ. Sci. Technol. **19** [1985] 726/30.

Unsworth, M. H., Harrison, R. M., Is Lead Killing German Forests?, Nature **317** [1985] 674.

Faulstich, H., Stournaras, C., Potentially Toxic Concentrations of Triethyl Lead in Black Forest Rainwater Samples, Nature **317** [1985] 714/5.

Harrison, R. M., Radojević, M., Hewitt, C. N., Measurements of Alkyllead Compounds in the Gas and Aerosol Phase in Urban and Rural Atmospheres, Sci. Total Environ. **44** [1985] 235/44.

Harrison, R. M., Hewitt, C. N., Radojevic, M., Environmental Pathways of Alkyllead Compounds, Heavy Met. Environ. 5th Intern. Conf., Athens 1985, Vol. 1, pp. 82/4; C.A. **106** [1987] No. 89402.

Van Cleuvenbergen, R., Chakraborti, D., Van Mol, W., Adams, F., Ionic Alkyllead Species in Atmospheric Fall-out and Surface Water, Heavy Met. Environ. 5th Intern. Conf., Athens 1985, Vol. 1, pp. 153/5; C.A. **106** [1987] No. 72628.

Hewitt, C. N., Harrison, R. M., Total Speciation of Gas-phase Alkyllead in the Atmosphere, Heavy Met. Environ. 5th Intern. Conf., Athens 1985, Vol. 1, pp. 171/3; C.A. **106** [1987] No. 55028.

Harrison, R. M., Chemical Speciation and Reaction Pathways of Metals in the Atmosphere, Advan. Environ. Sci. Technol. **17** [1986] 319/33.

Chau, Y. K., Occurrence and Speciation of Organometallic Compounds in Freshwater Systems, Sci. Total Environ. **49** [1986] 305/23.

Van Cleuvenbergen, R. J. A., Chakraborti, D., Adams, F. C., Occurrence of Tri- and Dialkyllead Species in Environmental Water, Environ. Sci. Technol. **20** [1986] 589/93.

De Jonghe, W. R. A., Adams, F. C., Biogeochemical Cycling of Organic Lead Compounds, Advan. Environ. Sci. Technol. **17** [1986] 561/94.

50

Hewitt, C. N., Organolead Compounds in the Environment, in: Craig, P. J., Organometallic Compounds in the Environment, Principles and Reactions, Longman, Burnt Mill, Harlow, Essex, 1986, pp. 160/97.

Elbaz-Poulichet, F., Holliger, P., Martin, J. M., Petit, D., Stable Lead Isotopes Ratios in Major French Rivers and Estuaries, Sci. Total Environ. **54** [1986] 61/76.

Radojevic, M., Harrison, R. M., Alkyllead Compounds in Surface and Potable Waters, Environ. Technol. Letters **7** [1986] 519/24.

Blais, J. S., Marshall, W. D., Determination of Alkyllead Salts in Runoff, Soils, and Street Dusts Containing High Levels of Lead, J. Environ. Qual. **15** [1986] 255/60.

Radojevic, M., Harrison, R. M., Alkyllead Compounds in Dust, Sediment and Soil Samples, Environ. Technol. Letters **7** [1986] 525/30.

Mukai, H., Organometallic Compounds in the Atmosphere, Kokuritsu Kogai Kenkyusho Kenkyu Hokoku No. 102 [1986] 109/20; C.A. **106** [1987] No. 107148.

Faulstich, H., Stournaras, C., Endres, K. P., Trialkyllead. Occurrence, Biological Interactions, and Possible Impact on Forest Decline, Experientia **43** [1987] 115/27.

Radojevic, M., Harrison, R. M., Concentrations and Pathways of Organolead Compounds in the Environment: A Review, Sci. Total Environ. **59** [1987] 157/80.

Røyset, O., Thomassen, Y., Presence of Alkyllead in Rural and Urban Air: Evaluation of Some Sources of Alkyllead Pollution of the Atmosphere in Norway, Atmos. Environ. **21** [1987] 655/8.

Hewitt, C. N., Harrison, R. M., Atmospheric Concentrations and Chemistry of Alkyllead Compounds and Environmental Alkylation of Lead, Environ. Sci. Technol. **21** [1987] 260/6.

Patents Related to the Use as Antiknock Agents

Calingaert, G., Ethyl Corp., Improvements in or Relating to Alkyllead Compounds or Compositions which are Stabilized against Decomposition, Brit. 670526 [1949/52]; C.A. **1953** 9346.

Jackson, R. G., Shell Research Ltd., Fuel Compositions, Brit. 829635 [1960]; C.A. **1960** 14675; Addn. Brit. 795067 [1958] 10733; C.A. **1959** 10733.

Calingaert, G., Wintringham, J. S., Ethyl Corp., Antiknock Compounds, Ger. 843038 [1949/52]; C. **1953** 316.

Graham, R., Lewis, A., Shell Refining & Marketing Co. Ltd., Gasolines, Swed. 135366 [1951/52]; C. **1953** 8262.

Bartholomew, E., Ethyl Corp., Antiknock Agent, U.S. 2398281-2 [1944/46]; C.A. **1946** 3889/90.

Luten, D. B., Jr., Shell Development Co., Stabilization of Gasolines, U.S. 2410829 [1944/46]; C. **1947** 698.

Partridge, W. A., Alty, H. J., Anglo-Iranian Oil Co., Ltd., Additive for Stabilizing Leaded Gasolines, U.S. 2452489 [1948]; C.A. **1949** 2423.

Orelup, J. W., Alkyl Amine Inhibitors in Fuels Containing Aromatic Amines, U.S. 2461917 [1949]; C.A. **1949** 4003.

Fischer, H. G. M., Standard Oil Development Co., Hydroquinone as Inhibitor against Lead Precipitation, U.S. 2461972 [1949]; C.A. **1949** 4003.

De Verter, P. L., Standard Oil Development Co., Antiknock Motor Fuel, U.S. 2479326 [1949]; C.A. **1949** 8667.

Calingaert, G., Wintringham, J. S., Ethyl Corp., Antiknock Mixtures, U.S. 2479900/2479903 [1949]; C.A. **1949** 8667.

Smith, H. G., Cantrell, T. L., Gulf Oil Corp., Stabilized Leaded Gasolines, U.S. 2560489 [1951]; C.A. **1951** 9256.

Bottoms, R. R., National Cylinder Gas Co., Combustion Catalyst Compositions, U.S. 2591503 [1952]; C.A. **1953** 296.

Ecke, G. G., Kolka, A. J., Ethyl Corp., Stabilized Antiknock Additive, U.S. 2836568 [1955/58]; C.A. **1958** 17692.

Lyben, R. G., Ethyl Corp., Antiknock Compositions, U.S. 2849302/2849304 [1958]; C.A. **1959** 2599.

Shepherd, C. C., Ethyl Corp., Synergistic Antioxidants for Alkyllead Antiknock Agents, U.S. 2865722 [1958]; C.A. **1959** 5660.

Scheule, H. J., E. I. du Pont de Nemours & Co., Stabilizing Leaded Gasolines, U.S. 3037851 [1956/62]; C.A. **57** [1962] 8810/1.

Cook, S. E., Shapiro, H., Ethyl Corp., Thermal Stabilization of Alkyllead Compounds, U.S. 3038916/3038919 [1960/62]; C.A. **57** [1962] 8810.

Niedzielski, E. L., E. I. du Pont de Nemours & Co., Antiknock Mixtures, U.S. 3074788 [1960/63]; C. **1966** No. 5-2652.

Cook, S. E., Ethyl Corp., Stabilizing Alkyllead Compounds, U.S. 3081326 [1961/63]; C.A. **58** [1963] 13688.

Cook, S. E., Thomas, W. H., Ethyl Corp., Stable Lead Alkyl Compositions and a Method for Preparing the Same, U.S. 3133098/3133099 [1962/64]; C.A. **61** [1964] 1695/6.

Miller, M. R., E. I. du Pont de Nemours & Co., Tetraalkyllead Compositions, U.S. 3175982 [1960/65]; C.A. **63** [1965] 1642.

Thomas, W. H., Cook, S. E., Ethyl Corp., Stabilization of Lead Alkyls, U.S. 3197492 [1962/64/65]; C.A. **63** [1965] 9730.

Whitehurst, D. D., Mobil Oil Corp., Sorbent for Heavy Metals, U.S. 3856664 [1972/74]; C.A. **82** [1975] No. 127370.

Crisman, R. W., Deangelis, D., Mobil Oil Corp., Fuel Sludge Treatment Method, Especially for Leaded Gasoline Storage-Tank Sludge, U.S. 4447332 [1983/84]; C.A. **101** [1984] No. 75888.

Bass, D. R. W., Associated Octel Co. Ltd., Elimination of Organolead Ions from Effluents, Fr. Demande 2543128 [1984]; C.A. **102** [1985] No. 66948.

Aleksandrov, Yu. A., Spiridonova, M. N., Tanaseichuk, B. S., Tikhonova, L. G., Gorki Scientific-Research Institute of Chemistry, Stabilization of Tetraalkyllead Compounds, U.S.S.R. 382636 [1971/73]; C.A. **79** [1973] No. 137292.

Patented Uses as Components of Polymerization Catalysts

Solvay & Cie., Polymerization Catalyst, Belg. 547618 [1956]; C.A. **1960** 16917.

Solvay & Cie., Ethylene Polymerization Catalyst, Belg. 553839 [1957], Addn. Belg. 547618 [1957]; C.A. **1959** 18550.

Solvay & Cie., Catalysts for Polymerization of Propylene, Belg. 586754 [1960]; C.A. **1960** 26003.

Solvay & Cie., Catalysts for Olefin Polymerization, Belg. 620081 [1962/63]; C.A. **58** [1963] 14138.

Solvay & Cie., Catalysts for Polymerization of Olefins, Belg. 620387 [1962/63]; C.A. **58** [1963] 14137.

Badische Anilin- & Soda-Fabrik, A.-G., Polymerization of Olefinically Unsaturated Hydrocarbons, Brit. 798447 [1958]; C.A. **1959** 2690.

Wood, D. G. M., Imperial Chemical Industries Ltd., Polymers, Brit. 808132 [1959]; C.A. **1960** 21850.

Crawford, J. W. C., Wood, D. G. M., Imperial Chemical Industries Ltd., Crystalline Polymers or Copolymers of Styrene, Brit. 812176 [1959]; C.A. **1959** 15649.

Jones, J. F., Mital, A. J., B. F. Goodrich Co., Water-Insoluble Carboxyl-Containing Interpolymers, Brit. 836755 [1960]; C.A. **1960** 26015.

Raum, A. L. J., Distillers Co. Ltd., Polymerization of α-Olefins, Especially Ethylene, Brit. 841527 [1960]; C.A. **1961** 1083.

Petrochemicals Ltd., Olefin Polymerization Catalyst, Brit. 841872 [1960].

Kluiber, R. W., Carrick, W. L., Union Carbide Corp., Process for Producing Ethylene-1-Olefine Copolymers, Brit. 856859 [1959/60]; C.A. **1961** 13924; U.S. 3073809 [1958/63]; C. **1964** No. 49-2580; Australian 234822 [1959/61]; C. **1964** No. 20-2627.

Solvay & Cie., Polymerization of Propylene, Brit. 887544 [1959/62]; C.A. **56** [1962] 15678.

Hanford, W. E., Harmon, J., E. I. du Pont de Nemours & Co., Higher Alkyl Chlorides from Ethylene and Hydrogen Chloride, U.S. 2418832 [1947]; C.A. **1948** 581.

Ray, R. L., Sistrunk, T. O., Ethyl Corp., Polymerization of Olefins, U.S. 2868772 [1959]; C.A. **1959** 8711.

Isbenjian, H., Aries Associates, Inc., Catalysts for Ethylene Polymerization, U.S. 2898330 [1959]; C.A. **1960** 966.

Aries, R. S., Aluminium Halide Catalysts for Ethylene Polymerization, U.S. 2900374 [1959]; C.A. **1959** 23099.

Stuart, A. P., Sun Oil Co., Polymerization of Ethylene with Cerium Acetylacetonate-Metal Alkyl Catalyst, U.S. 2921060 [1960]; C.A. **1960** 12657.

Jones, J. F., Mital, A. J., B. F. Goodrich Co., Cross-Linked Interpolymers, U.S. 2985631 [1961]; C.A. **1961** 22924.

Arnold, L. F., B. F. Goodrich Co., Cross-linked Nitrile Polymers, U.S. 2991276 [1961]; C.A. **1961** 26535.

Loeb, W. E., Union Carbide Corp., Polymerization of α-Monoolefins in an Aqueous Diluent, U.S. 3166547 [1959/65]; C.A. **62** [1965] 9259.

Finkelshtein, E. Sh., Strel'chik, B. S., Kotov, S. V., Smagin, V. M., Chernykh, S. P., Vdovin, V. M., Catalyst for Disproportionating Unsaturated Hydrocarbons, U.S.S.R. 1171087 [1981/85]; C.A. **104** [1986] No. 40618.

Willis, C. L., Shell Internationale Research Maatschappij B.V., Process for the Preparation of Block Polymers, Eur. 131322 [1984/85].

1 Mononuclear Organolead Compounds

1.1 Tetraorganolead Compounds

Tetraorganolead compounds in general exhibit lower thermal stability and greater reactivity than the corresponding compounds of the lighter Group 14 elements. The thermal stability is strongly dependent on the nature of the organic group bound to lead.

The tetraorganolead molecules have a tetrahedral configuration around the central lead atom, and therefore the lead electrons can be considered to be sp^3 hybridized. The lead-to-carbon bond has a high degree of covalent character.

1.1.1 Compounds of the Type PbR_4

Symmetrical tetraalkyllead compounds are colorless. The compounds with low molecular mass are liquids at room temperature, and they are soluble in common organic solvents, such as chloroform, ether, ethanol, and hydrocarbons. The tetra-n-alkyllead compounds are usually thermally stable up to about 100°C. They cannot be distilled at atmospheric pressure without decomposition, except $Pb(CH_3)_4$; but as a precaution, even this compound is best vacuum or steam distilled. Tetra-sec-alkyllead compounds are somewhat less stable than the straight-chain compounds. For these and the higher tetraalkyllead compounds, distillation under reduced pressure or steam distillation is required; but also under these conditions, caution is necessary. Compounds, with $R = n-C_{14}H_{29}$ and higher analogues, are low melting solids and are soluble in aromatic hydrocarbons and chloroform. In other organic solvents, only slight solubility is often observed.

Symmetrical tetraaryllead compounds are solids, which are usually colorless, having melting points in the range of about 120 to 250°C. They are soluble in chloroform, acetone, and aromatic hydrocarbons; they are much less or largely insoluble in ethers, alcohols, and aliphatic hydrocarbons. They generally are thermally more stable than the tetraalkyllead compounds. Often decomposition occurs only at, or slightly above, the melting point.

PbR_4-type compounds usually are not sensitive to air, insoluble in water, and very stable to hydrolysis. However, prolonged contact with air causes slow degradation, especially in the presence of light. They react readily with halogens, hydrogen halides, nonmetal halides, a series of salts (mainly metal halides), and with aqueous solutions of strong acids. They are much less sensitive to bases.

General Literature

Preparation, Reactions

Yakubovich, A. Ya., Ginsburg, V. A., The Diazo Method of Synthesis of Heteroorganic Compounds of the Aliphatic Series, Usp. Khim. **20** [1951] 734/58.

Seyferth, D., The Preparation of Organometallic and Organometalloidal Compounds by the Diazoalkane Method, Chem. Rev. **55** [1955] 1155/77.

Kay, J., Rowland, F. S., Formation of Volatile Compounds by ^{212}Pb Recoiling from Alpha Decay, J. Am. Chem. Soc. **80** [1958] 3165.

Shushunov, V. A., Sokolov, N. A., The Kinetic Method of Physico-Chemical Analysis. The Reaction of Some Liquid Alkyl Halides with the Alloys Na + Pb and K + Na + Pb, Tr. Khim. Khim. Tekhnol. **1** [1958] 265/9; C.A. **1960** 6277.

Shapiro, H., Preparation of Tetraalkyllead Compounds from Lead or Its Alloys, Advan. Chem. Ser. No. 23 [1959] 290/8.

Garrett, A. B., The Flash of Genius 5, Lead Tetraethyl: Thomas Midgley Jr., T. A. Boyd, C. A. Hochwalt, J. Chem. Educ. **39** [1962] 414/5.

Ziegler, K., Die elektrochemische Synthese von Metallalkylen, Chem. Ing. Tech. **35** [1963] 325/31.

Treichel, P. M., Stone, F. G. A., Fluorocarbon Derivatives of Metals, Advan. Organometal. Chem. **1** [1964] 143/220.

Sittig, M., Lead Compounds, in: Chemical Process Monograph No. 20, Organometallics, Noyes Development Corp., Park Ridge, N. J., 1966, pp. 67/104.

Vyazankin, N. S., Kruglaya, O. A., Homolytic Reactions of Organometallic Compounds in Liquid Phase, Tr. Khim. Khim. Tekhnol. **1966** No. 1, pp. 3/16; C.A. **67** [1967] No. 100168.

Chambers, R. D., Chivers, T., Pentafluorophenyl-Metal Compounds, Organometal. Chem. Rev. **1** [1966] 279/304.

Abraham, M. H., Hill, J. A., Organometallic Compounds. Mechanisms of Electrophilic Substitution of Metal Alkyls, J. Organometal. Chem. **7** [1967] 11/21.

Huber, F., Horn, H., Haupt, H. J., Zur Acidolyse von Tetraorganoplumbanen, Z. Naturforsch. **22b** [1967] 918/21.

Huber, F., Horn, H., Bade, V., Acidolyse von Organobleiverbindungen, Angew. Chem. **79** [1967] 996; Angew. Chem. Intern. Ed. Engl. **6** [1967] 976.

Mantell, C. L., Tetra Alkyl Leads by Electrolysis, Commercial Plant, Electro-Organic Chemical Processing, Chemical Process Review No. 14, Noyes Development Corp., Park Ridge, N.J., 1968, pp. 165/70.

Moedritzer, K., Redistribution Equilibria of Organometallic Compounds, Advan. Organometal. Chem. **6** [1968] 171/271.

Huber, F., Bade, V., Kinetics and Mechanisms of MR$_4$ and R$_3$MX Acidolysis — A Coordination Problem?, Proc. 12th Intern. Conf. Coord. Chem., Sydney 1969, p. 31.

Moedritzer, K., The Redistribution Reaction, Organometal. React. **2** [1971] 1/115.

Davidson, P. J., Lappert, M. F., Pearce, R., Stable Homoleptic Metal Alkyls, Accounts Chem. Res. **7** [1974] 209/17.

Lowenheim, F. A., Moran, M. K., Lead Alkyls, in: Faith, W. L., Keyes, D. B., R. L. Clark's Industrial Chemicals, 4th Ed., Wiley, New York 1975, pp. 502/8.

Butin, K. P., Shishkin, V. N., Beletskaya, I. P., Reutov, O. A., Equilibria of Redistribution Reactions in Group IVB Organometallic Compounds, J. Organometal. Chem. **93** [1975] 139/71.

Tedoradze, G. A., Electrochemical Synthesis of Organometallic Compounds, J. Organometal. Chem. **88** [1975] 1/36.

Beletskaya, I. P., Artamkina, G. A., Reutov, O. A., The Interaction of Organometallic Derivatives with Organic Halides, Usp. Khim. **45** [1976] 661/94; Russ. Chem. Rev. **45** [1976] 330/47.

Davidson, P. J., Lappert, M. F., Pearce, R., Metal-σ-Hydrocarbyls, MR$_n$. Stoichiometry, Structures, Stabilities, and Thermal Decomposition Pathways, Chem. Rev. **76** [1976] 219/29.

Tuck, D. G., Direct Electrochemical Synthesis of Inorganic and Organometallic Compounds, Pure Appl. Chem. **51** [1979] 2005/18.

Owen, A. J., Electrochemistry of Organolead Compounds, in: Kuhn, A. T., The Electrochemistry of Lead, Academic, London 1979, pp. 163/97.

Mengoli, G., Recent Developments in the Electrochemical Synthesis of Lead and Tin Alkyl Derivatives, Rev. Silicon Germanium Tin Lead Compounds **4** [1979] 59/89.

Krasavin, V. P., Grinberg, E. E., Fetisov, Yu. M., Efremov, A. A., Egurnov, V. Ya., Liquid-Vapor Equilibrium of Binary Systems of Trimethylaluminum with Methyl Derivatives of Zinc, Gallium, Tin, and Lead, Tr. Vses. Nauchn. Issled. Inst. Khim. Reaktivov Osobo Chist. Khim. Veshchestv No. 41 [1979] 69/73; C.A. **93** [1980] No. 192977.

Chernykh, I. N., Tomilov, A. P., Electrosynthesis of Organometallic Compounds During Cathodic Processes, in: Feoktistov, L. G., Elektrosint. Monomerov, Nauka, Moscow 1980, pp. 190/208.

Patents

de Mahler, E. R. W., Simultaneous Production of Organic Lead and Tin Compounds, Belg. 418670 [1936]; C.A. **1937** 5816.

Shappirio, S., Organometallic Derivatives, Can. 371890 [1936/38]; C.A. **1938** 2545.

Sullivan Jr., F. W., Standard Oil Co. of Indiana, Process of Making Lead Alkyls, U.S. 2148138 [1936/39]; C.A. **1939** 4012.

Downing, F. B., Linch, A. L., E. I. du Pont de Nemours & Co., Stabilizing Sludges in Tetraalkyllead Manufacture, U.S. 2407261 [1942/46]; C.A. **1946** 7229.

Linch, A. L., E. I. du Pont de Nemours & Co., Stabilizing Sludges in Tetraalkyllead Manufacture, U.S. 2407262 [1942/46]; C.A. **1946** 7230.

Linch, A. L., E. I. du Pont de Nemours & Co., Stabilizing Sludges in Tetraalkyllead Manufacture, U.S. 2407263 [1942/46]; C.A. **1946** 7230.

Linch, A. L., E. I. du Pont de Nemours & Co., Process for Stabilizing or Deactivating Sludges, Precipitates, and Residues Occurring or Used in the Manufacture of Tetraalkylleads, U.S. 2407307 [1942/46]; C.A. **1946** 7230.

Linch, A. L., Inhibition of Spontaneous Ignition of Tetraalkyl Lead Compounds Absorbed on Sludges, U.S. 2432321 [1947]; C.A. **1949** 7676.

1.1.1.1 Tetramethyllead, Pb(CH₃)₄

CAS Registry Number: *[75-74-1]*

1.1.1.1.1 Formation and Preparation

From Grignard Compounds. $Pb(CH_3)_4$ is readily prepared by the reaction of $PbCl_2$ with Grignard reagents in ether according to the general equation

$$2 \, PbX_2 + 4 \, CH_3MgX \rightarrow Pb(CH_3)_4 + Pb + 4 \, MgX_2$$

Use of CH_3MgCl is advantageous [1 to 5].

Thus, dry $PbCl_2$ (1.8 mol) is added slowly to a solution of excess CH_3MgCl (4 mol) in absolute diethyl ether under constant shaking. The reaction mixture is refluxed for 2 [6] to 5 h [5] on a water bath, cooled in ice and decomposed with water, but not acidified. After extracting with diethyl ether, the solvent is removed by distillation [4, 5, 6]. The product, 150 to 160 g (about 65%) [1], requires for usual purposes no further purification other than a fractional distillation [3 to 6]. Solutions of CH_3MgBr [7, 8] or CH_3MgI [9, 10] in diethyl ether can also be used for the reaction with $PbCl_2$. Also, $Pb(NO_3)_2$ was reacted with CH_3MgI to obtain $Pb(CH_3)_4$ [11].

According to [1], the product from the reaction of $PbCl_2$ and CH_3MgCl is pure; however, it was later observed that $Pb(CH_3)_4$, obtained from Grignard compounds, contains $Pb_2(CH_3)_6$ [3, 4, 7, 12, 13]. It was also reported that, on using CH_3MgBr as the methylating agent, $Pb_2(CH_3)_6$ is primarily produced, which gives $Pb(CH_3)_4$ and elemental lead on standing for 4 d [8]. For reactions of CH_3MgBr and CH_3MgI with $PbCl_2$ giving $Pb_2(CH_3)_6$ as the major product and $Pb(CH_3)_4$ as the minor product, see [12, 14]. $Pb_2(CH_3)_6$ decomposes as mentioned above even at room temperature [8] or on slight warming of the ether solution or during evaporation of the solvent [3, 4, 13]. $Pb_2(CH_3)_6$ can also be removed by brominating the ether solution and then adding CH_3MgBr to obtain $Pb(CH_3)_4$ [7]. The amount of elemental lead formed during the reaction of the Grignard compound and lead(II) halide is significantly reduced, and the yield of $Pb(CH_3)_4$ is appreciably increased when the reaction is carried out in the presence of methyl halide. Reaction of CH_3MgCl or CH_3MgI with $PbCl_2$ or PbI_2 and CH_3I (1:3:3.3 mole ratio) in diethyl ether gives $Pb(CH_3)_4$ with a yield of about 70% [15, 16]; a 93% yield of $Pb(CH_3)_4$ was obtained from CH_3MgCl and $PbCl_2$ (1:3 mole ratio) with excess CH_3Cl in THF [17, 18, 19]. This reaction can also be carried out in two steps [19]: $(CH_3)_3PbMgCl$ is first prepared in THF at 5°C by adding $PbCl_2$ or $Pb(OOCCH_3)_2$ to CH_3MgCl (1:3 mole ratio). Then, excess CH_3Cl at ambient temperature is added to give a 94% yield of $Pb(CH_3)_4$ [19]. (For conversion of elemental lead, obtained from the technical reaction of lead-sodium alloy and alkyl halide to $Pb(CH_3)_4$ with CH_3MgCl, see Subsection "From Alloys", p. 58.) A mixture of CH_3MgCl and $CH_2{=}CHMgCl$ in THF reacts with $PbCl_2$ to give $Pb(CH_3)_4$, $Pb(CH{=}CH_2)_4$, and the appropriate mixed species [20]. Reaction of $PbCl_2$ with a mixture of CH_3MgCl, $CH_2{=}CHMgCl$, and CH_3Cl (1:3.2:3.2:5 mole ratio) in THF at 5°C produces a mixture of $Pb(CH_3)_4$ (49%), $(CH_3)_3PbCH{=}CH_2$ (41%), and $(CH_3)_2Pb(CH{=}CH_2)_2$ (10%) with a total of 94% yield, whereas reaction with a mixture of CH_3MgCl and allyl chloride mainly gives $(CH_3)_3PbC_3H_5$ and only little $Pb(CH_3)_4$ [17]. A Grignard synthesis of $Pb(CH_3)_4$ and other tetraalkyllead compounds in gasoline is patented [21]. In this procedure, CH_3Cl and other alkyl halides are added to gasoline, or are obtained by in situ halogenation of lower alkanes in gasoline, and the reaction is then performed with magnesium and lead halide [21].

According to [2], pure $Pb(CH_3)_4$ is obtained from $(CH_3)_3PbBr$ and CH_3MgCl. Reaction of $(CH_3)_3PbBr$ and $t{-}C_4H_9MgCl$ gives $Pb(CH_3)_4$ and Pb; whereas with $(CH_3)_3PbBr$ or $(CH_3)_3PbI$ and $s{-}C_4H_9MgCl$, a mixture of $Pb(CH_3)_4$ and $(CH_3)_3PbC_4H_9{-}s$ is formed in 23 and 19% yield, respectively [12].

On addition of $Pb(OOCCH_3)_4$ to CH_3MgCl in THF at 5°C, $Pb(CH_3)_4$ is obtained in a yield of 89% [22, 23] or 80% [24]. $PbCl_4$ or $(NH_4)_2[PbCl_6]$ are unfavorable educts for the Grignard reaction [1]. The Grignard method also converts milligram amounts of lead(II) halides into $Pb(CH_3)_4$, e.g., for mass spectrometric analysis [25, 26, 27]; see also [28]. The minimum quantity of lead(II) halide for successfully performing the Grignard reaction is as low as 10 mg when i-amyl ether is used as solvent [26]. For transformation of Pb^{2+} from aqueous solution into $Pb(CH_3)_4$ by the Grignard reaction for analytical purposes, see [29].

$Pb(CD_3)_4$ is prepared from $PbCl_2$ or PbI_2 and CD_3MgBr [30, 31] or CD_3MgI [32, 33, 34]; see also [11]. In a similar way, $Pb(CH_3)_4$ enriched with [14]C [35] or [13]C [36], or labelled with [203]Pb [37], [210]Pb [38 to 41], or [212]Pb [41, 42] can be synthesized. The synthesis of ^{14}C-labelled $Pb(CH_3)_4$ from $(CH_3)_3PbCl$ and labelled CH_3MgCl on a microscale was less efficient than was an electrochemical procedure [629].

No isotopic effect was observed in the reaction of CH_3MgI and PbI_2 [25, 27]. $Pb(CH_3)_4$, prepared from CH_3MgI, may contain traces of iodine; these are removed by treating with freshly prepared Ag_2O for several days [10, 43]; purification by distillation is described in [5, 10, 44]. For a discussion of the mechanism of the reaction of Grignard reagents and lead(II) compounds, see [19] and the appropriate literature in Section 1.1.1.2.

From Metal Alkyls. Reaction of CH_3Li with PbI_2 in ether gives quantitative yields of $Pb(CH_3)_4$ in the presence of CH_3I. Thus, an ether solution of CH_3Li (0.15 mol) is added dropwise to a suspension of PbI_2 (0.04 mol) in dry ether containing CH_3I (0.07 mol). The elemental lead formed in the first stage of the reaction dissolves during addition of the second half of the CH_3Li solution. After washing with dilute HCl and water, the ether solution is dried with $CaCl_2$ and worked up as usual [5, 15]; see also [16]. This type of reaction allows synthesis of $Pb(CH_3)_4$ on a small scale, e.g., for isotopic analysis of $PbCl_2$ obtained by leaching minerals containing >1% lead with hydrochloric acid [28]; see also [25]. Silicones have been identified as impurities in $Pb(CH_3)_4$ prepared by this method [45, 46]. A procedure for analysis of Pb^{2+} in aqueous samples involves the extraction into $CHCl_3$ as a dithiocarbamate complex, solvent evaporation, and methylation of the residue with CH_3Li to form $Pb(CH_3)_4$, which is then determined by atomic absorption spectrometry [29].

Methylaluminium compounds react with lead(II) compounds to give $Pb(CH_3)_4$. It is produced from $(CH_3)_3Al_2Cl_3$ and $PbCl_2$ in the presence of alkali metal halides in xylene or in the absence of solvents at temperatures not higher than 130°C [47, 48]. In the presence of NaF, a yield of 85% has been claimed [48]. CH_3AlCl_2 and $(CH_3)_2AlCl$ can also be used as methylating agents [48]. $Pb(OOCCH_3)_2$, $Pb(OOCH)_2$, lead(II) naphthenate, and their double salts with PbO and PbS can be methylated by stoichiometric amounts of $Al(CH_3)_3$ or $M[M'(CH_3)_4]$ (M = alkali ion, M' = Al, Ga, In) in an inert solvent [49]. However, it also was reported that CH_3AlX_2 (X = Cl, Br) and PbO react to give PbX_2 and no $Pb(CH_3)_4$ [630]. Reaction of PbO and $Al(CH_3)_3$ yields about 26 to 28% $Pb(CH_3)_4$ depending on the reaction conditions (8 h/135°C or 140 h/80°C). Lewis bases inhibit the synthesis of $Pb(CH_3)_4$ [50, 51]. The reaction can be performed in refluxing toluene [52]. $Pb(CH_3)_4$ is obtained by reaction of PbO and methylaluminium sesquihalides or mixtures of $(CH_3)_2AlCl$ and CH_3AlCl_2, either in the absence or in the presence of nonpolar solvents, like benzene or toluene, at temperatures between 80 and 170°C [53]. Reaction of PbO and $(CH_3)_2AlCl$ in nonpolar solvents, such as C_6H_6, xylene, and decane at 60 to 140°C gives $Pb(CH_3)_4$ in the presence of alkali metal halides [54]; addition of o-xylene as a thermal stabilizer to the reaction mixture before distillation is recommended [55]. A continuous method for synthesis of $Pb(CH_3)_4$ from PbO and $(CH_3)_2AlCl$ in xylene was investigated in [56]. $Pb(CH_3)_4$ is obtained (up to 25%, 135°C) from the reaction of $Al(CH_3)_3$ and PbS with elemental lead and $((CH_3)_2Al)_2S$ as side products [51, 57]. Other methylation agents for PbS are $(CH_3)_2AlCl$ (advantageous in the presence of Lewis bases) [51, 58], and $(CH_3)_3Al_2Cl_3$ or their mixtures in nonpolar solvents, such as C_6H_6, xylene, toluene, decane, or tetralin at 110 to 140°C. The effect of alkali metal halides was also studied [59]. A mixture of $(CH_3)_2AlCl$ and $(C_2H_5)_2AlCl$ in p-xylene or benzene reacts with PbO, PbS, or $PbCl_2$ in the presence of Lewis bases, e.g., benzonitrile, anisole, or NaCl or KCl to give $Pb(CH_3)_4$ besides $(CH_3)_{4-n}Pb(C_2H_5)_n$ (n = 1, 2, 3, 4), even at room temperature. Overall yields at 135°C with the presence of KCl after 1 h were 50, 21, and 11%, respectively [51, 60, 61]. $B(CH_3)_3$ can also be employed to methylate PbO to give $Pb(CH_3)_4$ [52]. Reaction of $Zn(CH_3)_2$ and $PbCl_2$ was described in the first report on $Pb(CH_3)_4$ as one method for its preparation [62, 63]; however, $[CH_3Hg]^+$ does not methylate Pb^{II} salts [631].

$(CH_3)_3PbCl$ reacts with $Sn_2(CH_3)_6$ finally to give $Pb(CH_3)_4$, $Sn(CH_3)_4$, and $PbCl_2$ [64]. Generally, methyl-transfer equilibria are observed with $(CH_3)_3PbCl$ and methyltin compounds. Thus, $Sn(CH_3)_4$ or $(CH_3)_3SnC_4H_9$-t give mixtures containing $Pb(CH_3)_4$ and $(CH_3)_3SnCl$ or

$(CH_3)_2Sn(Cl)C_4H_9$-t, respectively, in equilibrium. For rate and equilibrium constants, see [64]. With $((CH_3)_3Sn)_2CH_2$, 10 to 20% $Pb(CH_3)_4$ was formed at equilibrium [628].

For the reaction with methyl cobalamin (methyl vitamin B_{12}), see "From Lead(II) and Lead(IV) Compounds and CH_3I or Other Methylating Agents", p. 71.

From Alloys and Methyl Halides. One of the commercial procedures now in use for the manufacture of $Pb(CH_3)_4$ is based on the reaction of lead-sodium alloy with methyl halide. Actually, the first paper on $Pb(CH_3)_4$ already mentioned the preparation from CH_3I and a lead-sodium (5:1) alloy [62, 63].

On a laboratory scale, $Pb(CH_3)_4$ can be prepared when CH_3Cl is reacted with a lead-sodium alloy (22% Na) [65] in the presence of pyridine [66]. Addition of some water at hourly intervals increases the yield to 50% [66]. For details, see [65].

In industrial syntheses, CH_3Cl and alloys essentially having the composition PbNa are used. The commercial batch process established for the manufacture of $Pb(C_2H_5)_4$ from alkyl halide and lead-sodium alloy (see "Organolead Compounds", Vol. 2, Section 1.1.1.2, to be published) is unsatisfactory for manufacturing $Pb(CH_3)_4$ owing to the high pressure necessary to maintain CH_3Cl as a liquid, the slowness of the reaction [67, 68], and the poor yield (only 25% after 50 h at 150°C) [68]. The reaction, however, can be conducted with higher yields and at lower temperatures when catalysts, such as Lewis acids [68 to 71], and a hydrocarbon, such as toluene, or isooctane [68] are added. Ethers (such as a dialkyl or diaryl ether [72 to 75], THF [74, 76], low molecular weight ethylene glycol dialkyl ether, or polyethylene glycol [77, 78, 79]), or NH_3 [80 to 83], amines [84, 85, 86], or amides [87, 88] are used as cocatalysts. The hydrocarbons serve as diluents, lowering the pressure, and as thermal stabilizers, and ensure the safety of the process [68, 89 to 96]. The reaction is carried out under dry nitrogen and in autoclaves under the autogenous pressure of CH_3Cl. Only one-fourth of the lead in PbNa is transformed into $Pb(CH_3)_4$ according to the equation

$$4\,PbNa + 4\,CH_3Cl \rightarrow Pb(CH_3)_4 + 3\,Pb + 4\,NaCl$$

while three-fourth are converted into finely divided elemental lead. The lead can be recycled, or it can be reacted with methylating agents to produce $Pb(CH_3)_4$.

Various chemical and constructional modifications for industrial production are described in a series of patents listed in Table 1 and in the following discussion. Numerous patents covering the preparation of $Pb(C_2H_5)_4$ claim the applicability of the specific procedure for the

Table 1
Preparation of $Pb(CH_3)_4$ from PbNa Alloy (= 100 parts per weight) and CH_3Cl in Pressure Reactors.

No.	conditions (parts per weight CH_3Cl)	catalyst system (parts per weight) and remarks	$Pb(CH_3)_4$ (yield in %)	Ref.
1	85 to 90°C (200)	$AlCl_3$ (0.2), $C_6H_5CH_3$ (20), continuous process, autogenous pressure	75 to 80	[68]
2	110°C, 15 min (77)	$AlCl_3$ (0.62), $Pb(CH_3)_4$ (2.5), graphite (2.5)	86.7	[93]
3	like No. 2	$AlCl_3$ (0.62), graphite (2.5)	52	[93]

Table 1 (continued)

No.	conditions (parts per weight CH$_3$Cl)	catalyst system (parts per weight) and remarks	Pb(CH$_3$)$_4$ (yield in %)	Ref.
4	like No. 2	AlCl$_3$ (0.62), Pb(C$_2$H$_5$)$_4$ (2.5), graphite (2.5)	61	[93]
5	90°C (45)	AlCl$_3$ (0.71), Pb(CH$_3$)$_4$ (7) as 38% solution in C$_6$H$_5$CH$_3$, (thickness of PbNa flakes 0.4 to 1.1 mm)		[97]
6	110°C, 1 h (77.5)	AlCl$_3$ (0.625), Li[AlH$_4$] (0.625), graphite (2.5)	75	[98]
7	like No. 6	AlCl$_3$ (0.63), graphite (2.5)	52	[98]
8	like No. 6	AlCl$_3$ (1.25), graphite (2.5)	80	[98]
9	like No. 6	Li[AlH$_4$] (1.25), graphite (2.5)	71	[98]
10	like No. 6	AlCl$_3$ (0.25), Li[AlH$_4$] (0.25), graphite (2.5)	84	[98]
11	50°C, 4 h (77.5)	AlCl$_3$ (0.63), Al(C$_4$H$_9$-i)$_3$ (1.25), graphite (2.5)	51	[71]
12	like No. 11	AlCl$_3$ (0.63), graphite (2.5)	30	[71]
13	110°C, 1.5 h (77.5)	AlCl$_3$ (0.63), Al(C$_4$H$_9$-i)$_3$ (0.63), graphite (2.5)	88	[71]
14	like No. 13	AlCl$_3$ (0.63), graphite (2.5)	<80	[71]
15	like No. 13	Al(C$_4$H$_9$-i)$_3$ (0.63), graphite (2.5)	<80	[71]
16	85°C, 4 h (77.5 CH$_3$Br)	AlCl$_3$ (0.63), Al(C$_4$H$_9$-i)$_3$ (0.63)	76	[71]
17	110°C, 1.5 h (90)	Al alloy (0.21), graphite (3.4), Br$_2$ (0.17), Br$_2$ and CH$_3$Cl introduced at −78°C before heating	99.5	[99]
18	85 to 110°C	Group 2 or 13 metal alkyl or chloride in C$_6$H$_5$CH$_3$	not reported	[68]
19	100 to 113°C, 6 h, 14 atm	Al(CH$_3$)$_3$ (0.2), C$_6$H$_5$CH$_3$ (10)	70	[68]
20	100 to 113°C, 6 h, 25 atm	Al(CH$_3$)$_3$ (0.2)	poor	[68]
21	85 to 110°C, 1 to 7 h	Al(C$_2$H$_5$)$_3$, CH$_3$O(CH$_2$CH$_2$O)$_2$CH$_3$ (1:0.75 mole ratio) in C$_6$H$_5$CH$_3$	93	[77, 78, 79]
22	like No. 21	Al(C$_2$H$_5$)$_2$H or (CH$_3$)$_3$Al$_2$Cl$_3$, CH$_3$O(CH$_2$CH$_2$O)$_2$CH$_3$ (1:0.75 mole ratio)	81.2 to 94.3	[79]

Table 1 (continued)

No.	conditions (parts per weight CH$_3$Cl)	catalyst system (parts per weight) and remarks	Pb(CH$_3$)$_4$ (yield in %)	Ref.
23	like No. 21	(CH$_3$)$_3$Al$_2$Cl$_3$, C$_2$H$_5$O(CH$_2$CH$_2$O)$_2$C$_2$H$_5$ (1:0.53 mole ratio)	90.1	[79]
24	like No. 21	(CH$_3$)$_3$Al$_2$Cl$_3$, CH$_3$O(CH$_2$CH$_2$O)$_4$CH$_3$ or CH$_3$O(CH$_2$CH$_2$O)$_2$C$_2$H$_5$ (1:0.53 mole ratio)	88.5 to 88.8	[79]
25	like No. 21 (ninefold excess CH$_3$Cl)	Al(C$_2$H$_5$)$_3$ (1.1), C$_2$H$_5$O(CH$_2$CH$_2$O)$_3$Na (1.27)	93	[100]
26	like No. 21	Al, AlCl$_3$, AlR$_3$, ether (e.g., (C$_6$H$_5$)$_2$O)	not reported	[73]
27	95 to 110°C	Al(C$_2$H$_5$)$_3$, CH$_3$OCH$_2$OCH$_3$ (1:1.26 mole ratio) in C$_6$H$_5$CH$_3$; see also use of other organoaluminium compounds and ethers in [72 to 75]	not reported	[72, 75]
28	110°C, 2 h (37)	Al(C$_2$H$_5$)$_3$ (0.24 wt% Al), C$_6$H$_5$CH$_3$ (5.4); see also in the presence of (CH$_3$O)$_2$CH$_2$	77.9 to 78.3	[72, 74]
29	110°C, 35 min (37)	Al(C$_2$H$_5$)$_3$ (0.11 wt% Al), (C$_6$H$_5$)$_2$O (1:0.6 mole ratio) in toluene	73	[74]
30	95 to 110°C, 2 h (37)	(CH$_3$)$_3$Al$_2$Cl$_3$ (0.93 wt%), C$_6$H$_5$CH$_3$ (5.4)	76.8	[101, 102]
31	like No. 30	(CH$_3$)$_3$Al$_2$Cl$_3$ (0.46 wt%), metadioxane (1.03 wt%), C$_6$H$_5$CH$_3$ (5.4); see also other organoaluminium compounds and ethers	91.1	[101,102]
32	like No. 30	(CH$_3$)$_3$Al$_2$Cl$_3$ (0.46 wt%), N(C$_4$H$_9$)$_3$ (1:1 mole ratio), C$_6$H$_5$CH$_3$ (5.4)	70.5	[84]
33	120°C	(CH$_3$)$_3$Al$_2$Cl$_3$, NaF or CaF$_2$ promotors	79.4 to 88.9	[103]
34	80 to 95°C, 12.7 atm	(CH$_3$)$_3$Al$_2$Cl$_3$ (0.2), C$_6$H$_5$CH$_3$ (10), amorphous carbon (0.5 to 1.2)	not reported	[104]
35	95 to 110°C (69)	NH$_3$, C$_6$H$_5$CH$_3$ (4.3), 60 ppm H$_2$O in CH$_3$Cl; (CH$_3$OH also improves yield)	up to 95	[80]
36	25 to 30°C, 2 h, then 50°C, 0.5 h (100)	NH$_3$ (1 to 150), H$_2$O or CH$_3$OH (0.014 to 2.5), C$_6$H$_5$CH$_3$, graphite; general range of reaction temperature −20 to 120°C; CH$_3$I impurity in CH$_3$Cl improves yield	not reported	[81, 82]
37	50°C (41.3)	NH$_3$, CH$_3$CN or C$_3$H$_7$CN or t-C$_4$H$_9$CN (10:1 mole ratio), C$_6$H$_5$CH$_3$ (4.15), 25 ppm H$_2$O in CH$_3$Cl	90.7	[83]

Table 1 (continued)

No.	conditions (parts per weight CH₃Cl)	catalyst system (parts per weight) and remarks	Pb(CH₃)₄ (yield in %)	Ref.
38	50 to 90°C (130 to 600)	CH_3NH_2 (10 to 50), H_2O (0.04 to 1)	not reported	[85, 86]
39	25 to 60°C (100 to 600)	$CH_3OCH_2CH_2NH_2$ (10 to 20), H_2O (0.3 to 3), C_1 to C_{18} alcohol (0.3 to 3); see also the use of other amines	86	[85, 86]
40	0°C (130)	$CH_3CON(CH_3)_2$ (0.5), 9 ppm H_2O in CH_3Cl (no yield with 25 ppm H_2O)	7	[87]
41	0°C (130)	$CH_3CON(CH_3)_2$ (0.5), H_2O (2:1 mole ratio); see also the use of other amines and ROH compounds	71	[87]
42	50°C, 1 h (130)	CH_3ONH_2, $[HONH_3]Cl$, $[i-C_3H_7NH_3]Cl$ as catalysts and CH_3OH, CH_3I as promotors	76 to 94	[88]
43	120°C, 2 h (770)	CH_3OH (13), diglyme (15), anthracene (4.4 or 0)	54	[105, 106]
44	85°C, 2 h (165)	$CH_3OCH_2CH_2OCH_3$ (9.8), H_2O (~20 ppm)	not reported	[107]
45	15 to 25°C	THF heating at 120 to 130°C	not reported	[76]
46	not reported	$ClCH_2CH_2Cl$ (78) in $C_6H_5CH_3$, $C_6H_5C_2H_5$, i-octane, or o-xylene (0 to 10)	not reported	[108]
47	~50°C, ~12 atm	$NaNH_2$, cumene; cyclopentadiene or $C_6H_5C≡CH$ was also used in the catalytic system	82	[109]
48	not reported	onium salts or Lewis bases; also with PbK and RX (X = Br, I)	not reported	[110]
49	80 to 90°C, 3 h (100)	$K[(CH_3)_2AlCl_2]$ (0.2), $C_6H_5CH_3$ (6)	94	[111]
50	100 to 120°C, 2 to 4 h	Al amalgam (2.17% Hg) in CH_3OH, C_6H_{14}, or C_6H_6	66 to 81	[112]
51	100°C, 3 h	Al amalgam (1:1)	73 to 82	[113]
52	not reported	LiAl alloy (1.1:1 mole ratio)	not reported	[114]
53	120°C, 3 h	MgAl alloy (1:2 mole ratio)	70 to 79	[115]

62

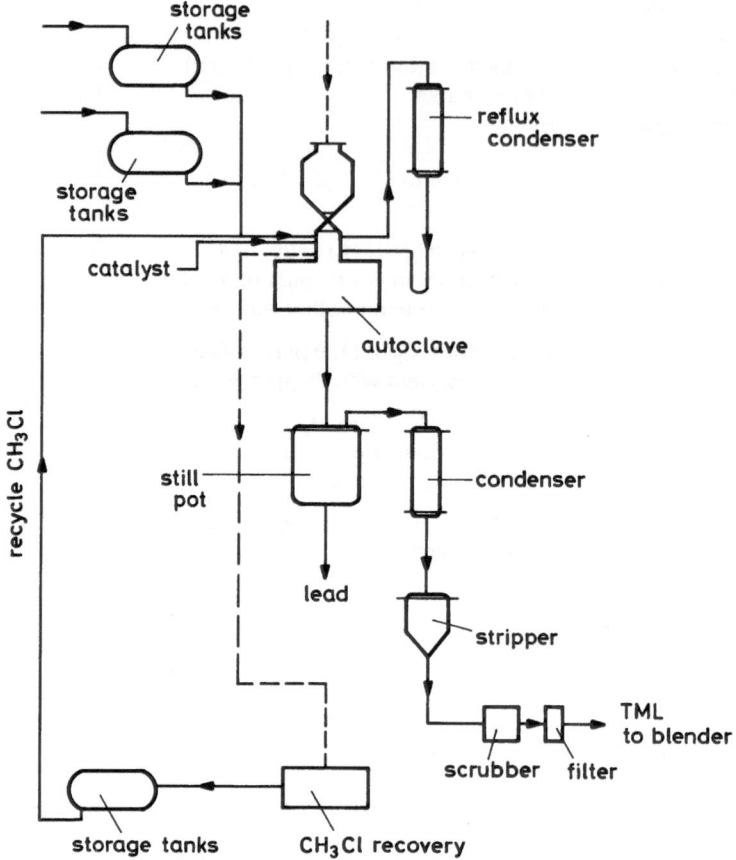

Fig. 1. Flowsheet for the process of synthesis of tetramethyllead (TML) from lead-sodium alloy and methyl chloride (according to [117]).

synthesis of $Pb(CH_3)_4$. For technical details of the production of $Pb(CH_3)_4$, see the relevant patents and a review [116], which though concentrating on the production of $Pb(C_2H_5)_4$ is informative regarding techniques applied in producing $Pb(CH_3)_4$. For descriptions of the commercial procedure, see [117, 118] and literature on the similarly performed synthesis of $Pb(C_2H_5)_4$ in Section 1.1.1.2; see also [119]. The principle of the process for the production of $Pb(CH_3)_4$ is shown in a flow chart in **Fig. 1** [117]. For sequential batch manufacture, see [120].

Reaction of PbNa alloy with CH_3Br at 85°C in the presence of $AlCl_3$ and $Al(C_4H_9-i)_3$ gives a 76% yield of $Pb(CH_3)_4$ in 4 h [71]. Also ternary alloys, such as lead-sodium-potassium alloy, containing 1 mole lead per 1 mole alkali metal [121], or lead-sodium-magnesium alloy in the presence of ethers [122 to 125] can be used for producing $Pb(CH_3)_4$ from CH_3Cl at elevated pressure; see also [126]. $PbMg_2$ reacts with CH_3Cl at 94°C in the presence of THF to give a 60.8% yield of $Pb(CH_3)_4$ [125]. Also reaction with CH_3Br [127] and CH_3I [128] can be realized in the presence of catalysts. When the lead-sodium alloy used in the reaction with alkyl halide contains 0.05 to 1% Mg, higher yields of the tetraalkyllead compound are obtained, and fewer problems are encountered during its recovery [129]. From sodium amalgam, containing about

3.5% Pb and CH$_3$Cl, only traces ($\sim$1%) of Pb(CH$_3$)$_4$ are obtained in addition to Hg(CH$_3$)$_2$ [130]. Dimethyl sulfate or a methyl phosphate, like trimethyl phosphate, can be used instead of methyl halide to methylate lead-sodium alloys at 110 to 150°C [131].

Production of Pb(CH$_3$)$_4$ by reaction of PbNa or alkaline earth-metal-lead alloys with CH$_3$X and CH$_3$MgX (X = halide) in ether is also possible under pressure and at elevated temperatures [132, 133]. A procedure to synthesize Pb(CH$_3$)$_4$ or other tetraalkyllead compounds in gasoline by reaction of lead-sodium alloy with CH$_3$Cl or alkyl halides, which are introduced or which are eventually obtained by in situ halogenation of lower alkanes in the gasoline, is described in [21].

Pb(CH$_3$)$_4$ is produced, together with Pb(C$_2$H$_5$)$_4$ and the mixed tetraalkyllead compounds (CH$_3$)$_n$Pb(C$_2$H$_5$)$_{4-n}$ (n = 1, 2, 3), by alkylation of PbNa alloy with a mixture of CH$_3$Cl and C$_2$H$_5$Cl in the presence of a catalyst, such as Al(CH$_3$)$_3$ [94, 95, 96], AlCl$_3$ [70], or an aluminium alloy, preferably at about 100°C for 4 h [69, 70]. A similar mixture, containing about 5 mol% Pb(CH$_3$)$_4$, is obtained by heating NaPb, CH$_3$Cl, and Pb(C$_2$H$_5$)$_4$ at 100 to 114°C for 125 min [134].

A series of procedures has been developed to convert the elemental lead obtained during manufacture of tetraalkyllead from lead-sodium alloy and alkyl halide into Pb(CH$_3$)$_4$. Due to the great activity of this finely divided material, it reacts with CH$_3$Cl in ether to give Pb(CH$_3$)$_4$. Reaction can be achieved in a two-stage process by introducing a solution of CH$_3$Cl in ether after an initial reaction period into the reacting mixture of CH$_3$Cl and the PbNa alloy to which Mg had been added. The reaction is performed in a temperature range of 45 to 85°C and 75 lb/in^2. In a similar one-stage reaction, all CH$_3$Cl and ether are added simultaneously during the reaction period [135]. Pb(CH$_3$)$_4$ and (CH$_3$)$_{4-n}$Pb(C$_2$H$_5$)$_n$ (n = 1 to 4) are obtained by a similar procedure, in which PbNa and CH$_3$Cl are reacted in the first stage and the precipitated lead and C$_2$H$_5$MgCl in the second stage [132]. Appreciable amounts of Pb(CH$_3$)$_4$ are obtained in addition to Pb(C$_2$H$_5$)$_4$ on treating the mixture resulting from the reaction of PbNa alloy with C$_2$H$_5$Cl in the presence of tricresyl phosphate at 80°C with CH$_3$MgCl·2 (C$_2$H$_5$)$_2$O [136]. Finely divided lead, to which up to 30 parts Mg per 100 parts Pb had been added, reacts with CH$_3$Cl in the presence of a catalyst, such as an alkyl ether or tertiary amines or tetraalkylammonium iodide, to give Pb(CH$_3$)$_4$ in high yields compared to the 25% conversion of lead in the methylation of PbNa. The reaction can be carried out as well in conjunction with the synthesis of Pb(CH$_3$)$_4$ from lead-sodium alloy and CH$_3$Cl, either in one stage or in two stages, the magnesium in the form of chips, turnings, etc., being charged to the reaction vessel together with the alloy [89 to 92]. Magnesium-lead alloys, preferentially Mg$_2$Pb, react in the presence of catalysts like aliphatic ethers, tertiary amines, pyridine, and tetraorganoammonium iodides at 80 to 120°C with pressures up to several hundred lb/in^2 with CH$_3$Br [127] or CH$_3$I [128] to give Pb(CH$_3$)$_4$. Another procedure starts from finely powdered Pb, Mg, and CH$_3$Cl, and is operated at 0 to 100°C in the presence of [N(C$_4$H$_9$)$_4$]I and (C$_2$H$_5$OCH$_2$CH$_2$)$_2$O [137]. Finely divided lead is also methylated by methyl sulfate or trimethyl phosphate to Pb(CH$_3$)$_4$ in good yield, especially in the presence of a catalyst, like an iodide and at temperatures ranging from 100 to 200°C [138]. Separation of Pb(CH$_3$)$_4$ from the reaction mixture is accomplished by distillation or extraction [71, 93, 98, 101]. A major problem is the handling of the slurry containing the elemental lead produced during the reaction. The organic phase containing Pb(CH$_3$)$_4$ can be separated after addition of water and acid to bring the pH to <7 [139]. Steam distillation is carried out [104, 111, 116]. Anticoagulating agents are used to prevent the lead slurry from agglomerating. Suitable agents are acidic mixtures of a polysulfide, like Na$_2$S$_2$, and a hydrolyzable Fe or Al compound like FeCl$_3$ or Al$_2$(SO$_4$)$_3$. Emulsifiers, like alkylbenzene sulfonates facilitate the separation [140]. Addition of a thermal stabilizing agent like toluene or a variety of other types of compounds (vide infra) is recommended during steam distillation

64

[94, 141]. Separation of Pb(CH$_3$)$_4$ by distillation is improved when THF is used as the solvent [76]. For the separation of Pb(CH$_3$)$_4$, the same yield is reported for steam distillation or extraction [93]. Extraction can be accomplished with toluene [80, 85, 87, 88, 99, 107]. An anhydrous method of recovery involves the vaporization of Pb(CH$_3$)$_4$ with a preheated inert gas or CH$_3$Cl [142]. Pb(CH$_3$)$_4$ can be recovered from the gas phase and vent gases with an organic scrubbing solution [143, 144].

Effluents from the manufacture of Pb(CH$_3$)$_4$ can be treated with an alkali metal borohydride, e.g., NaBH$_4$ at pH 8 to 11 to substantially reduce the level of dissolved lead compounds, like [Pb(CH$_3$)$_3$]$^+$ [145, 146]. Zn can also be used [147]. Liquid NH$_3$ and toluene are used to remove solid NH$_4$Cl from the apparatus for producing Pb(CH$_3$)$_4$ by the NH$_3$- or amine-catalyzed reaction of CH$_3$Cl with a PbNa alloy [148]. Stabilization of Pb(CH$_3$)$_4$, and of antiknock fluids containing Pb(CH$_3$)$_4$, is accomplished by addition of compounds, like toluene [149], xylene [141, 150], styrenes [151], naphthalenes [149, 151, 152], anthracenes [152], substituted phenols [141, 152 to 155], olefinic hydrocarbons [152], alcohols [141, 152], amines [155], hydroquinones [156], ethers [141], saturated or unsaturated carboxylic acids [141, 152], esters of phosphoric acid [152], or of sulfuric acid [157], or of sulfonic acids [141], imidazoles [158], alkyl halides and alkyl thiocyanates [141], or tall oil [159]; see also "Organolead Compounds", Vol. 2, Section 1.1.1.2, to be published.

By Electrolysis. Manufacture of Pb(CH$_3$)$_4$ is also accomplished by electrolysis of methyl-metal compounds at sacrificial lead anodes. The "NALCO" procedure uses Grignard compounds in a polyether solvent system:

anode process: $\quad$ Pb $\;+ 4\,CH_3MgX \rightarrow Pb(CH_3)_4 + 4\,MgX^+ + 4\,e^-$
cathode process: $\quad 4\,e^- + 4\,MgX^+ \;\rightarrow 2\,Mg + 2\,MgX_2$
$\qquad\qquad\qquad\quad 2\,Mg + 2\,CH_3X \;\rightarrow 2\,CH_3MgX$

overall reaction: $\quad$ Pb $\;+ 2\,CH_3MgX + 2\,CH_3X \rightarrow Pb(CH_3)_4 + 2\,MgX_2$

General descriptions of the technical procedure are given in [160 to 169]. A scheme of the procedure is shown in **Fig. 2**. The first chemical step of the process consists of the preparation of CH$_3$MgCl from Mg turnings and CH$_3$Cl in a solvent system preferentially containing a polyether (see Table 2) [170 to 186]. The conductivity of the Grignard solution is sufficiently high so that addition of a supporting electrolyte is not required. An advantageous cell consists of a tube reactor. The cell case is a stainless steel tube, which also acts as the cathode. In the center of the tube, a lead rod is arranged to serve as the connection to the positive electrical lead. The tube is filled with lead shot or pellets, which form the sacrificial anode, and can be continuously refilled through a feeder at the top of the tube. The inner tube wall is insulated and separated from the anode by a fine-mesh plastic netting [180, 187, 188], a perforated polymer, or a diaphragm of ceramic or other electrically nonconducting material [177, 189, 190, 191]. The Grignard solution is pumped through the tube, which is cooled to working temperature, in the NALCO process to 40 to 50°C. A number of such steel tube cells are mounted in parallel in a tube-and-shell reactor arrangement, equipped with a heat exchanger; see also [192]. One electrolyzer design contains a plurality of insulated cathode assemblies mounted within the vessel in a bed of lead granules that serve as the anode [193].

At the cathode, magnesium (theoretically 50% of the initial amount of magnesium in the Grignard solution) is deposited. It reacts with excess CH$_3$Cl added to the electrolyte to regenerate CH$_3$MgCl. The amount of magnesium removed as MgCl$_2$ is replaced by addition of CH$_3$MgCl. The yield is more than 96% [161, 163, 164, 165, 167]. The process can be used to make Pb(CH$_3$)$_4$, Pb(C$_2$H$_5$)$_4$, mixtures of both, or the mixed compounds (CH$_3$)$_{4-n}$Pb(C$_2$H$_5$)$_n$

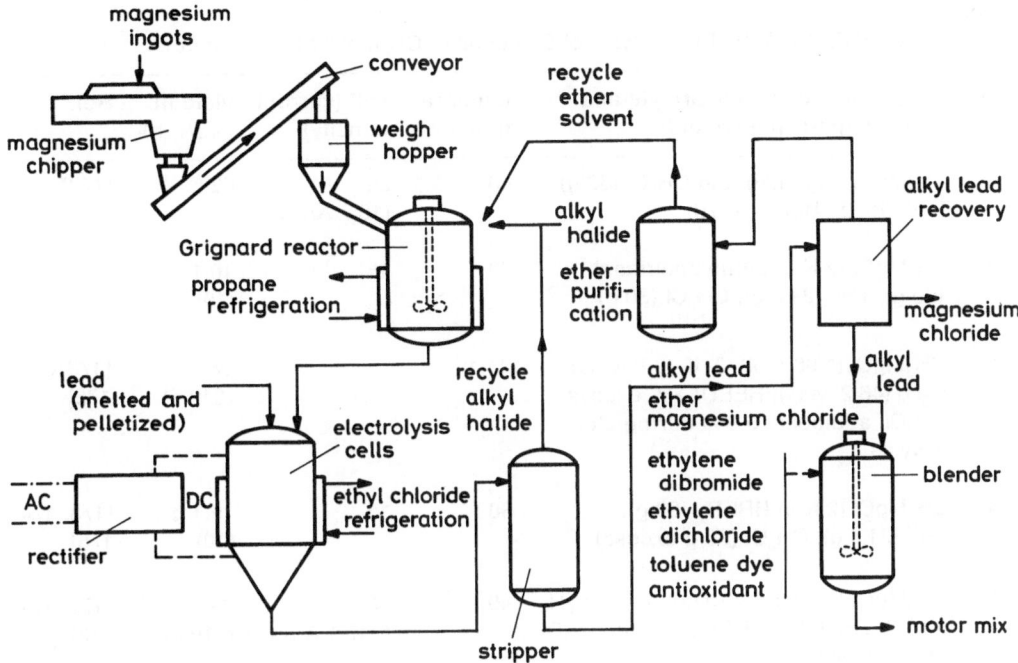

Fig. 2. Flowsheet for the NALCO (Nalco Chemical Co.) process for the electrolytic production of tetramethyllead (according to [161, 163]).

(n = 1, 2, 3) without major modification of the equipment [161]. Electrolytes and electrolysis conditions are compiled in Table 2 [170 to 186]. For technical details, see the relevant patents summarized in Table 2 and [187, 189, 190, 191, 194, 195]; see also [116]. The yield of Pb(CH$_3$)$_4$ can still be increased if the residual solution, in which only a small amount of Grignard reagent remains, is treated with PbCl$_2$ (1:1 mole ratio) at about 100°C [172]. For preparation of Pb(CH$_3$)$_4$ by electrolysis of a Grignard solution where electrochemical control of the concentration of the Grignard reagent is regulated by a coupled half-cell, see [196].

Explanations for Table 2: The yields given in column 5 refer to the % conversion of CH$_3$MgCl to Pb(CH$_3$)$_4$, if not stated otherwise, after a quantity of electricity (ampere hour (A·h)) is applied. The polyethers used in the solvent system of the Grignard solution are abbreviated as follows:

BEET = benzyl ethyl ether of triethylene glycol
HEED = hexyl ethyl ether of diethylene glycol
BBED = dibutyl ether of diethylene glycol
HEET = hexyl ethyl ether of triethylene glycol
HEEG = hexyl ethyl ether of ethylene glycol
EEET = diethyl ether of tetraethylene glycol
THFE = tetrahydrofurfuryl ethyl ether
DEED = diethyl ether of diethylene glycol

Table 2
Preparation of $Pb(CH_3)_4$ by Electrolysis of Solutions of CH_3MgCl at Lead Anodes.

No.	composition of the electrolyte and remarks (parts per weight)	tempera- ture in °C	volt (current density)	yield in % (A·h)	Ref.
1	CH_3MgCl (Mg (425) and CH_3Cl (885)), THF (5220), BEET (3470)	46	28 (16.7 A/ft^2)	82.5	[170]
2	CH_3MgCl (2 N in dibutylmethanol, 190 g), THF (21.4 g), CH_3Cl (30 g)	30	15	10.1 g (10)	[171, 638]
3	CH_3MgCl (1.98 mol), THF (9.8 wt%), C_6H_6 (46.2 wt%), HEED; 2% excess CH_3Cl, additional 5.3% during elec- trolysis	41.1		95.3 (219.6)	[172 to 175]
4	CH_3MgCl (2 N in BBED, 190 g), THF (21.4 g), CH_3Cl (30 g excess)	30	15	19.1 g (10)	[173, 174, 175]
5	CH_3MgCl (1.30 mmol/g solution), C_6H_6 (45%), THF (10%), HEED (45%), CH_3Cl (2% excess)	40	30 (247.1 A/dm^2)	99.1 (219.9)	[173, 174, 175]
6	CH_3MgCl (0.95 mol), C_6H_6 (4.5 mol), THF (1 mol), HEED (1 mol); additional CH_3Cl at the beginning and during electrolysis	23	27.7 (2.72 A/dm^2)	81.2 (2443)	[176, 177]
7	like No. 6; initial CH_3Cl and CH_3MgCl concentrations (0.03:1 mole ratio)	30	30 (20.2 A/ft^2)	91.1 (2515)	[177]
8	like No. 6; with polarity reversals to remove Pb precipitated between anode and cathode	30	25 to 30 (25.3 A/ft^2)	96.4 (22 h)	[177]
9	CH_3MgCl (1.30 mmol/g solution), C_6H_6 (45 wt%), THF (10 wt%), HEED (45 wt%), CH_3Cl (2% excess)	40		99.1 (219)	[173, 174, 175]
10	CH_3MgCl (1 mol), BBED (1 mol), THF, C_6H_6 (0.19 mol/mol CH_3MgCl), CH_3Cl (0.12 mol excess during electrolysis)	29.2	27.6	90.3 (204)	[178, 179]
11	CH_3MgCl (1.98 mmol/g solution), HEED [180] or HEET [181] or HEEG [182], C_6H_6, THF (1:2:0.77 mole ratio); CH_3Cl addition during electrolysis	39.5	25.5	97.9 (216)	[180, 181, 182]

Table 2 (continued)

No.	composition of the electrolyte and remarks (parts per weight)	tempera-ture in °C	volt (current density)	yield in % (A·h)	Ref.
12	CH_3MgCl (1.30 mmol/g solution), CH_3Cl, THF, C_6H_6 (1:0.3:1.02:4.41 mole ratio) in HEED [180], HEET [181], or HEEG [182]	40.2	22.2 to 22.4	99.1 (11.5)	[180, 181, 182]
13	CH_3MgCl (1.05 mmol/g solution), CH_3Cl, THF, C_6H_6 (1:0.63:0.04:5.53 mole ratio) in HEED [180], HEET [181], or HEEG [182]	49.4	30.1	94.5 (9.75)	[180, 181, 182]
14	CH_3MgCl (1.54 mmol/g solution), C_6H_6, xylene, CH_3Cl (1:4.07:0.12:0.3 mole ratio) in HEED [180], HEET [181], or HEEG [182]	40.1	29.7	94.6 (24 h)	[180, 181, 182]
15	CH_3MgCl (1 M solution in dibutylmethanol, 2680 g), THF (551 g), C_6H_6 (1739 g), CH_3Cl (224 g, maintained during electrolysis)	30	30 (140 A/m²)	91.8 (220)	[183]
16	like No. 15, without C_6H_6	30	30 (56 A/m²)	82.8 (173)	[183]
17	Mg, CH_3Cl (99 g), BBED, THF, C_6H_6 (0.95:0.98:1:0.77:2 mole ratio) additional CH_3Cl (264 g) during electrolysis	30	30 (120 A/m²)	92.1 (192)	[183]
18	like No. 17	40	30 (183 A/m²)	97 (192)	[183]
19	like No. 17; HEED instead of BBED, additional CH_3Cl (300 g), additional C_6H_6 during electrolysis to reach 45.6 wt%	40	30 (287 A/m²)	99.1 (216)	[183]
20	like No. 19; without THF, C_6H_6 replaced by xylene (46 wt%)	40.2	29.5 (129 A/m²)	94.7 (210)	[183]
21	like No. 20; xylene conc. initially 58%	44.2	29.6 (64 A/m²)	90.5 (156)	[183]
22	Mg, CH_3Cl, HEED, C_6H_6 (0.5:0.8:1:2.75 mole ratio), additional C_6H_6 and CH_3Cl during electrolysis	40 to 50	30 (258 A/m²)	94.5 (227)	[183]

Table 2 (continued)

No.	composition of the electrolyte and remarks (parts per weight)	temperature in °C	volt (current density)	yield in % (A · h)	Ref.
23	like No. 22; initial C_6H_6 conc. 37.9%, addition of more CH_3Cl during electrolysis	50	30.7 (261 A/m^2)	94 (217)	[183]
24	CH_3MgCl (1.44 mmol/g solution), THF (45 wt%), EEET (55 wt%), CH_3Cl (0.15 mol excess per mol Grignard solution)	43	25.6 (14.8 A/ft^2)	98 (255)	[184, 185]
25	CH_3MgCl (1.49 mmol/g solution), THF (35 wt%), EEET (65 wt%), CH_3Cl like No. 24	35	26.4	80 (246)	[184, 185]
26	CH_3MgCl (1.34 mmol/g solution), THF (50 wt%), THFE (50 wt%), CH_3Cl like No. 24	42	20.5 (28.2 A/ft^2)	91.2 (250.2)	[184, 185]
27	CH_3MgCl (1.16 mmol/g solution), THF (50 wt%), THFE (50 wt%), CH_3Cl (0.3 mol excess per mol Grignard solution)	34	22.2 (17.3 A/ft^2)	99.2 (250.2)	[184, 185]
28	CH_3MgCl (1.32 mmol/g solution), THF (75 wt%), THFE (25 wt%), CH_3Cl like No. 24	45	26.8 (15.9 A/ft^2)	92.6 (217)	[184, 185]
29	CH_3MgCl (1.48 mmol/g solution), THF (50 wt%), DEED (50 wt%), CH_3Cl like No. 24	51	19 (26.3 A/ft^2)	99 (385.4)	[184, 185]
30	CH_3MgCl in THF (60 wt%), BEET (40 wt%)	no further conditions and yield reported			[186]

Addition of CH_3Cl and C_2H_5Cl during electrolysis to the initial solution of CH_3MgCl in hexyl ethyl ether of ethylene glycol [182], or of diethylene glycol [180], or of triethylene glycol [181], and addition of C_6H_6 and THF gives $Pb(CH_3)_4$, $Pb(C_2H_5)_4$, and the mixed tetraalkyllead compounds [180, 181, 182]. A similar procedure is described for the preparation of a mixture of $Pb(CH_3)_4$ and tert-butyllead compounds [180, 181, 182]. $Pb(CH_3)_4$ and mixed tetraorganolead compounds, containing methyl and phenyl groups, are obtained when a solution of CH_3MgCl in dibutyl ether of diethylene glycol or THF and/or C_6H_6, to which C_6H_5Cl has been added in a 2:1 mole ratio, is circulated through a steel pipe electrolyzer filled with lead pellets with an applied potential of 30 V [197, 198]. Small amounts of $Pb(CH_3)_4$ and $Pb(CH=CH_2)_4$ are obtained in addition to the mixed compounds during electrolysis of $CH_2=CHMgCl$ in a THF-polyether-C_6H_6 solvent at lead anodes at 5.3 V, when CH_3Cl is added during the procedure [199]. Recovery of $Pb(CH_3)_4$ and solvent components from electrolysis mixtures is accomplished by steam or azeotropic distillation or by extraction methods, which are described in the patents listed in Table 2, and in [200, 201]. For recovery of $MgCl_2$ in the electrolytic production of

$Pb(CH_3)_4$, see [202], and for the recovery of CH_3Cl from the electrolyte containing $Pb(CH_3)_4$, see [203]. For methods of monitoring the concentration of the Grignard electrolyte during electrolysis, see [196, 204, 205].

$Pb(CH_3)_4$ can also be prepared by electrolyzing a solution of CH_3MgCl in THF using 60 Hz a.c. at two lead electrodes at 30 to 31 °C with 59 V at 1 A (current density 10 A/dm^2), while CH_3Cl is bubbled through the agitated solution. The current yield is 69% [206]. An alternating current electrolytic cell is described in [207]. A patent describes the preparation of $Pb(CH_3)_4$ in 89% yield, based on loss of lead, by electrolyzing CH_3I in the presence of $NaClO_4$ using a zinc cathode and a lead anode [208].

Another anodic procedure for preparing $Pb(CH_3)_4$ uses solutions of organoaluminium compounds as electrolytes under inert gas. Whereas molten $Na[Al(C_2H_5)_4]$ (m.p. 124 °C) can be used as anolyte for the preparation of $Pb(C_2H_5)_4$ (see "Organolead Compounds", Vol. 2, Section 1.1.1.2, to be published), the electrolysis of $Na[Al(CH_3)_4]$ at lead anodes to gain $Pb(CH_3)_4$ has to be carried out in a solvent, due to the 238 °C melting point of pure $Na[Al(CH_3)_4]$ [130, 209 to 212]; see also [213, 214]. THF and the dimethyl ether of diethylene glycol are the most favorable solvents [130, 212], but dioxane and other polyethers, like diglyme or triglyme, can also be used [209, 210, 211]. A rotating lead disk electrode as anode is used, and mercury serves as the cathode. Sodium, deposited during electrolysis, dissolves in the mercury, which is circulated, so that the amalgam reaches a concentration of 0.3 to 0.4% Na [130]. $Al(CH_3)_3$ forms a high-boiling adduct with the ether used. Some elemental lead precipitates which presumably is a product of redistribution of lower lead methyls formed at the anode [130, 209, 210, 211, 213]. Thus, 200 g $Na[Al(CH_3)_4]$ and 400 g THF with 5.5 V/3.8 A at 90 °C (7 h) gave 60 g $Pb(CH_3)_4$ (90% based on current supplied), which was removed along with THF by distillation and purified by steam distillation [209, 210, 211]. Using $CH_3O(CH_2CH_2O)_2CH_3$ as the solvent, $Pb(CH_3)_4$ can be removed by distillation at 50 to 80 °C and 50 to 100 Torr. $Na[Al(CH_3)_4]$ is regenerated by reaction of the appropriate $Al(CH_3)_3$-ether adduct with sodium amalgam and CH_3Cl [215]. It also can be recovered by reacting $Al(CH_3)_3$ with $Na[Al(CH_3)_3OR]$ (R = C_4H_9-n, C_6H_{13}-n) [216]. The overall reactions, therefore, correspond to the equations:

$$4\,Na[Al(CH_3)_4] + 4x\,Hg\,(cathode) + Pb\,(anode) \rightarrow 4\,Na(Hg)_x + Pb(CH_3)_4 + 4\,Al(CH_3)_3$$
$$4\,Al(CH_3)_3 + 4\,CH_3Cl + 8\,Na(Hg)_x \rightarrow 4\,Na[Al(CH_3)_4] + 4\,NaCl + 8x\,Hg$$

$$4\,Na(Hg)_x + 4\,CH_3Cl + Pb \rightarrow Pb(CH_3)_4 + 4\,NaCl + 4x\,Hg$$

The direct reaction is not feasible, the maximum yield of $Pb(CH_3)_4$ being only about 1% [130]. Since sodium amalgam is a product of the "Mülheimer Verfahren" to produce $Pb(C_2H_5)_4$ (see "Organolead Compounds", Vol. 2, Section 1.1.1.2, to be published), both procedures can be combined through a common amalgam circuit to gain $Pb(CH_3)_4$ and $Pb(C_2H_5)_4$ in equimolar amounts [130, 212].

Other organoaluminium complexes have been proposed for use as anolytes in the procedure: $K[Al(CH_3)_4]$ and $M[Al(CH_3)_3OR]$ (M = Na, K; R = C_4H_9-n, C_6H_{13}-n, C_8H_{17}) [209, 210, 211, 216, 217]. An electrolytic cell especially suited for this procedure is described in [218]. $Pb(CH_3)_4$ is produced in a similar way from a solution of $Na[B(CH_3)_4]$, $K[B(CH_3)_4]$, or $Na[(CH_3)_3BOCH_3]$ in THF or other ethers at 4.5 V and 90 °C [209, 210, 211]; see also [219].

$Pb(CH_3)_4$ is one of the products of electrolyzing a mixture of $Mg(CH=CH_2)_2$ and $Al(CH_3)_3$ or $B(CH_3)_3$ in $CH_3O(CH_2CH_2O)_2CH_3$ at 50 to 75 °C or a mixture of $Mg(CH=CH_2)_2$ and $(CH_3)_2AlOCH_3$ or $(CH_2=CH)_2AlCH_3$ in THF-$CH_3O(CH_2CH_2O)_2CH_3$ at 65 °C at a lead anode. If CH_3Cl is introduced into the mixture during electrolysis, $Pb(CH_3)_4$ is also produced when methyl-free aluminium compounds like $Al(C_6H_5)_3$ or $Al(OC_6H_5)_3$ are used [220].

Mixtures containing equimolar amounts of $Na[Al(CH_3)_4]$ and $Na[Al(C_2H_5)_4]$ can be electrolyzed without using a solvent at about $100°C$ in a cell with axially positioned lead anode and steel or copper cathodes; 3.6 V are applied (current density $0.25 A/cm^2$) [221, 222]. The only product isolated was $Pb(C_2H_5)_4$ [130, 221, 222]. In the patents, it is claimed that $Pb(CH_3)_4$ can be prepared similarly [221, 222]. $Pb(CH_3)_4$ can be prepared even from an aqueous system. Thus, a solution of 28.3 g $Na[B(CH_3)_4]$ in 31 g H_2O with a lead anode at 1.5 V (current density $3.5 A/dm^2$) produces 9.2 g $Pb(CH_3)_4$ with a current yield of 78% [223].

$Pb(CH_3)_4$ is also formed at a lead cathode during electrolysis of methyl halides. Reduction of CH_3Br or CH_3I at a lead electrode in dimethylformamide containing NaBr or NaI gives $Pb(CH_3)_4$ with a current yield of 85% (bromide system) and 90% (iodide system), respectively [224, 225]. Similarly, small amounts of $Pb(CH_3)_4$ are obtained in addition to $(CH_3)_{4-n}Pb(C_2H_5)_n$ (n = 1 to 4) by reduction of mixtures of CH_3I and C_2H_5I in dimethylformamide containing $NaClO_4$ [225]. Regarding the electrode process, see also [612]. Electrolysis of a catholyte containing 46.7 g CH_3Br, 36 g H_2O, 25.2 g $[N(C_2H_5)_4]Br$, and 260 g acetonitrile at a lead cathode at 45°C produces in 220 min (current density $0.06 A/cm^2$, platinum anode) an 84% yield of $Pb(CH_3)_4$, based on consumption of current [226, 227, 228]. In the absence of water the yield was only 73%, and replacement of acetonitrile by ethanol reduced the yield to 35% [226, 227, 228]. Other catholytes used for cathodic production of $Pb(CH_3)_4$ contain 1,2-dimethoxyethane or acetonitrile as solvent, CH_3X (X = Cl, Br, I) for delivering methyl groups, and $[R_4N]Br$ (R = C_2H_5, C_4H_9) or $[S(CH_3)_3]I$ for increasing the conductivity. Solutions of $[N(C_2H_5)_4]Br$ in acetonitrile or of Na_2CO_3 in water are employed as anolytes [229 to 232]. In a similar procedure, CH_3Br, water, $[N(C_4H_9)_4]Br$ or $[P(C_4H_9)_4]Br$, and acetone, THF, acetonitrile, or propionitrile are components of the catholyte; the anolyte is an aqueous solution of Na_2CO_3, $(NH_4)_2CO_3$, or $[N(C_4H_9)_4]Br$. $Pb(CH_3)_4$ is produced at the lead cathode in high yield [233, 234]. For recovery of $Pb(CH_3)_4$ from acetonitrile, see [227, 235]. Addition of a thermal stabilizer, like toluene, during distillation of the azeotrope, which is formed by $Pb(CH_3)_4$ and acetonitrile, is recommended [236, 237]. Similarly, reduction of $[S(CH_3)_3]I$ in dimethylformamide with $[N(C_2H_5)_4]Br$ as support electrolyte gives $Pb(CH_3)_4$ [238]. For cathodic synthesis of $Pb(CH_3)_4$, in addition to $Pb(C_2H_5)_4$, and the mixed tetraalkyllead compounds, see [224, 225, 229, 230]. Electrolysis of an aqueous solution containing 2 to 30% of $(CH_3)_3PbCl$ at 1 to 30 V with current densities of 0.2 to 5 A/in^2 is reported to give very high yields of $Pb(CH_3)_4$. The electrolysis can be performed at room temperature or up to $\sim 93°C$ using carbon anodes and lead cathodes; the starting pH should be 6 to 8 [239].

[14]C-labelled $Pb(CH_3)_4$ is obtained on a microscale by electrochemical reduction of labelled CH_3I at a sacrificial lead cathode using anhydrous dimethylformamide as solvent and $NaClO_4$ as supporting electrolyte [629].

Polarographic reduction of CH_3Br at a lead electrode in a mixture of $[R_4N]Br$ (R = C_2H_5, C_4H_9), acetonitrile, and water gives $Pb(CH_3)_4$ in yields approaching 100% [241, 242]. With CH_3I similar results are obtained [240]. Direct formation of the Pb-C bonds is inferred from the efficiency of the cathodic synthesis of $Pb(CH_3)_4$. Stepwise methylation of lead during the reduction of the methyl halides at the cathodes is assumed. The last alkylation step by CH_3Br is sufficiently in favor of formation of $Pb(CH_3)_4$, precluding formation of $Pb_2(CH_3)_6$ [242]; see also [243]. However, $Pb(CH_3)_4$ is assumed to be a dismutation product of the radical $Pb(CH_3)_3^{\cdot}$ in the electrochemical reduction of $[Pb(CH_3)_3]^+$ at a mercury electrode [244, 245]. For reviews of the electrosynthesis of $Pb(CH_3)_4$, see [243, 246 to 251].

From Lead and Methyl Halides. Ordinary forms of lead are not capable of reacting with organic halides to form $Pb(CH_3)_4$. However, finely divided lead, e. g., that formed in the reaction of CH_3Li with lead(II) halides, reacts with CH_3I giving a quantitative yield of $Pb(CH_3)_4$ [15, 252];

see also Subsection "From Metal Alkyls" p. 57. $Pb(CH_3)_4$ was also proposed as the product of refluxing pyrophoric lead (obtained by pyrolysis of lead citrate) and CH_3I for 8 h in a polar solvent (5% yield) [253, 254]. Lead powder undergoes methylation by CH_3I even in aqueous media in aerobic and anaerobic systems, giving detectable quantities of methyllead salts and $Pb(CH_3)_4$ within 24 h [255, 256]. Similar results were reported for the reaction of elemental lead with $[S(CH_3)_3]I$ [257]. However, in another analogous experiment no $Pb(CH_3)_4$ was detected [256], and also reactions of elemental lead with $S(CH_3)_2$ or $[(CH_3)_3N^+CH_2COO^-]$ gave no $Pb(CH_3)_4$ [256, 257]. It was detected by gas chromatography when elemental lead and CH_3I were kept in water in the dark at room temperature for up to 2 weeks [257, 258, 259]. The reaction of Pb^{2+} with CH_3I in the presence of Al proceeds to produce $Pb(CH_3)_4$ only after reduction of Pb^{2+} to Pb [260].

On reaction of excess CH_3Cl under Ar with finely divided lead and excess Li containing Na and K impurities, which are necessary for the reaction, 85.5% of the lead is converted in 4 h into $Pb(CH_3)_4$ [261]. Presence of ethers eliminates the formation of some byproducts [262]. It also was reported, that the preparation of $Pb(CH_3)_4$ from finely divided lead and CH_3Cl at elevated temperatures and pressure is improved by activating the lead by addition of sodium to give thin surfaces of sodium-lead alloy [263, 264]. Synthesis of $Pb(CH_3)_4$ can also be accomplished by heating finely divided lead with CH_3Cl or CH_3Br in the presence of I_2 or iodides, e.g., PbI_2, NaI, KI, HgI_2, etc. as catalysts; yields are increased by addition of $AlCl_3$ [265]. Elemental lead, produced by the reaction of PbNa and C_2H_5Cl, is methylated with dimethyl sulfate in the presence of PbI_2 as catalyst, or with CH_3I in the presence of iodine as catalyst with a yield of 65 to 70% $Pb(CH_3)_4$ [67]. For methylation of elemental lead, obtained from the industrial reaction of lead-sodium alloy and alkyl halide, see Subsection "From Alloys and Methyl Halides", p. 58.

From Lead and Methyl Radicals. Methyl radicals deplete a carbon-free lead mirror [10], presumably forming $Pb(CH_3)_4$ [43], although formation of $Pb(CH_3)_2$ has also been suggested [266]. Doubts about the role of CH_3 radicals in these experiments [267] have been rejected [268, 269]. The results of Paneth and Hofeditz [43] are experimentally confirmed in [270]. The mechanism of the reaction of Pb and CH_3 radicals was studied in [271, 272]. H_2, He, or N_2 was used as the carrier gas [271, 272]. The reaction was employed to prove the existence of CH_3 radicals in the pyrolysis of CH_3CHO [273], and in the decomposition of $CuCH_3$ [274]. Determination of $Pb(CH_3)_4$ produced from radicals obtained by thermolysis or photolysis of $O(CH_3)_2$ or CH_3COCH_3 and a lead mirror is used to estimate the number of radicals [275]. For methods to measure the rate at which lead mirrors are removed by methyl radicals, see [266, 273, 275]; see also [38, 271, 276].

A free radical procedure for preparing $Pb(CH_3)_4$ and mixed tetraalkyllead compounds by passing CH_3Cl and other alkyl chlorides over a heated mixture of lead and an initiator metal, like copper, is described in [277]. Reaction of lead deposited as a mirror on the inside of a quartz tube with CH_3 radicals formed by decomposition of di-tert-butylperoxide is employed to prepare $Pb(CH_3)_4$ from small amounts of lead, e.g., from minerals for mass spectrometric analysis [278, 279]. $^{212}Pb(CH_3)_4$ is formed by reaction of CH_3 radicals with recoil products of ^{224}Ra [280, 281, 282]; see also [283].

From Lead(II) and Lead(IV) Compounds and CH_3I or Other Methylating Agents. CH_3I was reported to methylate lead(II) in aqueous solutions of $Pb(OOCCH_3)_2 \cdot 3H_2O$, $Pb(NO_3)_2$, $Pb(ClO_4)_2 \cdot 3H_2O$, or Pb^{II} oxalate producing small amounts of $Pb(CH_3)_4$ [284]. The yield of $Pb(CH_3)_4$ was increased under highly basic conditions [284]; see also [285, 286]. Mixtures of CH_3I and C_2H_5I, reacting with Pb^{II} compounds at pH 13, gave only little $Pb(CH_3)_4$, and the

72

amount of tetraalkyllead produced increases with increasing ethyl substitution [284]. $[S(CH_3)_3]I$, $[(CH_3)_3SO]I$, and $[O(CH_3)_3][SbCl_6]$ can also be used to methylate lead(II) compounds in aqueous solution to give $Pb(CH_3)_4$, but only with very low yields [284]. In other experiments, no methylation of lead(II) by CH_3I [257], $[S(CH_3)_3]I$ [256, 257, 287], $[(CH_3)_3SO]I$, D,L-methionine-S-methylsulfonium chloride [287], $[(CH_3)_3N^+CH_2COO^-]$ [256, 257], $CH_3As(OH)_2O$ or $(CH_3)_2As(OH)O$ [631] was observed. Conflicting papers report that lead(II) salts are methylated by CH_3I only to the $[Pb(CH_3)_3]^+$ stage, and not to $Pb(CH_3)_4$ [255, 256] or that the reaction could not be reproduced, whether it was performed with or without the presence of elemental magnesium [259]. The methylation of lead(II) salts by CH_3I is enhanced by the presence of bulk metals, e.g., Al, Zn [285], or Mg [257, 285]. The reaction was later ascribed to the methylating action of CH_3I on elemental lead, which was produced by reduction of Pb^{2+} by elemental aluminium being present as covering foil [260]; see also [255]. It was also reported, that $Pb(CH_3)_4$ is not formed by direct reaction of PbS [288], $Pb(NO_3)_2$ [287, 289], or of lead(II) salts [260] with CH_3I in aqueous media.

The question as to whether or not there are biological pathways for the formation of $Pb(CH_3)_4$ has inspired various experiments to methylate lead(II) and lead(IV) compounds with the biologically relevant methyl donor methyl vitamin B_{12} (methyl cobalamin) and appropriate model compounds. Demethylation of methyl cobalamin marked with $^{14}CH_3$ occurs with $Pb(OOCCH_3)_4$, PbO_2, and Pb_3O_4. Although volatilization of ^{14}C was observed, no methyllead compounds have been identified [258, 290, 291]. However, in later kinetic studies of the interaction of PbO_2 and methyl cobalamin, traces of $Pb(CH_3)_4$ have been identified [292]. Methyl transfer was not observed to take place from methyl cobalamin to $(CH_3)_3PbOOCCH_3$, $(CH_3)_3PbCl$, and $(CH_3)_2PbCl_2$ [293]. In contrast, slow demethylation of methyl cobalamin by dialkyllead compounds [294] and production of $Pb(CH_3)_4$ under anaerobic [295] or under aerobic conditions [296] by reaction of methyl cobalamin with trimethyl- and dimethyllead compounds was reported. The major route to produce $Pb(CH_3)_4$ in this reaction, however, is not demethylation of methyl cobalamin, but redistribution [295]. Methylation of $[Pb(CH_3)_3]^+$ and $[Pb(CH_3)_2]^{2+}$ to $Pb(CH_3)_4$ is also accomplished by a dimethyl cobalt complex $[(CH_3)_2\text{-}CoN_4]ClO_4 \cdot H_2O$ (N_4 = 2,3,9,10-tetramethyl-1,4,8,11-tetraazacyclotetradeca-1,3,8,10-tetraene, Formula I) in acetonitrile or in aqueous media. The yields are 33 to 65% [289, 297]. The same complex methylates lead(II) compounds in acetonitrile or in aqueous media to $Pb(CH_3)_4$ when a 1:2 mole ratio is used [287, 289, 297] with a yield of 0.05 to 0.18%, and elemental lead is also produced [297]. Yields are lower in the presence of MnO_2, and are not enhanced on addition of CH_3I [287, 289]; see also [298]. $Pb(CH_3)_4$ was also obtained in very low yield from $Pb(NO_3)_2$ and complex I in a sediment system [632].

I

II

No methyl transfer from methyl cobalamin to Pb^{2+} was observed [287, 290, 291, 293, 295, 299, 632]; see also [294, 300, 301]. Lead(II) is also unreactive towards methyl pyridinatocobaloxime [302], and formation of $Pb(CH_3)_4$ was not observed in the reaction of the trans-dimethylcobalt complex II with Pb^{2+} [303].

Other Methods. A procedure to synthesize $Pb(CH_3)_4$ and other tetraalkyllead compounds by reacting vapors of pyrolyzed hydrocarbon compounds, e.g., acetone, ether, or heptane, and finely divided lead is described in a patent [304]. In electric discharges using lead electrodes in hydrogen containing hydrocarbon compounds, no $Pb(CH_3)_4$ is produced [305]. Reduction of $(CH_3)_3PbI$ with sodium in liquid ammonia gives $Pb(CH_3)_4$ as well as elemental lead and some $Pb_2(CH_3)_6$ [12].

A great number of methyllead compounds show a more or less explicit tendency to redistribute, and formation of $Pb(CH_3)_4$ depends on the temperature and other conditions. Disproportionation of trimethyllead compounds $(CH_3)_3PbX$ gives $Pb(CH_3)_4$ and $(CH_3)_2PbX_2$; the reaction is reversible [306, 307]; see also [308]. $(CH_3)_2PbX_2$ can redistribute irreversibly into CH_3X, PbX_2, and $(CH_3)_3PbX$, which again can produce $Pb(CH_3)_4$ [307]. Formation of $Pb(CH_3)_4$ has been observed during the disproportionation of $(CH_3)_3PbCl$ on heating at 112°C for 3 d [309], at 70°C in $CDCl_3$ [310], or at 60°C in pyridine [311]; see also [6]. $Pb(CH_3)_4$ is also formed during the disproportionation of $(CH_3)_3PbH$ and $(CH_3)_2PbH_2$ [312, 313, 314], dimethyl-lead dioxinate at 70°C in C_6D_6 [310], $(CH_3)_3PbC_6H_4C_6H_5$-4 [315], $((CH_3)_3Pb)_2S$ [316], $(CH_3)_3PbSCH_2COOPb(CH_3)_3$, and $(CH_3)_3PbSCH_2CH_2COOPb(CH_3)_3$ [317], $[(CH_3)_3Pb]_2SO_4$ [318], $(CH_3)_3PbOS(O)CH_3$ [319], $(CH_3)_3PbMn(CO)_5$ [320, 321, 322], and $(CH_3)_3PbRe(CO)_5$ [321, 322, 324]. Lead powder promotes the disproportionation of $(CH_3)_3PbCl$ [255] and so does sulfide in aqueous systems [316]. During equilibration in the system $(CH_3)_3PbCl$-$(C_2H_5)_3PbCl$ [325], formation of $Pb(CH_3)_4$ has also been considered in addition to alkyl-alkyl exchange [308]. $Pb(CH_3)_4$ was observed as a result of disproportionation of $(CH_3)_3PbCl$ due to elevated temperature in the injector part of a gas chromatographic system [326]. $Pb(CH_3)_4$ is one of the products of the redistribution reaction of $Pb(C_2H_5)_4$ with CH_3X (X = Cl, Br, I) in the presence of AlX_3 and I_2 or alkyl iodide at a temperature of preferably 80 to 120°C without solvent or in an aprotic polar solvent like N,N-dimethylformamide or hexamethylphosphortriamide [327]. CH_3I, $AlCl_3$, and $Pb(C_2H_5)_4$ undergo the exchange reaction even at −20°C [327]. $Pb(CH_3)_4$ labelled with ^{14}C can be obtained by vapor-phase exchange of unlabelled $Pb(CH_3)_4$ and CH_3I enriched with ^{14}C [616]. $Pb(CH_3)_4$ is a product in the random equilibrium mixture formed during the redistribution reaction of single organolead compounds containing CH_3 groups bonded to lead, e.g., $(CH_3)_3PbC_2H_5$ or $(CH_3)_2Pb(C_2H_5)_2$ or $(CH_3)_2Pb(C_4H_9$-i$)_2$ in hexane [328], or of mixtures of methylalkyllead compounds, e.g., $(CH_3)_3PbC_2H_5$ and $CH_3Pb(C_2H_5)_3$ or $(CH_3)_3Pb(C_3H_7$-i) and $(CH_3)_2Pb(C_3H_7$-i$)_2$ without solvent [328]. The redistribution is catalyzed by $AlCl_3$ [328]. The analysis of random equilibrium mixtures of $(CH_3)_{4-n}Pb(C_2H_5)_n$ (n = 0 to 4) is described in [329]. Further, $Pb(CH_3)_4$ is present in the equilibrated mixtures of methylmetal compounds and organolead compounds, e.g., $Sn(CH_3)_4$ and $Pb(C_2H_5)_4$ [330] or $Hg(CH_3)_2$ and $Pb(C_2H_5)_4$ [331]. The proportion of $Pb(CH_3)_4$ in the final mixture depends on the specific affinities of the metals for CH_3 and the other organic groups [331, 332]. The major route in the production of $Pb(CH_3)_4$ from methyl cobalamin and trimethyllead or dimethyllead compounds is not demethylation, but redistribution [295]. Alkyl exchange between $(CH_3)_3PbC_4H_9$-t and $(CH_3)_3PbCl$ gives $Pb(CH_3)_4$ and $(CH_3)_2Pb(Cl)C_4H_9$-t, the equilibrium being slightly in favor of the reactants [34]. $Pb_2(CH_3)_6$ and $(CH_3)_3PbCl$ (1:2 mole ratio) react to give $Pb(CH_3)_4$ and $PbCl_2$ (2:1 mole ratio); the same stoichiometry is valid when other trimethyllead compounds $(CH_3)_3PbX$ (X = NO_3, OH, OCH_3, CN) are used. For kinetics, see [34]. Reaction of $Pb_2(CH_3)_6$ with $(CH_3)_3SnCl$ in the same mole ratio gives $Pb(CH_3)_4$, $Sn(CH_3)_4$, and $PbCl_2$ (1:2:1 mole ratio) with first-order rate constants [34]. $Pb(CH_3)_4$ is formed during the reactions of $(CH_3)_3PbBr$ and $[N(CH_3)_4]Br$ [333] and of $(CH_3)_3PbH$ and NH_3 or $N(CH_3)_3$ [334]. The product of the reaction of $(CH_3)_3PbCl$ and $K[BH_4]$ in NH_3, presumably $(CH_3)_3PbBH_4 \cdot n\,NH_3$ and subsequently $(CH_3)_3PbH$, decomposes to form $Pb(CH_3)_4$ [334, 335]. Reaction of $(CH_3)_3PbCl$ and $Na[BH_4]$ in methanol gives $(CH_3)_3PbH$, then $(CH_3)_3PbOCH_3$, and finally produces $Pb(CH_3)_4$ [14]. $[Pb(CH_3)_3]^+$ reacts with sulfide between pH 6 and 7 to produce $Pb(CH_3)_4$ [258]. $Pb(CH_3)_4$ is formed in the early

74

stages of the reaction of solutions of $(CH_3)_3PbCl$ and $(CH_3)_2PbCl_2$ in water, more rapidly in 5% NaCl solution, slowly in methanol, or in moist pyridine with elemental zinc [147]. Reaction of a mixture of $(C_2H_5)_3PbCl$ and $(CH_3)_3PbCl$ in 5% NaCl solution with elemental zinc at 40°C gives $Pb(CH_3)_4$, $Pb(C_2H_5)_4$, CH_4, C_2H_6, and $Pb_2(C_2H_5)_6$ [336].

$Pb(CH_3)_4$ is produced on decomposition of $(CH_3)_3PbOOCCH_3$ [337], 2-methyl-2-(tri-methylplumbyl)-1,3-dithiane [338], $(CH_3)_2Pb(SC_6H_5)_2$ [339], and of 1,3-dithiolatobenzene- and 1,3-dithiolato-4-chlorobenzenedimethyllead [340]. Some $Pb(CH_3)_4$ is produced on heating $(CH_3)_2Pb(C(Si(CH_3)_3)_3)OOCH$ with HCOOH in C_6H_6 [341], and on heating $(CH_3)_3PbCF_3$ and $Hg(CF_3)_2$ [342]. Trace amounts of $Pb(CH_3)_4$ are formed when $(CH_3)_3PbF$ is attacked by moisture [342]. Thermal and photochemical decomposition of $(\eta^5\text{-}C_5H_5)Fe(CO)_2Pb(CH_3)_3$ [343] and of $(\eta^5\text{-}C_5H_5)M(CO)_3Pb(CH_3)_3$ (M = Cr, Mo, W) leads to $Pb(CH_3)_4$ [344]. It is also formed in addition to elemental lead during the reaction of $(CH_3)_2Pt(bipyridine)$ and $(CH_3)_3PbCl$ [345], and it is produced during the reaction of $(CH_3)_3PbOOCCH_3$ and $(\eta^5\text{-}C_5H_5)_2MH_2$ (M = Mo, W), probably from the decomposition of an intermediate [346].

$Pb(CH_3)_4$ is a product of the reaction of $Pb_2(CH_3)_6$ with $FeCl_3$, Hg_2Cl_2, $CuCl_2$, $CuCl$, CH_3HgCl, $HgCl_2$, or $AgNO_3$ [347]. $Pb_2(CH_3)_6$ is converted into $Pb(CH_3)_4$ and elemental lead by pyrolysis [348], and at 95°C [334] in the presence of silica gel and other silica-type catalysts [349]. Thermal decomposition of a mixture of $Pb_2(CH_3)_6$ and $Pb_2(C_2H_5)_6$ leads to $Pb(CH_3)_4$ and the four other possible tetraalkyllead compounds [350]. The decomposition of $Pb_2(CH_3)_6$ to form $Pb(CH_3)_4$ is very rapid when dissolved in strong donor solvents and slow in C_6H_6 or acetone [351]. $Pb(CH_3)_4$ and elemental lead, essentially in a 3:1 mole ratio, are products of the thermolysis of $Pb_2(CH_3)_6$ in C_6H_6 or toluene [14, 352], C_6H_5Cl, $CHCl_3$, CH_2Cl_2, dioxane, $P(N(CH_3)_2)_3$, and acetone [14]. $Pb(CH_3)_4$ is also produced in CCl_4 solutions of $Pb_2(CH_3)_6$ and is also a product of $Pb_2(CH_3)_6$ methanolysis [14]. $Pb_2(CH_3)_6$ reacts with 7,7,8,8-tetracyanoquinodimethane to give $Pb(CH_3)_4$ and the appropriate lead(II) salt [353]. Traces of $Pb(CH_3)_4$ are produced in addition to $(CH_3)_{4-n}Pb(C_2H_5)_n$ (n = 1 to 3) from solutions of $Pb(C_2H_5)_4$ in sea water irradiated with sun light [354, 355].

By Biomethylation. Anaerobic bacterial cultures which had been incubated with freshwater lake sediment, cultures from surface water of a lake, or an aerated aquarium, to which $(CH_3)_3PbOOCCH_3$, $(CH_3)_3PbCl$, or dimethyllead compounds like $(CH_3)_2PbCl_2$ had been added, evolve $Pb(CH_3)_4$ [307, 356 to 364] with an average yield of 3 to 5% [363], but nearly quantitative conversion of $(CH_3)_3PbOOCCH_3$ was also reported [364]. Bacterial cultures spiked with $(CH_3)_3PbCl$ were found to evolve $Pb(CH_3)_4$ only as a minor product [307]; see also [365]. Transformation of $(CH_3)_3PbOOCCH_3$ into $Pb(CH_3)_4$ is accomplished by indigenous microorganisms [360, 366] or by Aeromonas sp. [360, 362], Pseudomonas, Alcaligenes, Acinetobacter, Flavobacterium [361]. Maximum production was at a pH near 5, and it decreased with increasing pH, concurrent with the rate of bacterial growth [362]. Later work showed that formation of $Pb(CH_3)_4$ from $(CH_3)_3PbOOCCH_3$ in lake sediments increased with the pH value [367].

The stoichiometry of the redistribution of trimethyllead compounds observed in aqueous solutions [307] is different from that in bacterial cultures. The increased amounts of $Pb(CH_3)_4$, and the lower amounts of lead(II), are explained by the occurrence of biomethylation in addition to the usual redistribution [307, 357]. Formation of $Pb(CH_3)_4$ in bacterial cultures containing trimethyllead compounds occurs by both chemical and biological mechanisms [307, 357, 358, 361, 366, 367] over the pH range of 3.5 to 7.5 [367]. The amount produced biologically was found to be 15 to 20% [307, 358, 361, 366], but 50 to 76% (pH range 3.5 to 7.5) was also reported [367]. Production of $Pb(CH_3)_4$ from sediments containing $(CH_3)_3PbOOCCH_3$

was approximately threefold greater under anaerobic conditions than in air [368]. UV irradiation did not cause further chemical conversion of $(CH_3)_3PbOOCCH_3$ to $Pb(CH_3)_4$ in the absence of microorganisms [356, 362]. From experiments with labelled compounds giving only unlabelled $Pb(CH_3)_4$, it is inferred that only chemical and no biological methylation occurs [369, 370]; see also [298].

Conversion of $(CH_3)_3PbOOCCH_3$ to $Pb(CH_3)_4$ in a sediment system was assumed to occur only by chemical redistribution [293, 316, 363]. $((CH_3)_3Pb)_2S$ was supposed to be the intermediate in a sulfide-containing medium [293, 316]; see also [368, 369]. The observations reported in [239] are explained on the basis that both redistribution and biomethylation occur [357, 361].

Bacteria from a natural lake can methylate Pb^{2+} under anaerobic conditions to give $Pb(CH_3)_4$ [357, 359, 361, 366]. $Pb(CH_3)_4$ production from sediment samples was increased on addition of $Pb(NO_3)_2$ or $PbCl_2$; however, in some cases, this effect has failed [356]; see also [371]. To achieve biomethylation of Pb^{2+} to $Pb(CH_3)_4$, the concentration of Pb^{2+}, the content of sulfur, the age of the inoculum, the type of the nutrient, and the pH have to be controlled [307, 366].

About 0.03% of lead added as $Pb(NO_3)_2$ to marine sediments is biomethylated to $Pb(CH_3)_4$ [364]. In various experiments, conversion into $Pb(CH_3)_4$ was inconsistent [356, 362, 368] and time-independent [368]. Bacterial cultures spiked with $(C_2H_5)_3PbCl$ produce some $Pb(CH_3)_4$ (from Pb^{2+}, resulting from redistribution of $(C_2H_5)_3PbCl$) in addition to $Pb(C_2H_5)_4$ and $(C_2H_5)_3PbCH_3$ [307, 361]. Formation of a volatile organolead compound (presumably $Pb(CH_3)_4$) was observed upon aerobic incubation of marine sediment with $Pb(OOCCH_3)_2$, but conversion did not occur under anaerobic conditions or when sterilized sediment was used [372]. In other work, no transformation of Pb^{2+} to $Pb(CH_3)_4$ by bacterial isolates [362], and using a variety of sediments [368, 371] and culture systems [371], has been observed. Also attempts failed to detect $Pb(CH_3)_4$ as a product of transmethylation of lead(II) salts by methylarsenic compounds produced by the mould *Scopulariopsis brevicaulis* [371]. From experiments with labelled and unlabelled methyl donors and Pb^{2+} compounds, in which no $Pb(CH_3)_4$ was produced, it is concluded that biological methylation of Pb^{2+} does not occur [369, 370, 373]. In a series of papers, general doubt has been raised regarding the biological production of $Pb(CH_3)_4$ from Pb^{2+} [374, 375], and the question whether or not biomethylation of lead(II) compounds occurs is at present still open [376]; for reviews, see [376, 377, 378]; see also [632]. A correlation between standard reduction potentials of lead(II) compounds and the mechanism of biomethylation is discussed in [374]. For environmental aspects of the generation of $Pb(CH_3)_4$ from sediments and bacterial cultures, see Section 1.1.1.1.9.

Purification and Analysis. $Pb(CH_3)_4$ is purified by trap-to-trap distillation in vacuum [8, 28], by steam distillation [2], by fractionating in a rectifying column described in [379] at 10.5°C [380], by gas chromatography [381, 382], or by fractional crystallization by zone melting [383]. Purification of very small amounts ($< 1\ \mu L$) is accomplished by gas chromatography [278]. Separation of toluene is achieved by crystallization [384, 385], or on a column of silica gel (mean pore diameter 60 Å) using acetone as eluent [386], and separation of $Pb(C_2H_5)_4$ and other tetraalkyllead compounds is accomplished by fractional distillation [330, 387, 388]. A Standard Test Method for separation of $Pb(CH_3)_4$ and $Pb(C_2H_5)_4$ present in gasoline blend by distillation and a qualitative test for presence or absence of these compounds is given in [389]. Gas chromatography is also applied to purify $Pb(CD_3)_4$ [33]. For spectral determination of impurities in high-purity $Pb(CH_3)_4$, see [390]. Bismuth impurities are removed from 35 to 95% solutions of $Pb(CH_3)_4$ in stabilizer solvents, like toluene and 1,2-dichloroethane, by treating with an alkane peroxocarboxylic acid [391], with ozone in air, or in an inert carrier gas [392].

Analysis of random equilibrium mixtures of $(CH_3)_{4-n}Pb(C_2H_5)_n$ (n = 0 to 4) was performed by fractional vacuum distillation before the advent of gas chromatography [329]. Gas chromatography using different types of columns and detectors is employed for analytical identification [29, 381, 393, 394], for determination in solutions [395 to 398], and for separation from other lead tetraalkyls [60, 256, 355, 397, 399 to 412], especially in gasoline or antiknock mixtures [400, 410, 413 to 427]; see also [428]. $Pb(CH_3)_4$ is determined along with other lead tetraalkyls with a combination of gas chromatography and atomic absorption spectroscopy in gasolines [411, 424, 425, 429 to 442], water [443], blood [444], and other samples [29, 368, 369, 371, 427, 435, 445 to 458, 629]; see also [378, 459, 460]. In the analysis of organolead in the presence of alkyl halides, care is required using a gas chromatography-atomic absorption combination since lead deposited in the transfer line from previous analyses can be re-volatilized by CH_3I as it elutes from the gas chromatographic column, and it can produce a false $Pb(CH_3)_4$ peak [255]. Determination is also feasible with gas chromatography combined with flame photometry [461], mass spectroscopy [256], inductively coupled plasma emission spectrometry [462, 463], inductively coupled plasma mass spectrometry [463], atomic fluorescence spectrometry [633], or with a combination of high-pressure liquid chromatography and atomic absorption spectrometry [464, 634], or Zeeman atomic absorption spectrometry [465]; see also [466, 467]. Reversed-phase high-pressure liquid chromatography coupled to an atomic absorption spectrometer [468] or to a chemical reaction detector allows separation and determination of $Pb(CH_3)_4$, Pb^{2+}, and other organolead compounds [469, 613, 614]. High-pressure liquid chromatography with UV detection can be used for determination of $Pb(CH_3)_4$ alone [470] and for separate determination of $Pb(CH_3)_4$ and $Pb(C_2H_5)_4$ [471], but not for quantitative determination in mixtures of $(CH_3)_{4-n}Pb(C_2H_5)_n$ (n = 0 to 4) [470, 471]. This can, however, be accomplished with detection by inductively coupled plasma [471]. Successful separation of such mixtures by high-pressure liquid chromatography is described in [472]; see also [473]. Reversed-phase high-pressure liquid chromatography with electrochemical detection at a mercury electrode with acetonitrile, containing 0.05 M $[N(C_2H_5)_4]ClO_4$ as the eluent, also provides a specific and sensitive method for determination of $Pb(CH_3)_4$ and $Pb(C_2H_5)_4$ [615]. Other methods investigated for determination of $Pb(CH_3)_4$, mainly in gasolines, are emission spectrometry [474], mass spectrometry [475, 476], coulometric or amperometric titration [477, 478, 479], X-ray fluorescence [474, 480 to 483], and X-ray absorption [484]. For a possible interference in mass spectrometric determination of $Pb(CH_3)_4$, see [46]. Standard Test Methods by X-ray procedures are described in [483, 485]. Detection with a galvanic cell of trace amounts of $Pb(CH_3)_4$ in gasolines is described in [486, 487, 488]. $Pb(CH_3)_4$ and $Pb(C_2H_5)_4$ in gasolines are separated by distillation and estimated by flame photometry [489]; see also [388, 481]. Atomic absorption spectrometry is applied for determination of $Pb(CH_3)_4$ in gasolines after appropriate sample pretreatment [457, 474, 481, 490 to 509, 635, 636]. A Standard Method is described in [510]; see also [501]. Different absorbance values of $Pb(CH_3)_4$ and $Pb(C_2H_5)_4$ have to be considered [465, 481, 490, 491, 497, 499, 500, 504, 511, 512]. Variation in response is overcome by reaction with iodine [493] and addition of a liquid anion exchanger [492, 501, 502, 507]. An alternative procedure is offered by a flow-injection technique with atomic absorption spectrometric detection [513]. Determination in gasolines [471] and in aqueous and alcoholic solution [514] is also accomplished by reversed-phase high-pressure liquid chromatography and detection with inductively coupled plasma emission. For an analogous microdetermination, see [515].

$Pb(CH_3)_4$ is determined in air or other gas samples by gas chromatography after appropriate sampling using flame ionization [394, 409], electron capture detectors [394, 406], a photoionization detector [395], mass spectrometric detection [402], or detection by atomic absorption spectrometric procedures [427, 435, 436, 438, 445, 446, 449, 455, 458, 462, 463, 503, 516 to 521]. A gas chromatographic procedure for monitoring $Pb(CH_3)_4$ is described in [395].

Determination by atomic absorption spectrometry after sampling and appropriate pretreatment is described in [449, 490, 522 to 533]. Rapid determination of a trace amount of $Pb(CH_3)_4$ or other lead compounds is accomplished by direct supply of air samples to the burner of an atomic absorption spectrophotometer for monitoring purposes [534]. Determination in air and other gas samples is also accomplished by gas chromatographic-mass spectrometric isotope dilution analysis [11], by degradating $Pb(CH_3)_4$ with a solid scrubber or a scrubber solution, and analyzing the reaction product with usual methods. As scrubber reagents are used, I_2 [413, 419, 535 to 540], ICl [419, 541 to 546], hydrochloric acid [547], activated [548, 637] or iodized carbon [419], and also condensation or absorption followed by reaction with nitric acid [549, 550] or with a mixture of nitric acid and H_2O_2 [551] are employed; see also [552]. For determination after trapping and separation by gas chromatography with a microwave plasma detector, see [553]. Procedures for sampling $Pb(CH_3)_4$ and other tetraalkyllead compounds in air are described in [11, 29, 419, 427, 449, 517, 518, 522 to 533, 535, 536, 538 to 543, 546, 548, 550, 551, 553, 554, 637]; see also [555]. Amberlite XAD-4 performed the best in a study of adsorption and desorption of $Pb(CH_3)_4$ on active carbon and different porous polymers [556], however, according to later work not under all conditions [637]. Monitoring of $Pb(CH_3)_4$ and $Pb(C_2H_5)_4$ by atomic absorption spectroscopy in air is described in [449, 534, 557]; see also [637]. For determination of $Pb(CH_3)_4$ and other tetraalkyllead compounds in air by a denuder diffusion technique, see [558]. Standard atmospheres containing $Pb(CH_3)_4$ are generated from constant temperature diffusion cells [409] or evaporators [538, 548]. For flammable gas detection in the presence of $Pb(CH_3)_4$ a pellistor catalytic sensor was developed [559]. For reviews on measurements of $Pb(CH_3)_4$ and other organolead compounds in air, see [560, 561].

For wet chemical analysis, $Pb(CH_3)_4$ is converted with bromine into $PbBr_2$ which is determined conventionally [275, 562 to 564]. The same reaction is used as a Standard Test Method for determining trace amounts of $Pb(CH_3)_4$ and other lead compounds in primary reference fuels [565]. For determination with iodine, see [538, 566]; with nitric acid, see [637]; with nitric acid and bromine, see [567]; and with a mixture of HCl, HNO_3, and $HClO_4$, see [568]. Iodometric titration is used in [569]. Determination of $Pb(CH_3)_4$ and other lead tetraalkyls in gasolines is accomplished by reaction with a solution of ICl in hydrochloric acid and titration or colorimetry of the extracted degradation products [544, 570]. This latter procedure is used in Standard Test Methods for determination of $Pb(CH_3)_4$ and other lead tetraalkyls [571, 572] and for determination of trace amounts in gasolines [573]. Similarly, lead tetraalkyls are determined in crude oil [574]. Treatment of gasoline with I_2 under irradiation and subsequent colorimetric titration of Pb^{2+} is also used for determination of $Pb(CH_3)_4$ and other tetraalkyllead compounds [575] and as a Standard Test Method for appropriate determinations in unleaded gasolines [576]. In a series of procedures, $Pb(CH_3)_4$ or other lead tetraalkyls in gasolines is first converted with concentrated hydrochloric acid to $PbCl_2$, which is extracted and determined by atomic absorption spectrometry [577 to 580], polarographically [581], complexometrically [582, 583], as $PbCrO_4$ by gravimetry [584 to 587], or by volumetric iodometric titration [588 to 591]. The reaction with hydrochloric acid, together with different subsequent determination procedures, is employed for Standard Test Methods [578 to 583, 586, 587, 589, 590, 591]; see also [592]. $Pb(CH_3)_4$ and $Pb(C_2H_5)_4$, present in gasoline, may be converted with a mixture of mercaptoacetic acid and nitrous acid in the presence of hydrochloric acid into water-soluble species; lead is then determined by atomic absorption spectrometry [593]. Determination can also be accomplished by reaction with Ag^+ [594, 595], using photoelectric [594], or flameless atomic absorption spectrophotometric indication [595]. Monitoring of $Pb(CH_3)_4$ and $Pb(C_2H_5)_4$ in air by reaction with I_2, and indication with an optical sensor [540] or by wet analytical methods, is described in [419, 538, 541, 542, 543, 557]; see also [555].

$Pb(CH_3)_4$ in street dust is extracted with ammoniacal methanol and reacted with I_2; the extracted lead is determined by atomic absorption spectroscopy [596]; see also [597].

Paper chromatography is used in the determination of $Pb(CH_3)_4$ in gasolines in addition to $Pb(C_2H_5)_4$ and other lead tetraalkyls [598], and in microdeterminations besides $(CH_3)_3PbCl$, $(CH_3)_2PbCl_2$, and Pb^{2+} [599]. For determination by thin-layer chromatography, see [408].

Procedures for determination of $Pb(CH_3)_4$ present in mixtures with $(CH_3)_{4-n}Pb(C_2H_5)_n$ (n = 1 to 4), $[PbR_3]^+$, $[PbR_2]^{2+}$ (R = CH_3, C_2H_5), and Pb^{2+} in biological or water samples are based on extraction, coupled with propylation [600] or n-butylation [597, 601, 602] of the ionic species to the appropriate tetraalkyllead compounds, and gas chromatography in combination with atomic absorption spectroscopy [600, 601, 602] or coupled with differential pulse electro-chemical techniques [245]; see also [378, 456, 603]. Simultaneous determination of $Pb(CH_3)_4$, $Pb(C_2H_5)_4$, $[Pb(CH_3)_3]^+$, and Pb^{2+} is accomplished by atomic absorption spectrometry after hydrogenating the latter two species with $NaBH_4$ and separating the volatile compounds gas chromatographically [442]; see also [635].

Determination in natural waters, sea water, waste water, marine sediments, and biological samples is accomplished after extraction by different analytical procedures, mostly micromethods [245, 404, 407, 408, 427, 443, 450, 452, 455, 463, 503, 549, 567, 568, 602, 604 to 609]; see also [597, 610]. $Pb(CH_3)_4$ is adsorbed onto the walls of glass sampling bottles, and it is recommended that the extraction of aqueous samples be carried out inside the sampling bottle [597]; see also [397]. A sampler for collecting evolved $Pb(CH_3)_4$ from sediment is described in [611]. For the determination of microamounts of $Pb(CH_3)_4$ by wet methods in the presence of $Pb(C_2H_5)_4$ and degradation products of these compounds (µg/L or nanomol range), see [567, 568, 599, 605, 606].

Thermodynamic Data of Formation. The enthalpy of formation from the elements (graphite, $H_{2(gas)}$, $Pb_{(metal)}$) ΔH_f° in kcal/mol at standard conditions for gaseous $Pb(CH_3)_4$ determined by rotating bomb calorimetry is $\Delta H_{f,298.16}^{\circ} = 32.6$ [383]; a recalculated value of 32.4 ± 0.5 is given in [619]; see also [617, 618, 625]. The corresponding values for liquid $Pb(CH_3)_4$ are 23.5 [383] and 23.3 ± 0.5 [619]; see also [617, 622, 623]. Earlier values determined by the less reliable stationary bomb calorimetry [619, 621] deviate appreciably and are reported in [624]. The value $\Delta H_f^{\circ} = 26.12$ kcal/mol was calculated by the MNDO method [626]. For a correlation of the dissociation energy $D(M-CH_3)$ with $\Delta H_{f(g)}^{\circ}$ for $M(CH_3)_4$ (M = Si, Ge, Sn, Pb), see [627].

The Gibbs energy of formation from the elements at standard conditions for gaseous $Pb(CH_3)_4$ is $\Delta G_{f,298.16}^{\circ} = 64.7$ kcal/mol [383]. Other less reliable values are reported in [624]. The Gibbs energy for liquid $Pb(CH_3)_4$ is $\Delta G_{f,298.16}^{\circ} = 62.8$ kcal/mol [383].

The entropy of formation ΔS_f° in $cal \cdot mol^{-1} \cdot K^{-1}$ at standard conditions for gaseous $Pb(CH_3)_4$ is given as $\Delta S_{f,298.16}^{\circ} = -107.7$ and for liquid $Pb(CH_3)_4$ $\Delta S_{f,298.16}^{\circ} = -131.7$ [383]. Other less reliable values for gaseous $Pb(CH_3)_4$ are reported in [624]. The molal entropy of $Pb(CH_3)_4$, S° (ideal gas at 1 atm), is 100.48 ± 0.20 cal/K at 298.16 K. The logarithm of the equilibrium constant of formation for gaseous $Pb(CH_3)_4$ is $\log K_{f,298.16} = -47.4$, whereas $\log K_{f,298.16}$ for liquid $Pb(CH_3)_4 = -46.0$ [383]; see also Section 1.1.1.1.3.4, p. 120.

Molal thermodynamic properties of $Pb(CH_3)_4$ in the ideal gas state comprising S°, Cp°, ΔH_f°, ΔG_f°, and $\log K_f$ are calculated for temperatures between 0 and 1000 K. The calculations are based on rigid-rotator, harmonic oscillator, and independent-internal-rotator approximations [620].

References:

[1] Grüttner, G., Krause, E. (Ber. Deut. Chem. Ges. **49** [1916] 1415/28).
[2] Krause, E. (Diss. Friedrich Wilhelm-Universität Berlin 1917).
[3] Krause, E. (Ber. Deut. Chem. Ges. **62** [1929] 1877/8).

[4] Krause, E., v. Grosse, A. (Die Chemie der metall-organischen Verbindungen, Bornträger, Berlin 1937, p. 389).

[5] Baudler, M. (in: Brauer, G., Handbuch der Präparativen Anorganischen Chemie, 3rd Ed., Enke, Stuttgart 1978, Vol. 2, pp. 781/2).

[6] Calingaert, G. (Chem. Rev. **2** [1925] 43/83).

[7] Hein, F., Heuser, E. (Z. Anorg. Allgem. Chem. **254** [1947] 138/50).

[8] Singh, G. (J. Organometal. Chem. **11** [1968] 133/43).

[9] Jones, L. W., Werner, L. (J. Am. Chem. Soc. **40** [1918] 1257/75).

[10] Simons, J. H., McNamee, R. W., Hurd, C. D. (J. Phys. Chem. **36** [1932] 939/48).

[11] Nielsen, T., Egsgaard, H., Larsen, E., Schroll, G. (Anal. Chim. Acta **124** [1981] 1/13).

[12] Calingaert, G., Soroos, H. (J. Org. Chem. **2** [1938] 535/9).

[13] Tagliavini, G., Schiavon, G., Belluco, U. (Gazz. Chim. Ital. **88** [1958] 746/54).

[14] Arnold, D. P., Wells, P. R. (J. Organometal. Chem. **111** [1976] 269/83).

[15] Gilman, H., Jones, R. G. (J. Am. Chem. Soc. **72** [1950] 1760/1).

[16] Gilman, H. (Advan. Organometal. Chem. **7** [1968] 1/52).

[17] Gorsich, R. D., Robbins, R. O. (J. Organometal. Chem. **19** [1969] 444/6).

[18] Gorsich, R. D., Ethyl Corp. (U.S. 3444223 [1964/68/69]; C.A. **71** [1969] No. 39179).

[19] Williams, K. C. (J. Organometal. Chem. **22** [1970] 141/8).

[20] Ethyl Corp. (Neth. Appl. 6505907 [1964/65]; C.A. **64** [1966] 14005).

[21] Shappirio, S. (U.S. 2012356 [1932/35]; C.A. **1935** 6752).

[22] Williams, K. C. (J. Org. Chem. **32** [1967] 4062/3).

[23] Williams, K. C., Ethyl Corp. (U.S. 3488369 [1967/70]; C.A. **72** [1970] No. 55658).

[24] Barbieri, G., Benassi, R. (Atti Soc. Nat. Mat. Modena **105** [1974] 39/46).

[25] Ducheylard, G., Lazard, B., Roth, E. (J. Chim. Phys. **50** [1953] 497/500).

[26] Collins, C. B., Farquhar, R. M., Russell, R. D. (Bull. Geol. Soc. Am. **65** [1954] 1/22).

[27] Bate, G. L., Miller, D. S., Kulp, J. L. (Anal. Chem. **29** [1957] 84/8).

[28] Richards, J. R. (Mikrochim. Acta **1962** 620/7).

[29] Brueggemeyer, T. W., Caruso, J. A. (Anal. Chem. **54** [1982] 872/5).

[30] Cook, C. L., Clouston, J. G. (Nature **177** [1956] 1178/9).

[31] Clouston, J. G., Cook, C. L. (Trans. Faraday Soc. **54** [1958] 1001/7).

[32] Costa, G., de Alti, G. (Atti Accad. Nazl. Lincei Classe Sci. Fis. Mat. Nat. Rend. [8] **28** [1960] 627/31).

[33] Lacey, M. J., Macdonald, C. G., Pross, A., Shannon, J. S., Sternhell, S. (Australian J. Chem. **23** [1970] 1421/9).

[34] Arnold, D. P., Wells, P. R. (J. Organometal. Chem. **111** [1976] 285/96).

[35] Bambynek, W. (Z. Physik. Chem. [Frankfurt] **25** [1960] 403/14).

[36] Dreeskamp, H., Stegmeier, G. (Z. Naturforsch. **22a** [1967] 1458/64).

[37] Heard, M. J., Wells, A. C., Newton, D., Chamberlain, A. C. (Manage. & Control Heavy Met. Environ. Intern. Conf., London 1979, pp. 103/8).

[38] Leighton, P. A., Mortensen, R. A. (J. Am. Chem. Soc. **58** [1936] 448/54).

[39] Edwards, R. R., Coryell, C. D. (AECU-50 [1948] 63/73; N.S.A. **2** [1949] No. 570).

[40] Edwards, R. R., Day, J. M., Overman, R. F. (J. Chem. Phys. **21** [1953] 1555/8).

[41] Duncan, J. F., Thomas, F. G. (J. Inorg. Nucl. Chem. **29** [1967] 869/90).

[42] Mortensen, R. A., Leighton, P. A. (J. Am. Chem. Soc. **56** [1934] 2397/8).

[43] Paneth, F., Hofeditz, W. (Ber. Deut. Chem. Ges. **62** [1929] 1335/47).

[44] Chaudhry, A. U., Gowenlock, B. G. (J. Organometal. Chem. **16** [1969] 221/6).

[45] Slawson, W. F., Russell, R. D. (Mikrochim. Ichnoanal. Acta **1963** 165/8).

[46] Richards, J. R. (Vacuum **16** [1966] 310/1).

[47] Pasynkiewicz, S., Malinowski, S., Bitter, J. (Przemysl Chem. **44** [1965] 500/3).

80

[48] Pasynkiewicz, S., Malinowski, S., Bitter, J., Sliwa, E., Politechnika Warszawska (Pol. 54578 [1965/68]; C.A. **70** [1969] No. 29061).

[49] Blitzer, S. M., Pearson, T. H., Ethyl Corp. (Fr. 1168218 [1956/58]; C. **1959** 17583; Belg. 553653 [1956/60]).

[50] Boleslawski, M., Pasynkiewicz, S. (J. Organometal. Chem. **43** [1972] 81/93).

[51] Boleslawski, M. (Prace Nauk. Politech. Warsz. Chem. No. 18 [1977] 1/124).

[52] Blitzer, S. M., Pearson, T. H., Ethyl Corp. (U.S. 2989558 [1958/61]; C.A. **1961** 23345).

[53] Pasynkiewicz, S., Boleslawski, M., Kunicki, A., Jaworski, K., Politechnika Warszawska (Pol. 82544 [1972/75]; C.A. **89** [1978] No. 109973).

[54] Pasynkiewicz, S., Boleslawski, M., Dichter, M., Politechnika Warszawska (Pol. 65480 [1968/72]; C.A. **79** [1973] No. 66581).

[55] Pasynkiewicz, S., Boleslawski, M., Jaworski, K., Politechnika Warszawska (Pol. 78697 [1973/75]; C.A. **85** [1976] No. 94483).

[56] Boleslawski, M., Pasynkiewicz, S., Synoradzki, L. (Przemysl Chem. **55** [1976] 80/2).

[57] Boleslawski, M., Pasynkiewicz, S., Kunicki, A. (J. Organometal. Chem. **73** [1974] 193/8).

[58] Boleslawski, M., Pasynkiewicz, S., Kunicki, A. (Przemysl Chem. **51** [1972] 446/9).

[59] Pasynkiewicz, S., Boleslawski, M., Kunicki, R., Politechnika Warszawska (Pol. 66224 [1969/72]; C.A. **79** [1973] No. 53576).

[60] Jaworski, K., Wilkanowicz, L., Kunicki, A. (J. Organometal. Chem. **102** [1975] 431/6).

[61] Pasynkiewicz, S., Kunicki, A., Jaworski, K., Wilkanowicz, L., Boleslawski, M., Politechnika Warszawska (Pol. 99990 [1975/78]; C.A. **91** [1979] No. 125912).

[62] Cahours, A. (Ann. Chim. Phys. [3] **62** [1861] 257/350).

[63] Cahours, A. (Liebigs Ann. Chem. **122** [1862] 48/71).

[64] Arnold, D. P., Wells, P. R. (J. Organometal. Chem. **108** [1976] 345/52).

[65] Saunders, B. C., Stacey, G. J. (J. Chem. Soc. **1949** 919/25).

[66] Heap, R., Saunders, B. C. (J. Chem. Soc. **1949** 2983/8).

[67] Shapiro, H. (Advan. Chem. Ser. No. 23 [1959] 290/8).

[68] Cook, S. E., Sistrunk, T. O., Ethyl Corp. (U.S. 3049558 [1959/62]; C.A. **57** [1962] 16656).

[69] Calingaert, G., Beatty, H. A., Ethyl Gasoline Corp. (U.S. 2270109 [1938/42]; C.A. **1942** 3190).

[70] Ethyl Gasoline Corp. (Fr. 841535 [1939]; C.A. **1940** 4393).

[71] Tullio, V., E. I. du Pont de Nemours & Co. (U.S. 3072694 [1960/63]; C.A. **58** [1963] 13992).

[72] Beaird Jr., F. M., Kobetz, P., Ethyl Corp. (U.S. 3188333 [1963/65]; C.A. **63** [1965] 13316).

[73] Beaird Jr., F. M., Kobetz, P., Ethyl Corp. (U.S. 3226408 [1963/65]; C.A. **64** [1966] 9768).

[74] Kobetz, P., Beaird Jr., F. M., Ethyl Corp. (U.S. 3357928 [1965/67]; C.A. **68** [1968] No. 105365).

[75] Beaird Jr., F. M., Kobetz, P., Ethyl Corp. (U.S. 3391086 [1965/68]; C.A. **69** [1968] No. 77505).

[76] Gorina, F. A., Zhitareva, L. V., Samarin, K. M. (U.S.S.R. 706419 [1977/79]; C.A. **92** [1980] No. 146912).

[77] Kobetz, P., Beaird, F. M., Ethyl Corp. (Fr. 1372724 [1962/64]; C.A. **62** [1965] 586).

[78] Kobetz, P., Beaird, F. M., Ethyl Corp. (U.S. 3192240 [1963/65]).

[79] Kobetz, P., Beaird, F. M., Ethyl Corp. (Ger. 1210840 [1963/66]).

[80] E. I. du Pont de Nemours & Co. (Brit. 1015227 [1963/65]; C.A. **64** [1966] 8240).

[81] Pedrotti, R. L., Sandy, C. A., E. I. du Pont de Nemours & Co. (Fr. 1406132 [1963/65]; C.A. **63** [1965] 14904).

[82] Pedrotti, R. L., Sandy, C. A., E. I. du Pont de Nemours & Co. (U.S. 3281442 [1963/66]).

[83] Pedrotti, R. L., E. I. du Pont de Nemours & Co. (U.S. 3401187 [1965/68]; C.A. **70** [1969] No. 20224).

[84] Beaird Jr., F. M., Kobetz, P., Ethyl Corp. (U.S. 3188334 [1963/65]; C.A. **63** [1965] 13316).
[85] Sandy, C. A., Pedrotti, R. L., Tullio, V., E. I. du Pont de Nemours & Co. (Fr. 1480011 [1965/67]; C.A. **68** [1968] No. 114743), Sandy, C. A., E. I. du Pont de Nemours & Co. (U.S. 3400143 [1965/68]), Pedrotti, R. L., Sandy, C. A., Tullio, V., E. I. du Pont de Nemours & Co. (U.S. 3408375 [1965/68]).
[86] Pedrotti, R. L., Sandy, C. A., Tullio, V., E. I. du Pont de Nemours & Co. (Ger. 1287077 [1966/69]).
[87] Sandy, C. A., E. I. du Pont de Nemours & Co. (U.S. 3426056 [1965/69]; C.A. **70** [1969] No. 68525).
[88] Sandy, C. A., Tullio, V., E. I. du Pont de Nemours & Co. (U.S. 3401189 [1966/68]; C.A. **69** [1968] No. 96876).
[89] Calingaert, G., Shapiro, H., Ethyl Corp. (U.S. 2535190 [1950]; C.A. **1951** 3864).
[90] Calingaert, G., Shapiro, H., Ethyl Corp. (Brit. 673871 [1952]; C.A. **1952** 9121).

[91] Calingaert, G., Shapiro, H., Ethyl Corp. (Brit. 673997 [1952]; C.A. **1952** 11230).
[92] Calingaert, G., Shapiro, H., Ethyl Corp. (Ger. 888696 [1953]; C.A. **1955** 2483).
[93] Jarvie, J. M. S., Schuler, M. J., Sterling Jr., J. D., E. I. du Pont de Nemours & Co. (U.S. 3048610 [1960/62]; C.A. **58** [1963] 550).
[94] Ethyl Corp. (Neth. Appl. 6403049 [1964/65]; C.A. **64** [1966] 6694).
[95] Ethyl Corp. (Neth. Appl. 6412633 [1963/65]; C.A. **63** [1965] 11223).
[96] Ethyl Corp. (Brit. Amended 1088415 [1969]; C.A. **72** [1970] No. 81186).
[97] Luyben, W. L., Pedrotti, R. L., E. I. du Pont de Nemours & Co. (Ger. Offen. 1945429 [1968/70]; C.A. **73** [1970] No. 4033).
[98] Tullio, V., E. I. du Pont de Nemours & Co. (U.S. 3072695 [1960/63]; C.A. **58** [1963] 13993).
[99] Sandy, C. A., E. I. du Pont de Nemours & Co. (U.S. 3113955 [1961/63]; C.A. **60** [1964] 5550).
[100] Beaird Jr., F. M., Kobetz, P., Ethyl Corp. (U.S. 3226409 [1963/65]; C.A. **64** [1966] 9768).

[101] Kobetz, P., Beaird Jr., F. M., Ethyl Corp. (U.S. 3188332 [1963/65]; C.A. **63** [1965] 9986).
[102] Beaird Jr., F. M., Kobetz, P., Ethyl Corp. (U.S. 3338842 [1965/67]; C.A. **68** [1968] No. 49773).
[103] Thomas, W. H., Kobetz, P., Giraitis, A. P., Ethyl Corp. (U.S. 3661952 [1970/72]; C.A. **77** [1972] No. 88660).
[104] Cliver, L. G., Ethyl Corp. (U.S. 3452069 [1966/69]; C.A. **71** [1969] No. 70745).
[105] Newyear, E. G., PPG Industries, Inc. (U.S. 3642849 [1970/72]; C.A. **76** [1972] No. 99832).
[106] Newyear, E. G., PPG Industries, Inc. (U.S. 3956176 [1970/76]; C.A. **85** [1976] No. 108775).
[107] Sandy, C. A., E. I. du Pont de Nemours & Co. (U.S. 3401188 [1965/68]; C.A. **69** [1968] No. 96875).
[108] Scales, R. K., Ethyl Corp. (U.S. 3145224 [1963/64]; C.A. **61** [1964] 13345).
[109] Lamarche, D., Decarie, M., Miranda Inc. (U.S. 3636021 [1969/72]; C.A. **76** [1972] No. 127156).
[110] Montecatini Edison S.p.A. (Ital. 846023 [1968/69]; C.A. **75** [1971] No. 88766).

[111] Imura, S., Sakamoto, S., Toyo Ethyl Co., Ltd. (Japan. 70-36494 [1967/70]; C.A. **74** [1971] No. 76528).
[112] Briody, R. G., Cuevas, E. A., PPG Industries, Inc. (U.S. 3619176 [1969/71]; C.A. **76** [1972] No. 49125).
[113] Bartlett, R. S., PPG Industries, Inc. (U.S. 3637778 [1969/72]; C.A. **76** [1972] No. 85921).
[114] Bartlett, R. S., PPG Industries, Inc. (U.S. 3647839 [1969/72]; C.A. **76** [1972] No. 127159).
[115] Briody, R. G., Newyear, E. G., PPG Industries, Inc. (U.S. 3655706 [1969/72]; C.A. **76** [1972] No. 153932).

[116] Sittig, M. (Organometallics, Lead Compounds, Chemical Process Monograph No. 20, Noyes Development Corporation, Park Ridge, N.J., 1966, pp. 67/104).
[117] Anonymous (Erdöl Kohle Erdgas Petrochem. 19 [1966] 905/7).
[118] Anonymous (Chem. Age [London] 1966 977).
[119] Frey, F. W., Shapiro, H. (Fortschr. Chem. Forsch. 16 [1971] 243/97).
[120] Rich, W. W., E. I. du Pont de Nemours & Co. (U.S. 3440256 [1967/69]; C.A. 71 [1969] No. 39174).

[121] Associated Octel Co., Ltd. (Belg. 610109 [1960/62]; C.A. 57 [1962] 7024).
[122] Calingaert, G., Shapiro, H., Ethyl Corp. (U.S. 2535191 [1950]; C.A. 1951 3864).
[123] Calingaert, G., Shapiro, H., Ethyl Corp. (U.S. 2535192 [1950]; C.A. 1951 3865).
[124] Calingaert, G., Shapiro, H., Ethyl Corp. (Ger. 875355 [1953]; C.A. 1954 10763).
[125] Williams, K. C., Thomas, W. H., Ethyl Corp. (U.S. 3647838 [1969/72]; C.A. 76 [1972] No. 127160).
[126] Downing, F. B., Bake, L. S., E. I. du Pont de Nemours & Co. (U.S. 1979254 [1934]; C.A. 1935 111).
[127] Shapiro, H., Ethyl Corp. (U.S. 2535235 [1950]; C.A. 1951 3865).
[128] Shapiro, H., Ethyl Corp. (U.S. 2535237 [1950]; C.A. 1951 3865).
[129] E. I. du Pont de Nemours & Co. (Ger. 658566 [1935/38]).
[130] Lehmkuhl, H. (Chem. Ing. Tech. 36 [1964] 612/6).

[131] Shapiro, H., Krohn, I. T., Ethyl Corp. (U.S. 2688628 [1951/54]; C.A. 1955 14797).
[132] Calingaert, G., Shapiro, H., Ethyl Corp. (Ger. 848817 [1950/52]; C. 1953 3180).
[133] Ethyl Corp. (Brit. 673440 [1950/52]; C.A. 1953 3335).
[134] Barton, J. M., E. I. du Pont de Nemours & Co. (U.S. 3478072 [1967/69]; C.A. 72 [1970] No. 43872).
[135] Calingaert, G., Shapiro, H., Ethyl Corp. (U.S. 2535193 [1949/50]; C.A. 1951 3865).
[136] Pagliarini, P., Compagnia Italiana Petrolio S.p.A. (Brit. 918519 [1961/63]; C.A. 59 [1963] 5196).
[137] Galli, R., Giannaccari, B. M., Montecatini Edison S.p.A. (Ital. 874189 [1969/70]; C.A. 81 [1974] No. 136301).
[138] Krohn, I. T., Ethyl Corp. (U.S. 2727053 [1951/55]; C. 1957 1294).
[139] Ethyl Corp. (Brit. 724155 [1952/55]; C.A. 1956 8709).
[140] Gelius, R., VEB Chemiefaserwerk "Friedrich Engels" (Ger. 1203266 [1964/65]; C.A. 64 [1966] 6693).

[141] Shapiro, H., Neal, H. R., Ethyl Corp. (U.S. 2992250 to 2992261 [1959/61]; C.A. 1961 22799).
[142] Hopkins, F. M., Ethyl Corp. (Fr. 1441315 [1966]; C.A. 66 [1967] No. 30784).
[143] Feehs, R. H., E. I. du Pont de Nemours & Co. (U.S. 3413328 [1966/68]; C.A. 70 [1969] No. 58024).
[144] Jaasma, W. C., Ethyl Corp. (U.S. 3403495 [1967/68]; C.A. 70 [1969] No. 20222).
[145] Lores, C., Moore, R. B., E. I. du Pont de Nemours & Co. (U.S. 3770423 [1972/73]; C.A. 80 [1974] No. 137055).
[146] Anonymous (Res. Discl. No. 136 [1975] 51).
[147] Hitchen, M. H., Holliday, A. K., Puddephatt, R. J. (J. Organometal. Chem. 172 [1979] 427/44).
[148] Bryce-Smith, D., E. I. du Pont de Nemours & Co. (Ger. Offen. 1932705 [1970]; C.A. 72 [1970] No. 90637).
[149] Cook, S. E., Ethyl Corp. (U.S. 3081326 [1961/63]; C.A. 58 [1963] 13688).
[150] Cook, S. E., Thomas, W. H., Ethyl Corp. (U.S. 3133099 [1962/64]; C.A. 61 [1964] 1695).

[151] Calingaert, G., Ethyl Corp. (Brit. 670526 [1949/52]).
[152] Cook, S. E., Shapiro, H., Ethyl Corp. (U.S. 3038916 to 3038919 [1960/62]).
[153] Luten Jr., D. B., Shell Development Co. (U.S. 2410829 [1944/46]; C.A. **1947** 1090).
[154] Ecke, G. G., Kolka, A. J., Ethyl Corp. (U.S. 2836568 [1955/58]).
[155] Shepherd, C. C., Ethyl Corp. (U.S. 2865722 [1958]; C.A. **1959** 5660).
[156] Fischer, H. G. M., Standard Oil Development Co. (U.S. 2461972 [1949]; C.A. **1949** 4003).
[157] Partridge, W. A., Alty, H. J., Anglo-Iranian Oil Co., Ltd. (U.S. 2452489 [1948]; C.A. **1949** 2423).
[158] Aleksandrov, Yu. A., Spiridonova, M. N., Tanaseichuk, B. S., Tikhonova, L. G., Gorki Scientific-Research Institute of Chemistry (U.S.S.R. 382636 [1971/73]; C.A. **79** [1973] No. 137292).
[159] Cook, S. E., Thomas, W. H., Ethyl Corp. (U.S. 3133098 [1962/64]; C.A. **61** [1964] 1695).
[160] Anonymous (Chem. Eng. News **39** No. 38 [1961] 36).

[161] Anonymous (Chem. Eng. News **42** No. 49 [1964] 52/3).
[162] Anonymous (Chem. Week **1964** Dec. 12, pp. 77/9).
[163] Prescott, J. H. (Chem. Eng. [New York] **72** No. 21 [1965] 238/50).
[164] Bott, L. L. (Hydrocarbon Process. Petrol. Refiner **44** No. 1 [1965] 115/8; C.A. **62** [1965] 7788).
[165] Guccione, E. (Chem. Eng. [New York] **72** No. 13 [1965] 102/4).
[166] Sugino, K. (Yuki Gosei Kagaku Kyokaishi **24** [1966] 1170/82; C.A. **66** [1967] No. 51572).
[167] Mantell, C. L. (Electro-Organic Chemical Processing, Tetraalkyl Leads by Electrolysis. Commercial Plant, Chemical Process Review No. 14, Noyes Development Corporation, Park Ridge, N.J., 1968, pp. 165/70).
[168] Kato, M. (Kagaku Gijutsushi MOL **18** No. 9 [1980] 31/6; C.A. **94** [1981] No. 46250).
[169] Wagenknecht, J. H. (J. Chem. Educ. **60** [1983] 271/3).
[170] Braithwaite, D. G., Nalco Chemical Co. (U.S. 3312605 [1961/63/65/67]; C.A. **67** [1967] No. 11589).

[171] Linsk, J., Mayerle, E. A., Standard Oil Co., Indiana (U.S. 3155602 [1960/64]; C.A. **62** [1965] 2794).
[172] Linsk, J., Standard Oil Co., Indiana (U.S. 3116308 [1961/63]; C.A. **60** [1964] 6867).
[173] Coopersmith, J. M., Linsk, J., Field, E., Carl, R. W., Mayerle, E. A., Standard Oil Co., Indiana (Ger. 1157616 [1960/61/63]; C.A. **61** [1964] 1892).
[174] Linsk, J., Standard Oil Co., Indiana (U.S. 3298939 [1960/67]).
[175] Standard Oil Co., Indiana (Brit. 984421 [1961/65]).
[176] Braithwaite, D. G., Nalco Chemical Co. (Belg. 613892 [1961/62]; C.A. **59** [1963] 7559).
[177] Braithwaite, D. G., Nalco Chemical Co. (U.S. 3234112 [1961/66]).
[178] Braithwaite, D. G., Nalco Chemical Co. (Ger. 1202790 [1963/65]; C.A. **64** [1966] 3064).
[179] Braithwaite, D. G., Nalco Chemical Co. (U.S. 3256161 [1961/64/66]; C.A. **65** [1966] 6751).
[180] Braithwaite, D. G., Nalco Chemical Co. (Ger. 1226100 [1963/66]; C.A. **65** [1966] 19690).

[181] Braithwaite, D. G., Nalco Chemical Co. (Ger. 1231242 [1963/66]; C.A. **66** [1967] No. 95196).
[182] Braithwaite, D. G., Nalco Chemical Co. (Ger. 1231700 [1963/67]; C.A. **66** [1967] No. 76155).
[183] Braithwaite, D. G., Nalco Chemical Co. (Ger. 1216303 [1963/66]; C.A. **65** [1966] 8962).
[184] Nalco Chemical Co. (Neth. Appl. 6514238 [1964/66]; C.A. **65** [1966] 6751).
[185] Braithwaite, D. G., Bott, L. L., Phillips, K. G., Nalco Chemical Co. (U.S. 3380900 [1964/68]).
[186] Braithwaite, D. G., Bott, L. L., Nalco Chemical Co. (U.S. 3380899 [1964/68]; C.A. **69** [1968] No. 15414).

84

[187] Pearce, F. G., Wright, L. T., Birkness, H. A., Linsk, J., Standard Oil Co., Indiana (U.S. 3180810 [1961/65]; C.A. **63** [1965] 3902).

[188] Braithwaite, D. G., Nalco Chemical Co. (Ger. 1197086 [1962/65]; C.A. **63** [1965] 11023).

[189] Braithwaite, D. G., Hanzel, W., Nalco Chemical Co. (Belg. 611212 [1960/62]; C.A. **57** [1962] 16330).

[190] Braithwaite, D. G., D'Amico, J. S., Gross, P. L., Hanzel, W., Nalco Chemical Co. (U.S. 3141841 [1960/64]; C.A. **61** [1964] 10323).

[191] Braithwaite, D. G., D'Amico, J. S., Gross, P. L., Hanzel, W., Nalco Chemical Co. (U.S. 3287249 [1962/66]; C.A. **66** [1967] No. 51716).

[192] Pletcher, D. (Industrial Electrochemistry, Chapman and Hall, London 1982, pp. 84/5).

[193] Blackmar, G. E., Nalco Chemical Co. (U.S. 3573178 [1968/71]; C.A. **74** [1971] No. 150413).

[194] Braithwaite, D. G., Nalco Chemical Co. (Belg. 590453 [1960]).

[195] Braithwaite, D. G., D'Amico, J. S., Gross, P. L., Hanzel, W., Nalco Chemical Co. (Belg. 606111 [1961/62]).

[196] Baimbridge, C. L., Minderhout, J. R., Bearman, R. W., Carpenter, D. E., Nalco Chemical Co. (U.S. 3925169 [1973/75]; C.A. **84** [1976] No. 81745).

[197] Braithwaite, D. G., Nalco Chemical Co. (U.S. 3007858 [1959]; C.A. **56** [1962] 4526).

[198] Braithwaite, D. G., Nalco Chemical Co. (U.S. 3391066 [1961/68]; C.A. **69** [1968] No. 67537).

[199] Ethyl Corp. (Neth. Appl. 6507727 [1964/65]; C.A. **64** [1966] 17640).

[200] Braithwaite, D. G., Bott, L. L., Nalco Chemical Co. (U.S. 3359291 [1964/67]; C.A. **68** [1968] No. 95980).

[201] Braithwaite, D. G., Bott, L. L., Gross, P. L., Laubach, J. E., Altman, W. L., Hanzel, W., Nalco Chemical Co. (U.S. 3408273 [1964/68]; C.A. **70** [1969] No. 96952).

[202] Bott, L. L., Craig, R. L., Hunter, E. A., Mayerle, E. A., Nalco Chemical Co. (U.S. 3594120 [1968/71]; C.A. **75** [1971] No. 78717).

[203] Braithwaite, D. G., Nalco Chemical Co. (U.S. 3409518 [1966/68]; C.A. **70** [1969] No. 37180).

[204] Johnson, E. E., Nalco Chemical Co. (U.S. 3572932 [1969/71]; C.A. **75** [1971] No. 20582).

[205] Baimbridge, C. L., Minderhout, J. R., Bearman, R. W., Carpenter, D. E., Nalco Chemical Co. (U.S. 4002548 [1973/77]; C.A. **86** [1977] No. 80915).

[206] Ganci, J. B., Manos, P., E. I. du Pont de Nemours & Co. (U.S. 3630858 [1968/71]; C.A. **76** [1972] No. 80308).

[207] Shepard Jr., J. C., Wight, R. C., Kosted, P. W., Nalco Chemical Co. (U.S. 3928164 [1973/75]; C.A. **84** [1976] No. 81751).

[208] Consiglio Nazionale delle Ricerche (Neth. Appl. 7611475 [1977]; Ital. Appl. 75/84153 [1975]; C.A. **88** [1978] No. 89847).

[209] Ziegler, K. (Belg. 617628 [1961/62]; C.A. **60** [1964] 3008).

[210] Ziegler, K., Lehmkuhl, H. (Ger. 1220855 [1961/66]).

[211] Ziegler, K. (U.S. 3254008 [1962/66]).

[212] Lehmkuhl, H. (Ann. N.Y. Acad. Sci. **125** [1965] 124/36).

[213] Schäfer, R. (Diss. Aachen T. H. 1961).

[214] Lehmkuhl, H. (in: Baizer, M. M., Organic Electrochemistry, Dekker, New York 1973, pp. 621/76).

[215] Ziegler, K., Lehmkuhl, H., Schäfer, R. (Ger. 1174779 [1962/64]).

[216] Ziegler, K., Lehmkuhl, H. (U.S. 3254009 [1963/66]).

[217] Giraitis, A. P., Ethyl Corp. (U.S. 3177130 [1960/62/65]; C.A. **63** [1965] 1816).

[218] Ziegler, K., Eisenbach, W. (U.S. 3620954 [1967/71]; C.A. **77** [1972] No. 69463).

[219] Pinkerton, R. C., Ethyl Corp. (U.S. 3028325 [1959/62]; C.A. **57** [1962] 4471).

[220] Kobetz, P., Thomas, W. H., Ethyl Corp. (U.S. 3344048 [1964/67]; C.A. **68** [1968] No. 45633).

[221] McKay, T. W., Ethyl Corp. (U.S. 3088885 [1959/63]; C.A. **59** [1963] 10119).

[222] Kobetz, P., Pinkerton, R. C., Ethyl Corp. (U.S. 3028322 [1959/62]; C.A. **57** [1962] 11235).

[223] Ziegler, K., Steudel, O.-W. (Liebigs Ann. Chem. **652** [1962] 1/7).

[224] Fleischmann, M., Pletcher, D., Vance, C. J., Associated Octel Co., Ltd. (Brit. 1290211 [1970/72]; C.A. **77** [1972] No. 159570).

[225] Fleischmann, M., Pletcher, D., Vance, C. J. (J. Electroanal. Chem. **29** [1971] 325/34).

[226] E. I. du Pont de Nemours & Co. (Neth. Appl. 6508049 [1964/65]; C.A. **64** [1966] 17048).

[227] Smeltz, K. N., E. I. du Pont de Nemours & Co. (U.S. 3392093 [1964/68]).

[228] E. I. du Pont de Nemours & Co. (Brit. 1064081 [1965/67]).

[229] E. I. du Pont de Nemours & Co. (Brit. 949925 [1961/64]; C.A. **61** [1964] 3935).

[230] Silversmith, E. F., Sloan, W. J., E. I. du Pont de Nemours & Co. (U.S. 3197392 [1961/65]).

[231] Silversmith, E. F., Sloan, W. J., E. I. du Pont de Nemours & Co. (Ger. 1240082 [1962/67]).

[232] Silversmith, E. F., Sloan, W. J., E. I. du Pont de Nemours & Co. (Can. 690294 [1962/64]; C. **1966** No. 38-2525).

[233] Yang, K., Reedy, J. D., Johnson, M. A., Harwood, W. H., Continental Oil Co. (Ger. 1955201 [1969/70]; C.A. **73** [1970] No. 31079).

[234] Continental Oil Co. (Brit. 1285209 [1972]).

[235] Smeltz, K. C., E. I. du Pont de Nemours & Co. (U.S. 3442768 [1967/69]; C.A. **71** [1969] No. 39184).

[236] Hannan, J. F., E. I. du Pont de Nemours & Co. (U.S. 3362889 [1966/68]; C.A. **68** [1968] No. 51783).

[237] E. I. du Pont de Nemours & Co. (Neth. Appl. 6615216 [1966/67]; C.A. **67** [1967] No. 73686).

[238] Settineri, W. J., Wessling, R. A. (in: Kuhn, A. T., The Electrochemistry of Lead, Academic, London 1979, pp. 163/97).

[239] Mayerle, E. A., Minderhout, J. R., Nalco Chemical Co. (U.S. 3696009 [1971/72]; C.A. **77** [1972] No. 171997).

[240] Ulery, H. E. (J. Electrochem. Soc. **116** [1969] 1201/5).

[241] Schuler, M. (unpublished, cited in [240]).

[242] Kegelman, M. (unpublished, cited in [240]).

[243] Chernykh, I. N., Tomilov, A. P. (in: Feoktistov, L. G., Elektrosintes Monomerov, Izd. Nauka, Moscow 1980, pp. 190/208).

[244] Colombini, M. P., Fuoco, R., Papoff, P. (Ann. Chim. [Rome] **72** [1982] 547/66).

[245] Colombini, M. P., Fuoco, R., Papoff, P. (Sci. Total Environ. **37** [1984] 61/70).

[246] Marlett, E. M. (Ann. N.Y. Acad. Sci. **125** [1965] 12/24).

[247] Tomilov, A. P., Brago, I. N. (Progr. Elektrokhim. Org. Soedin. **1** [1969] 208/49).

[248] Danly, D. (in: Baizer, M. M., Organic Electrochemistry, Dekker, New York 1973, pp. 907/46).

[249] Tedoradze, G. A. (J. Organometal. Chem. **88** [1975] 1/36).

[250] Settineri, W. J., McKeever, L. D. (in: Weissberger, A., Techniques of Chemistry, Vol. 5, Pt. 2, Wiley, New York 1975, pp. 397/558).

[251] Owen, A. J. (in: Kuhn, A. T., The Electrochemistry of Lead, Academic, London 1979, pp. 163/97).

[252] Gilman, H., Jones, R. G. (J. Am. Chem. Soc. **68** [1946] 517/20).

[253] Willemsens, L. C. (Progr. Rept. No. 28 to International Lead Zinc Research Organization, Project LC-18, Jan. 1 — Dec. 31, 1966, p. 26).

86

[254] Anonymous (Lead Research Digest, No. 19, Pt. V, International Lead Zinc Research Organization, April 1967, pp. 1, 3).

[255] Jarvie, A. W. P., Whitmore, A. P. (Environ. Technol. Letters **2** [1981] 197/204).

[256] Craig, P. J., Rapsomanikis, S. (Environ. Sci. Technol. **19** [1985] 726/30).

[257] Craig, P. J., Moreton, P. A., Rapsomanikis, S. (Heavy Metals Environ. 4th Intern. Conf., Heidelberg 1983, pp. 788/92).

[258] Craig, P. J., Rapsomanikis, S. (NBS-SP-618 [1981] 54/64).

[259] Craig, P. J., Rapsomanikis, S. (J. Chem. Soc. Chem. Commun. **1982** 114).

[260] Snyder, L. J., Bentz, J. M. (Nature **296** [1982] 228/9).

[261] Cortez, H. V., Houston Chemical Corp. (Fr. 1544454 [1966/68]; C.A. **71** [1969] No. 91657).

[262] Cortez, H. V., Houston Chemical Corp. (Fr. 1544455 [1966/68]; C.A. **71** [1969] No. 113089).

[263] Gray, R. D., Mayer, S. E., Houston Chemical Corp. (U.S. 3442923 [1965/69]; C.A. **71** [1969] No. 39178).

[264] Gray, R. D., Mayer, S. E., PPG Industries, Inc. (U.S. 3472637 [1968/69]; C.A. **71** [1969] No. 126616).

[265] Pearsall, H. W., Ethyl Corp. (U.S. 2414058 [1947]; C.A. **1947** 2430).

[266] Miller, D. M., Winkler, C. A. (Can. J. Chem. **29** [1951] 537/43).

[267] Schultze, G., Müller, E. (Z. Physik. Chem. B **6** [1930] 267/71).

[268] Paneth, F. (Z. Physik. Chem. B **7** [1930] 155/6).

[269] Pearson, T. G., Robinson, P. L., Stoddart, E. M. (Proc. Roy. Soc. [London] A **142** [1933] 275/85).

[270] Rice, F. O., Johnston, W. R., Evering, B. L. (J. Am. Chem. Soc. **54** [1932] 3529/43).

[271] Paneth, F., Lautsch, W. (Ber. Deut. Chem. Ges. **64** [1931] 2708/18).

[272] Paneth, F., Herzfeld, K. (Z. Elektrochem. **37** [1931] 577/82).

[273] Burton, M., Ricci, J. E., Davis, T. W. (J. Am. Chem. Soc. **62** [1940] 265/7).

[274] Hurd, D. T., Rochow, E. G. (J. Am. Chem. Soc. **67** [1945] 1057/9).

[275] Feldman, M. H., Ricci, J. E., Burton, M. (J. Chem. Phys. **10** [1942] 618/23).

[276] Whittingham, G. (Nature **160** [1947] 671/2).

[277] Latham, K. G., Esso Research and Engineering Co. (Brit. 800609; C.A. **1959** 19880).

[278] Ulrych, T. J., Russell, R. D. (Geochim. Cosmochim. Acta **28** [1964] 455/69).

[279] Whittles, A. B. L., Slawson, W. F. (Geochim. Cosmochim. Acta **29** [1965] 142/3).

[280] Hoffmann, P., Bächmann, K., Klenk, H., Lieser, K. H. (Inorg. Nucl. Chem. Letters **7** [1971] 577/82).

[281] Hoffmann, P., Bächmann, K., Bögl, W., Klenk, H., Lieser, K. H. (Radiochim. Acta **16** [1971] 172/9).

[282] Hoffmann, P., Bächmann, K., Klenk, H., Trautmann, W., Lieser, K. H. (Z. Anal. Chem. **267** [1973] 277/80).

[283] Hoffmann, P., Bächmann, K., Klenk, H., Bögl, W., Lieser, K. H. (Angew. Chem. **83** [1971] 909; Angew. Chem. Intern. Ed. Engl. **10** [1971] 835).

[284] Ahmad, I., Chau, Y. K., Wong, P. T. S., Carty, A. J., Taylor, L. (Nature **287** [1980] 716/7).

[285] Chau, Y. K., Wong, P. T. S., Carty, A. J., Taylor, L. (10th Intern. Conf. Organometal. Chem., Toronto, Canada, 1981, Paper 1D 09).

[286] Kotulla, U. (Diss. Univ. Dortmund 1984).

[287] Rhode, S. F., Weber, J. H. (Environ. Technol. Letters **5** [1984] 63/8).

[288] Thayer, J. S., Olson, G. J., Brinckman, F. E. (Environ. Sci. Technol. **18** [1984] 726/9).

[289] Rapsomanikis, S., Ciejka, J. J., Weber, J. H. (Inorg. Chim. Acta **89** [1984] 179/83).

[290] Taylor, R. T., Hanna, M. L. (J. Environ. Sci. Health A **11** [1976] 201/11).

[291] Taylor, R. T. (PB-251553 [1976] 1/35; C.A. **87** [1977] No. 89872).

[292] Thayer, J. S. (J. Environ. Sci. Health A **18** [1983] 471/81).

[293] Jarvie, A. W. P., Markall, R. N., Potter, H. R. (Nature **255** [1975] 217/8).

[294] Ridley, W. P., Dizikes, L. J., Wood, J. M. (Science **197** [1977] 329/32).

[295] Craig, P. J., Rapsomanikis, S. (Inorg. Chim. Acta **107** [1985] 39/43).

[296] Anonymous (Lead Research Digest, No. 36, International Lead Zinc Research Organization, New York 1978, pp. 46/7).

[297] Dimmit, J. H., Weber, J. H. (Inorg. Chem. **21** [1982] 1554/7).

[298] Weber, J. H., Dimmit, J. H., Ciejka, J. J. (10th Intern. Conf. Organometal. Chem., Toronto, Canada, 1981, Paper 1D 04).

[299] Agnes, G., Bendle, S., Hill, H. A. O., Williams, F. R., Williams, R. J. P. (Chem. Commun. **1971** 850/1).

[300] Wood, J. M. (Recherche **7** [1976] 711/9).

[301] Wood, J. M. (Science **183** [1974] 1049/52).

[302] Lewis, J., Prince, R. H., Stotter, D. A. (J. Inorg. Nucl. Chem. **35** [1973] 341/51).

[303] Witman, M. W., Weber, J. H. (Inorg. Chem. **16** [1977] 2512/5).

[304] Rice, F. O. (Brit. 407 036 [1934]; C.A. **1934** 4847).

[305] Paneth, F., Matthies, M., Schmidt-Hebbel, E. (Ber. Deut. Chem. Ges. **55** [1922] 775/89).

[306] Huber, F., Gmehling, J., Pohl, U. (Proc. 16th Intern. Conf. Coord. Chem., Dublin 1974, Abstr. 3.26).

[307] Huber, F., Schmidt, U., Kirchmann, H. (ACS Symp. Ser. No. 82 [1978] 65/81).

[308] Moedritzer, K. (Organometal. Chem. Rev. **1** [1966] 179/278).

[309] Holliday, A. K., Jessop, G. N. (J. Chem. Soc. A **1967** 889/91).

[310] Glockling, F., Gowda, N. M. N. (J. Chem. Soc. Dalton Trans. **1982** 2191/5).

[311] Gmehling, J., Huber, F. (Z. Anorg. Allgem. Chem. **393** [1972] 131/5).

[312] Amberger, E. (Angew. Chem. **72** [1960] 494).

[313] Becker, W. E., Cook, S. E. (J. Am. Chem. Soc. **82** [1960] 6264/5).

[314] Duffy, R., Feeney, J., Holliday, A. K. (J. Chem. Soc. **1962** 1144/7).

[315] Curtis, M. D., Allred, A. L. (J. Am. Chem. Soc. **87** [1965] 2554/63).

[316] Jarvie, A. W. P., Whitmore, A. P., Markall, R. N., Potter, H. R. (Environ. Pollut. B **6** [1983] 69/79).

[317] Hager, C.-D., Huber, F. (Z. Naturforsch. **35b** [1980] 542/7).

[318] Gelius, R., Müller, R. (Z. Anorg. Allgem. Chem. **351** [1967] 42/7).

[319] Gelius, R. (Z. Anorg. Allgem. Chem. **349** [1967] 22/32).

[320] Haupt, H.-J., Schubert, W., Huber, F. (J. Organometal. Chem. **54** [1973] 231/8).

[321] Schubert, W. (Diss. Univ. Dortmund 1975).

[322] Ködel, W. (Diss. Univ. Dortmund 1978).

[323] Ködel, W., Huber, F., Haupt, H.-J. (Inorg. Chim. Acta **49** [1981] 209/12).

[324] Schubert, W., Haupt, H.-J., Huber, F. (Z. Naturforsch. **29b** [1974] 694/6).

[325] Calingaert, G., Soroos, H., Shapiro, H. (J. Am. Chem. Soc. **62** [1940] 1104/7).

[326] De Jonghe, W., Adams, F. (Z. Anal. Chem. **314** [1983] 552/4).

[327] Sandy, C. A., E. I. du Pont de Nemours & Co. (U.S. 4 069 237 [1976/78]; C.A. **88** [1978] No. 121 387).

[328] Calingaert, G., Beatty, H. A., Soroos, H. (J. Am. Chem. Soc. **62** [1940] 1099/104).

[329] Calingaert, G., Beatty, H. A., Neal, H. R. (J. Am. Chem. Soc. **61** [1939] 2755/8).

[330] Calingaert, G., Beatty, H. A. (J. Am. Chem. Soc. **61** [1939] 2748/54).

[331] Calingaert, G., Soroos, H., Thomson, G. W. (J. Am. Chem. Soc. **62** [1940] 1542/5).

[332] Calingaert, G., Soroos, H., Shapiro, H. (J. Am. Chem. Soc. **63** [1941] 947/8).

[333] Huber, F., Schönafinger, E. (Proc. 11th Intern. Conf. Coord. Chem., Haifa/Jerusalem 1968, pp. 409/10).

[334] Duffy, R., Holliday, A. K. (J. Chem. Soc. **1961** 1679/82).

[335] Duffy, R., Holliday, A. K. (Proc. Chem. Soc. **1959** 124/5).

[336] Hitchen, M. H., Holliday, A. K., Puddephatt, R. J. (J. Organometal. Chem. **184** [1980] 335/42).

[337] Hönigschmid-Grossich, R., Amberger, E. (Chem. Ber. **102** [1969] 3589/98).

[338] Drew, G. M., Kitching, W. (J. Org. Chem. **46** [1981] 558/63).

[339] Wieber, M., Baudis, U. (J. Organometal. Chem. **125** [1977] 199/207).

[340] Grätz, K., Huber, F., Silvestri, A., Barbieri, R. (J. Organometal. Chem. **273** [1984] 283/94).

[341] Glockling, F., Gowda, N. M. N. (Inorg. Chim. Acta **58** [1982] 149/53).

[342] Eujen, R., Lagow, R. J. (J. Chem. Soc. Dalton Trans. **1978** 541/4).

[343] Pannell, K. H. (J. Organometal. Chem. **198** [1980] 37/40).

[344] Pannell, K. H., Kapoor, R. N. (J. Organometal. Chem. **214** [1981] 47/52).

[345] Jawad, J. K., Puddephatt, R. J. (Inorg. Chim. Acta **31** [1978] L 391/L 392).

[346] Kubicki, M. M., Kergoat, R., Guerchais, J.-E., L'Haridon, P. (J. Chem. Soc. Dalton Trans. **1984** 1791/3).

[347] Arnold, D. P., Wells, P. R. (J. Organometal. Chem. **113** [1976] 311/9).

[348] Gilman, H., Bailie, J. C. (J. Am. Chem. Soc. **61** [1939] 731/8).

[349] McDyer, T. W., Closson, R. D., Ethyl Corp. (U.S. 2571987 [1951]; C.A. **1952** 3556).

[350] Calingaert, G., Soroos, H., Shapiro, H. (J. Am. Chem. Soc. **64** [1942] 462/3).

[351] Puddephatt, R. J., Thistlethwaite, G. H. (J. Organometal. Chem. **40** [1972] 143/50).

[352] Arnold, D. P., Wells, P. R. (J. Chem. Soc. Chem. Commun. **1975** 642/3).

[353] Dick, A. W. S., Holliday, A. K., Puddephatt, R. J. (J. Organometal. Chem. **96** [1975] C 41/C 42).

[354] Charlou, J. L. (CNEXO-ENSCR 78-5678 [1979] 1/15).

[355] Charlou, J. L. (CNEXO-ENSCR 79-5943 [1980] 1/30).

[356] Wong, P. T. S., Chau, Y. K., Luxon, P. L. (Nature **253** [1975] 263/4).

[357] Schmidt, U., Huber, F. (Nature **259** [1976] 157/8).

[358] Chau, Y. K., Wong, P. T. S. (Lead Marine Environ. Proc. Intern. Experts Discuss., Rovinj, Yugoslavia, 1977 [1980], pp. 225/31).

[359] Dumas, J.-P., Pazdernik, L., Belloncik, S., Bouchard, D., Vaillancourt, G. (Water Pollut. Res. Can. **12** [1977] 91/100).

[360] Silverberg, B. A., Wong, P. T. S., Chau, Y. K. (Arch. Environ. Contam. Toxicol. **5** [1977] 305/13).

[361] Huber, F., Schmidt, U. (Organometal. Coord. Chem. Germanium Tin Lead, 2nd Intern. Conf., Nottingham 1977, Paper A 7).

[362] Wong, P. T. S., Chau, Y. K. (Manage. & Control Heavy Met. Environ. Intern. Conf., London 1979, pp. 131/4).

[363] Craig, P. J. (Environ. Technol. Letters **1** [1980] 17/20).

[364] Thompson, J. A. J., Crerar, J. A. (Marine Pollut. Bull. **11** [1980] 251/3).

[365] Macaskie, L. E., Ainsworth, M. A., Dean, A. C. R. (Environ. Technol. Letters **6** [1985] 237/50).

[366] Schmidt, U. (Diss. Univ. Dortmund 1977).

[367] Baker, M. D., Wong, P. T. S., Chau, Y. K., Mayfield, C. I., Inniss, W. E. (Heavy Met. Environ. 3rd Intern. Conf., Amsterdam 1981, pp. 645/8).

[368] Thompson, J. A. J. (Heavy Met. Environ. 3rd Intern. Conf., Amsterdam 1981, pp. 653/6).

[369] Reisinger, K., Stoeppler, M., Nürnberg, H. W. (Nature **291** [1981] 228/30).

[370] Reisinger, K., Stoeppler, M., Nürnberg, H. W. (Heavy Met. Environ. 3rd Intern. Conf., Amsterdam 1981, pp. 649/52).

[371] Jarvie, A. W. P., Whitmore, A. P., Markall, R. N., Potter, H. R. (Environ. Pollut. B **6** [1983] 81/94).

[372] Berdicevsky, I., Shachar, M., Yannai, S. (Arch. Toxicol. Suppl. No. 6 [1983] 285/91).

[373] Rapsomanikis, S. (Diss. Leicester Polytechnic 1983).

[374] Wood, J. M. (Lead Marine Environ. Proc. Intern. Experts Discuss., Rovinj, Yugoslavia, 1977 [1980], pp. 299/303).

[375] Wood, J. M. (Chem. Eng. News **55** No. 27 [1977] 37).

[376] Craig, P. J. (Spec. Publ. Roy. Soc. Chem. No. 44 [1983] 277/322).

[377] Craig, P. J. (Handb. Environ. Chem. A **1** [1980] 169/227).

[378] Chau, Y. K. (Sci. Total Environ. **49** [1986] 305/23).

[379] Clusius, K., Riccoboni, L. (Z. Physik. Chem. B **38** [1937] 81/95).

[380] Staveley, L. A. K., Warren, J. B., Paget, H. P., Dowrick, D. J. (J. Chem. Soc. **1954** 1992/2001).

[381] Abel, E. W., Nickless, G., Pollard, F. H. (Proc. Chem. Soc. **1960** 288).

[382] Cook, C. L., Napier, I. M. (Australian J. Chem. **24** [1971] 179/82).

[383] Good, W. D., Scott, D. W., Lacina, J. L., McCullough, J. P. (J. Phys. Chem. **63** [1959] 1139/42).

[384] Jonas, A. E., Schweitzer, G. K., Grimm, F. A., Carlson, T. A. (J. Electron Spectrosc. Relat. Phenom. **1** [1972/73] 29/66).

[385] Oyamada, T., Iijima, T., Kimura, M. (Bull. Chem. Soc. Japan **44** [1971] 2638/42).

[386] Ferreira da Silva, D., Diehl, H. (Xenobiotica **15** [1985] 789/97).

[387] Anonymous, E. I. du Pont de Nemours & Co. (Petroleum Laboratory Method No. G 39-60-2, July 1960).

[388] Kuznetsova, L. N., Cherpak, A. G., Lebedev, S. R., Muratova, R. D. (Khim. Tekhnol. Topl. Masel **1985** No. 5, pp. 34/5).

[389] Anonymous (1985 Annual Book of ASTM Standards, Vol. 5.02, D 1949−79, Reapproved 1984, Am. Soc. Testing Mater., Philadelphia 1985, pp. 133/5).

[390] Zanozina, V. F., Tumanova, A. N., Emel'yanova, O. A. (Poluch. Anal. Chist. Veshchestv **1983** 41/3; C.A. **101** [1984] No. 47 921).

[391] Collier Jr., H. E., Hammond, G. S., E. I. du Pont de Nemours & Co. (U.S. 3187028 [1963/65]; C.A. **63** [1965] 17 770).

[392] Collier Jr., H. E., E. I. du Pont de Nemours & Co. (U.S. 3270042 [1963/66]; C.A. **65** [1966] 15 428).

[393] Boettner, E. A., Dallos, F. C. (J. Gas Chromatog. **1** [1965] 190/1).

[394] Tausch, H. (SGAE-2636 [1976] 1/11; C.A. **86** [1977] No. 95 082).

[395] Taylor, D. G. (NIOSH Manual of Analytical Methods, Vol. 4, 2nd Ed., Washington 1978, Method S384, pp. 1/10).

[396] Zabairova, R. A., Bortnikov, G. N., Gorina, F. A., Samarin, K. M. (Zavodsk. Lab. **47** [1981] 22; C.A. **94** [1981] No. 167 164).

[397] Jarvie, A. W. P., Markall, R. N., Potter, H. R. (Environ. Res. **25** [1981] 241/9).

[398] Zabairova, R. A., Bochkarev, V. N., Bortnikov, G. N., Gorina, F. A., Samarin, K. M. (Khim. Elementoorg. Soedin. **1982** 88/9).

[399] Prösch, U., Zöpfl, H.-J. (Z. Chem. [Leipzig] **3** [1963] 97/100).

[400] Henneberg, D., Schomburg, G. (Z. Anal. Chem. **215** [1965] 424/30).

[401] Pollard, F. H., Nickless, G., Uden, P. C. (J. Chromatog. **19** [1965] 28/56).

[402] Laveskog, A. (Proc. 2nd Intern. Clean Air Congr., Washington 1970 [1971], pp. 549/57).

[403] Laveskog, A. (by Nielsen, T. in: Grandjean, P., Grandjean, E. C., Biological Effects of Organolead Compounds, CRC Press, Boca Raton, Fla., 1984, p. 48, ref. 40).

[404] Hayakawa, K. (Nippon Eiseigaku Zasshi **26** [1971] 377/85).

[405] Zorin, A. D., Umilin, V. A., Vanchagova, V. K. (U.S.S.R. 519628 [1972/76]; C.A. **85** [1976] No. 153565).

[406] Panetsos, A. G., Kilikides, S. K., Psomas, I. E. (Chem. Chron. [Athens] **5** [1976] 199/205).

[407] Mor, E. D., Beccaria, A. M. (Lead Marine Environ Proc. Intern. Experts Discuss., Rovinj, Yugoslavia, 1977 [1980], pp. 53/9).

[408] Potter, H. R., Jarvie, A. W. P., Markall, R. N. (Water Pollut. Contr. **76** [1977] 123/8).

[409] Harrison, R. M., Laxen, D. P. H. (Environ. Sci. Technol. **12** [1978] 1384/92).

[410] Uden, P. C. (Anal. Proc. [London] **18** [1981] 189/92).

[411] Charlou, J. L., Caprais, M. P., Blanchard, G., Martin, G. (Environ. Technol. Letters **3** [1982] 415/24).

[412] Crippen, R. C. (GC/LC, Instruments, Derivatives in Identifying Pollutants and Unknowns, Pergamon, New York 1983, pp. 151/2).

[413] Parker, W. W., Smith, G. Z., Hudson, R. L. (Anal. Chem. **33** [1961] 1170/1).

[414] Barrall II, E. M., Ballinger, P. R. (J. Gas Chromatog. **1** [1963] 7/13).

[415] Parker, W. W., Hudson, R. L. (Anal. Chem. **35** [1963] 1334/5).

[416] Dawson Jr., H. J. (Anal. Chem. **35** [1963] 542/5).

[417] Bonelli, E. J., Hartmann, H. (Anal. Chem. **35** [1963] 1980/1).

[418] Kramer, K. (Erdöl Kohle **19** [1966] 182/5).

[419] Soulages, N. L. (Anal. Chem. **38** [1966] 28/33).

[420] Soulages, N. L. (Anal. Chem. **39** [1967] 1340/1).

[421] Soulages, N. L. (J. Gas Chromatog. **6** [1968] 356/60).

[422] Castello, G. (Chim. Ind. [Milan] **51** [1969] 700/4).

[423] Wilkowa, T. (Chem. Anal. [Warsaw] **19** [1974] 545/53).

[424] Coker, D. T. (Anal. Chem. **47** [1975] 386/9).

[425] Bye, R., Paus, P. E., Solberg, R., Thomassen, Y. (At. Absorption Newsletter **17** [1978] 131/4; C.A. **90** [1979] No. 57560).

[426] Frank, H. A. (J. Forensic Sci. Soc. **20** [1980] 285/92).

[427] Chakraborti, D., Jiang, S. G., Surkijn, P., De Jonghe, W., Adams, F. (Anal. Proc. [London] **18** [1981] 347/50).

[428] Ham, N. S., McAllister, T. (Australian J. Chem. **36** [1983] 1299/304).

[429] Kolb, B., Kemmner, G., Schleser, F. H., Wiedeking, E. (Z. Anal. Chem. **221** [1966] 166/75).

[430] Kolb, B., Kemmner, G., Schleser, F. H., Wiedeking, E. (Angew. Chem. **78** [1966] 719/20; Angew. Chem. Intern. Ed. Engl. **5** [1966] 678).

[431] Ballinger, P. R., Whittemore, I. M. (Am. Chem. Soc. Div. Petrol. Chem. Prepr. **13** [1968] 133/8).

[432] Segar, D. A. (Anal. Letters **7** [1974] 89/95; C.A. **81** [1974] No. 20613).

[433] Robinson, J. W., Vidaurreta, L. E., Wolcott, D. K., Goodbread, J. P., Kiesel, E. (Spectrosc. Letters **8** [1975] 491/507).

[434] Katou, T., Nakagawa, R. (Yokohama Kokuritsu Daigaku Kankyo Kagaku Kenkyu Senta Kiyo No. 1 [1975] 19/24).

[435] Chau, Y. K., Wong, P. T. S., Saitoh, H. (J. Chromatog. Sci. **14** [1976] 162/4).

[436] Robinson, J. W., Kiesel, E. L., Goodbread, J. P., Bliss, R., Marshall, R. (Anal. Chim. Acta **92** [1977] 321/8).

[437] De Jonghe, W., Chakraborti, D., Adams, F. (Anal. Chim. Acta **115** [1980] 89/101).

[438] De Jonghe, W. R. A., Chakraborti, D., Adams, F. (10th Ann. Symp. Anal. Chem. Pollut., Dortmund, FRG, 1980, Paper M 57).

[439] Chan, L. (Forensic Sci. Intern. **18** [1981] 57/62).
[440] Crippen, R. C. (GC/LC, Instruments, Derivatives in Identifying Pollutants and Unknowns, Pergamon, New York 1983, pp. 294/5).

[441] Bykhovskii, M. Ya., Braude, A. Yu., Gorovoi, B. M., Katsis, L. F. (Khim. Tekhnol. Topl. Masel **1984** No. 5, pp. 34/5; C.A. **101** [1984] No. 40713).
[442] Baussand, P., Foster, P., Besson, J., Laverlochere, J., Meinhrat, A. O. (Analusis **13** No. 2 [1985] 53/8).
[443] Bykhovskii, M. Ya., Braude, A. Yu., Gorovoi, B. M., Rotin, V. A. (Zh. Analit. Khim. **39** [1984] 2183/5; J. Anal. Chem. [USSR] **39** [1984] 1742/4).
[444] Andersson, K., Nilsson, C.-A., Nygren, O. (Scand. J. Work Environ. Health **10** [1984] 51/5).
[445] Chau, Y. K., Wong, P. T. S., Goulden, P. D. (Intern. Conf. Heavy Metals Environ. 1st Symp. Proc., Toronto 1975 [1977], Vol. 1, pp. 295/302, Paper D-84).
[446] Chau, Y. K., Wong, P. T. S., Goulden, P. D. (Anal. Chim. Acta **85** [1976] 421/4).
[447] Chau, Y. K., Wong, P. T. S. (NBS-SP-464 [1977] 485/90).
[448] Chau, Y. K., Wong, P. T. S., Goulden, P. D. (Lead Marine Environ. Proc. Intern. Experts Discuss., Rovinj, Yugoslavia, 1977 [1980], pp. 77/81).
[449] Coker, D. T. (Ann. Occup. Hyg. **21** [1978] 33/8).
[450] Chau, Y. K., Wong, P. T. S., Bengert, G. A., Kramar, O. (Anal. Chem. **51** [1979] 186/8).

[451] Vickrey, T. M., Howell, H. E., Harrison, G. V., Ramelow, G. J. (Anal. Chem. **52** [1980] 1743/6).
[452] Cruz, R. B., Lorouso, C., George, S., Thomassen, Y., Kinrade, J. D., Butler, L. R. P., Lye, J., Van Loon, J. C. (Spectrochim. Acta B **35** [1980] 775/83).
[453] Ebdon, L., Ward, R. W., Leathard, D. A. (Analyst **107** [1982] 129/43).
[454] Chau, Y. K., Wong, P. T. S., Kramar, O. (Anal. Chim. Acta **146** [1983] 211/7).
[455] Chau, Y. K., Wong, P. T. S. (NATO Conf. Ser. [1] **6** [1983] 87/103).
[456] Chakraborti, D., De Jonghe, W. R. A., Van Mol, W. E., Van Cleuvenbergen, R. J. A., Adams, F. C. (Anal. Chem. **56** [1984] 2692/7).
[457] de Mora, S. J., Hewitt, C. N., Harrison, R. M. (Anal. Proc. [London] **21** [1984] 415/8).
[458] Harrison, R. M., Hewitt, C. N. (Intern. J. Environ. Anal. Chem. **21** [1985] 89/104).
[459] Parris, G. E., Blair, W. R., Brinckman, F. E. (Anal. Chem. **49** [1977] 378/86).
[460] Radziuk, B., Thomassen, Y., Butler, L. R. P., Van Loon, J. C., Chau, Y. K. (Anal. Chim. Acta **108** [1979] 31/8).

[461] Mutsaars, P. M., Van Steen, J. E. (J. Inst. Petrol. **58** [1972] 102/7).
[462] Sommer, D., Ohls, K. (Z. Anal. Chem. **295** [1979] 337/41).
[463] Van Loon, J. C., Balilcki, M. R., Nimjee, M. C., Brzezinska, A., Douglas, D. (Heavy Metals Environ. 4th Intern. Conf., Heidelberg 1983, pp. 78/81).
[464] Botre, C., Cacace, F., Cozzani, R. (Anal. Letters **9** [1976] 825/30).
[465] Koizumi, H., McLaughlin, R. D., Hadeishi, T. (Anal. Chem. **51** [1979] 387/92).
[466] Vickrey, T. M., Harrison, G. V., Ramelow, G. J. (At. Spectrosc. **1** [1980] 116/9).
[467] Uden, P. C. (Chem. Anal. [N.Y.] **78** [1985] 229/84).
[468] Messman, J. D., Rains, T. C. (Anal. Chem. **53** [1981] 1632/6).
[469] Blaszkewicz, M., Neidhart, B. (Intern. J. Environ. Anal. Chem. **14** [1983] 11/21).
[470] Ruo, T. C. S., Selucky, M. L., Strausz, O. P. (Anal. Chem. **49** [1977] 1761/5).

[471] Ibrahim, M., Gilbert, T. W., Caruso, J. A. (J. Chromatog. Sci. **22** [1984] 111/5).
[472] Götze, H.-J., Telgheder, P. (Z. Anal. Chem. **320** [1985] 59/60).
[473] MacDonald, J. C. (Chem. Anal. [N.Y.] **78** [1985] 285/99).
[474] Boldrino, F., Guagliumi, M. (Riv. Combust. **24** [1970] 260/71).
[475] Howard, H. E., Ferguson, W. C., Snyder, L. R. (Anal. Chem. **32** [1960] 1814/5).

[476] Knof, H., Ewers, H., Albers, G. (Compend. Deut. Ges. Mineralölwiss. Kohlechem. **1974/75** Vol. 2, pp. 798/810; C.A. **83** [1975] No. 149935).
[477] Pilloni, G., Plazzogna, G. (Ric. Sci. Rend. A **4** [1964] 27/32).
[478] Pilloni, G. (Farmaco Ed. Prat. **22** [1967] 666/76).
[479] Hozman, R. (Sb. Praci Vyzk. Chem. Vyuziti Uhli Dehtu Ropy **15** [1978] 297/317; C.A. **89** [1978] No. 131999).
[480] Christofferson, G. D., Beach, J. Y. (9th Colloq. Spectros. Intern., Lyons 1961 [1962], Vol. 3, pp. 492/503; C.A. **58** [1963] 13670).

[481] Robinson, J. W. (Anal. Chim. Acta **24** [1961] 451/5).
[482] Anonymous (IP Standards for Petroleum and Its Products, Pt.1, Sect.2, IP 352/81, Institute of Petroleum, London 1977).
[483] Anonymous (Deutsche Norm, Prüfung von Mineralölerzeugnissen; Bestimmung des Bleigehaltes (Gesamtblei) von Ottokraftstoffen durch Röntgenfluoreszenzanalyse, DIN 51769, Pt. 6, Aug. 1982).
[484] Holle, B., Svajgl, O., Vitovec, V. (Sb. Praci Vyzk. Chem. Vyuziti Uhli Dehtu Ropy **16** [1979] 233/50).
[485] Anonymous (1985 Annual Book of ASTM Standards, Vol. 5.02, D 2599/81, Am. Soc. Testing Mater., Philadelphia 1985, pp. 533/6).
[486] Olson, D. C., Shell Oil Co. (U.S. 3960690 [1974/76]; C.A. **85** [1976] No. 145574).
[487] Olson, D. C., Shell Oil Co. (U.S. 4012290 [1974/77]; C.A. **86** [1977] No. 174115).
[488] Olson, D. C., Shell Oil Co. (U.S. 4153517 [1978/79]; C.A. **91** [1979] No. 94196).
[489] Van Rysselberge, J., Leysen, R. (Nature **189** [1961] 478).
[490] Thilliez, G. (Chim. Anal. [Paris] **46** [1964] 3/22).

[491] Trent, D. J. (At. Absorpt. Newsl. **4** [1965] 348/50).
[492] Lindemanis, E. (Direct Determination of Lead in Gasoline by Atomic Absorption, Analytical Method M 113-71, E. I. du Pont de Nemours & Co., Wilmington, Del., 1971).
[493] Kashiki, M., Yamazoe, S., Oshima, S. (Anal. Chim. Acta **53** [1971] 95/100).
[494] Kashiki, M., Oshima, S. (Bunseki Kagaku **20** [1971] 1398/405).
[495] Smirnov, B. V., Kyuregyan, S. K. (Zh. Prikl. Spektrosk. **41** [1984] 832/5; C.A. **102** [1985] No. 48357).
[496] Campbell, K., Palmer, J. M. (J. Inst. Petrol. **58** [1972] 193/200).
[497] Kashiki, M., Yamazoe, S., Ikeda, N., Oshima, S. (Anal. Letters **7** [1974] 53/64).
[498] Nishishita, T., Yamazoe, S., Mallett, W. R., Kashiki, M., Oshima, S. (Anal. Letters **8** [1975] 849/55).
[499] Miyagawa, H. (Nagoya Med. J. **20** [1975] 95/109).
[500] Lukasiewicz, R. J., Berens, P. H., Buell, B. E. (Anal. Chem. **47** [1975] 1045/9).

[501] Russell, T. J., Campbell, K. (in: Holding, S. T., Palmer, J. M., Analyst **109** [1984] 507/10, p. 507, ref. 16).
[502] Anonymous (CEN/TC 19 N.442, European Standard for Gasoline: Determination of Lead Content, Atomic Absorption Spectroscopy 1979).
[503] Van Loon, J. C. (Analytical Atomic Absorption Spectroscopy, Selected Methods, Academic, New York 1980, pp. 303/7, 310/1).
[504] Polo-Diez, L., Hernández-Méndez, J., Pedraz-Penalva, F. (Analyst **105** [1980] 37/42).
[505] Berenguer, V., de la Guardia, M., Guinon, J. L. (Anales Quim. B **78** [1982] 338/43; C.A. **98** [1983] No. 110059).
[506] Ng, K. C., Caruso, J. A. (Anal. Chem. **55** [1983] 2032/6).
[507] Holding, S. T., Palmer, J. M. (Analyst **109** [1984] 507/10).
[508] Aneva, Z., Iancheva, M. (Anal. Chim. Acta **167** [1985] 371/4).

[509] Ivovic, B., Milosevic, Z. (Nafta [Zagreb] **36** [1985] 55/9).

[510] Anonymous (1985 Annual Book of ASTM Standards, Vol. 5.03, D 3237-79, Reapproved 1984, Am. Soc. Testing Mater., Philadelphia 1985, 121/3).

[511] Buell, B. E. (Anal. Chem. **34** [1962] 635/40).

[512] Meszaros, J., Mandy, T., Gelencser, J. (Banyasz. Kohasz. Lapok Koolaj Foldgaz **11** [1978] 121/5; C.A. **90** [1979] No. 25727).

[513] Taylor, C. G., Trevaskis, J. M. (Anal. Chim. Acta **179** [1986] 491/6).

[514] Ibrahim, M., Nisamaneepong, W., Haas, D. L., Caruso, J. A. (Spectrochim. Acta B **40** [1985] 367/76).

[515] Ibrahim, M., Nisamaneepong, W., Caruso, J. (J. Chromatog. Sci. **23** [1985] 144/50).

[516] Robinson, J. W., Kiesel, E. L. (J. Environ. Sci. Health A **12** [1977] 411/22).

[517] Radziuk, B., Thomassen, Y., Van Loon, J. C., Chau, Y. K. (Anal. Chim. Acta **105** [1979] 255/62).

[518] De Jonghe, W. R. A., Chakraborti, D., Adams, F. C. (Anal. Chem. **52** [1980] 1974/7).

[519] Birnie, S. E., Noden, F. G. (Analyst **105** [1980] 110/8).

[520] Dmitriev, M. T., Braude, A. Yu., Bykhovskii, M. Ya., Emel'yanov, B. V., Pautova, L. F., Rotin, V. A. (Gig. Sanit. **1984** No. 9, pp. 55/7; C.A. **102** [1985] No. 31071).

[521] Hewitt, C. N., Harrison, R. M. (Anal. Chim. Acta **167** [1985] 277/87).

[522] Harrison, R. M., Perry, R. (Atmos. Environ. **11** [1977] 847/52).

[523] Harrison, R. M., Perry, R., Slater, D. H. (Atmos. Environ. **8** [1974] 1187/94).

[524] Harrison, R. M., Perry, R., Slater, D. H. (EUR-5360 [1974/75] Vol. 3, pp. 1783/8).

[525] Rohbock, E., Müller, J. (Mikrochim. Acta I **1979** 423/34).

[526] Rohbock, E., Georgii, H.-W., Müller, J. (Atmos. Environ. **14** [1980] 89/98).

[527] Robinson, J. W. (Atmos. Environ. **14** [1980] 1207).

[528] Rohbock, E., Georgii, H.-W., Müller, J. (Atmos. Environ. **14** [1980] 1207/8).

[529] De Jonghe, W. R. A., Chakraborti, D., Adams, F. C. (Atmos. Environ. **15** [1981] 421/2).

[530] Rohbock, E., Georgii, H.-W., Müller, J. (Atmos. Environ. **15** [1981] 422).

[531] Harrison, R. M., Laxen, D. P. H. (Atmos. Environ. **15** [1981] 422/3).

[532] Rohbock, E., Georgii, H.-W., Müller, J. (Atmos. Environ. **15** [1981] 423/4).

[533] Rohbock, E. (Atomspektrom. Spurenanal. Vortr. Kolloq. **1981/82** 267/74).

[534] Thilliez, G. (Anal. Chem. **39** [1967] 427/32).

[535] Snyder, L. J., Henderson, S. R. (Anal. Chem. **33** [1961] 1175/80).

[536] Snyder, L. J., Henderson, S. R., Ethyl Corp. (U.S. 3071446 [1962/63]; C.A. **58** [1963] No. 6200).

[537] Kehoe, R. A., Cholak, J., McIlhinney, J. G., Lofquist, G. A., Sterling, T. D. (Arch. Environ. Health **6** [1963] 255/72).

[538] Linch, A. L., Davis, R. B., Stalzer, R. F., Anzilotti, W. F. (Am. Ind. Hyg. Assoc. J. **25** [1964] 81/93).

[539] Zelaskowski, C. A., Carlisi, J. J., Mobil Oil Corp. (U.S. 3955927 [1973/76]; C.A. **85** [1976] No. 145572).

[540] Walker, A. O., Nalco Chem. Co. (U.S. 3870469 [1973/75]; C.A. **83** [1975] No. 136502).

[541] Moss, R., Browett, E. V. (Autom. Anal. Chem. Technicon Symp., New York and London 1965 [1966], pp. 285/90; C.A. **67** [1967] No. 57033).

[542] Moss, R., Browett, E. V. (Analyst **91** [1966] 428/38).

[543] Linch, A. L., Wiest, E. G., Carter, M. D. (Am. Ind. Hyg. Assoc. J. **31** [1970] 170/9).

[544] Snyder, L. J., Ethyl Corp. (U.S. 3912454 [1974/75]; C.A. **84** [1976] No. 62223).

[545] Hancock, S., Slater, A. (Analyst **100** [1975] 422/9).

[546] Birch, J., Harrison, R. M., Laxen, D. P. H. (Sci. Total Environ. **14** [1980] 31/42).

[547] Zuliani, G., Perin, G., Rausa, G. (Med. Lavoro **57** [1966] 771/80).

[548] Snyder, L. J. (Anal. Chem. **39** [1967] 591/5).

[549] Diehl, K.-H., Rosopulo, A., Kreuzer, W. (Z. Anal. Chem. **314** [1983] 755/7).

[550] Harrison, R. M., Radojević, M., Hewitt, C. N. (Sci. Total Environ. **44** [1985] 235/44).

[551] Jiang, S. G., Chakraborti, D., De Jonghe, W., Adams, F. (Z. Anal. Chem. **305** [1981] 177/80).

[552] Sawicki, C. R. (EPA-650-2-75-003 [1975] 1/14).

[553] Reamer, D. C., Zoller, W. H., O'Haver, T. C. (Anal. Chem. **50** [1978] 1449/53).

[554] Jiang, S. G., Ma, C. G., Liu, H. C., Ge, J. R., Li, M., Adams, F. C., Winchester, J. W. (Atmos. Environ. **18** [1984] 2553/6).

[555] Gunderson, E. C., Anderson, C. C. (DHHS [NIOSH] Publ. U.S. No. 80-133 [1980] 1/59; C.A. **94** [1981] No. 196841).

[556] Anonymous (Swedish Board of Occupational Safety and Health, Report 1981, 35f.).

[557] Blears, D. G., Coventry, R. J. (Inst. Chem. Eng. Symp. Ser. A No. 39 [1974] 322/37).

[558] Febo, A., Di Palo, V., Possanzini, M. (Sci. Total Environ. **48** [1986] 187/94).

[559] Jones, E. (Intern. Environ. Safety **1981** 52/3).

[560] De Jonghe, W. R. A., Adams, F. C. (Talanta **29** [1982] 1057/67).

[561] Berg, S., Jonsson, A. (in: Grandjean, P., Biological Effects of Organolead Compounds, CRC Press, Boca Raton, Fla., 1984, pp. 33/42).

[562] Ipatiew, W. N., Rasuwajew, G. A., Bogdanow, I. F. (Zh. Russ. Fiz. Khim. Obshchestva **61** [1929] 1791/9).

[563] Ipatiew, W. N., Rasuwajew, G. A., Bogdanow, I. F. (Ber. Deut. Chem. Ges. **63** [1930] 335/42).

[564] Ipatieff, V. N. (Catalytic Reactions at High Pressures and Temperatures, Macmillan, New York 1936, p. 353).

[565] Anonymous (1985 Annual Book of ASTM Standards, Vol. 05.01, D 1368-83, Am. Soc. Testing Mater., Philadelphia 1985, pp. 710/3).

[566] Henderson, S. R., Snyder, L. J. (Anal. Chem. **33** [1961] 1172/5).

[567] Brondi, M., Dall'Aglio, M., Ghiara, E., Mignuzzi, C., Tiravanti, G. (Sci. Total Environ. **19** [1981] 21/31).

[568] Beccaria, A. M., Mor, E. D., Poggi, G. (Ann. Chim. [Rome] **68** [1978] 607/17).

[569] Boleslawski, M., Pasynkiewicz, S., Jaworski, K. (J. Organometal. Chem. **30** [1971] 199/209).

[570] Moss, R., Campbell, K. (J. Inst. Petrol. **53** [1967] 89/93).

[571] Anonymous (Deutsche Norm, Prüfung von Mineralölerzeugnissen, Bestimmung des Bleigehaltes (Gesamtblei) von flüssigen Mineralöl-Kohlenwasserstoffen im Bereich von 50 bis 1000 µg/kg, Jodmonochloridaufschluß, Extraktion mit Dithizon und photometrische Endbestimmung des Bleidithizonats, DIN 51769, Pt. 9, Okt. 1981).

[572] Anonymous (1985 Annual Book of ASTM Standards, Vol. 5.03, D 3341-80, Am. Soc. Testing Mater., Philadelphia 1985, pp. 242/5).

[573] Anonymous (1985 Annual Book of ASTM Standards, Vol. 5.03, ANSI/ASTM D 3116-82, Am. Soc. Testing Mater., Philadelphia 1985, pp. 33/7).

[574] Campbell, K., Moss, R. (J. Inst. Petrol. **53** [1967] 194/200).

[575] Zeląskowski, C. A., Mobil Oil Corp. (U.S. 3934976 [1974/76]; C.A. **84** [1976] No. 167189).

[576] Anonymous (1985 Annual Book of ASTM Standards, Vol. 05.03, D 3348-85, Am. Soc. Testing Mater., Philadelphia 1985, pp. 260/6).

[577] Madec, M., La Villa, F. (Rev. Inst. Franc. Petrole **31** [1976] 687/701).

[578] Anonymous (Deutsche Norm, Prüfung von Mineralölerzeugnissen; Bestimmung des Bleigehaltes (Gesamtblei) von Ottokraftstoffen mit einem Bleigehalt über 25 mg/L; Direkte Bestimmung durch Atomabsorptionsspektroskopie, DIN 51769, Pt. 7, Okt. 1981).

[579] Anonymous (Deutsche Norm, Prüfung von Mineralölerzeugnissen; Bestimmung des Bleigehaltes (Gesamtblei) von Ottokraftstoffen mit einem Bleigehalt von 5 bis 25 mg/L; Direkte Bestimmung durch Atomabsorptionsspektroskopie, DIN 51769, Pt. 8, Okt. 1981).

[580] Anonymous (Deutsche Norm, Prüfung von Mineralölerzeugnissen; Bestimmung des Bleigehaltes (Gesamtblei) von Schmierstoffen; Direkte Bestimmung durch Atomabsorptionsspektroskopie, DIN 51769, Pt. 10, Okt. 1981).

[581] Anonymous (1973 Annual Book of ASTM Standards, Vol. 17, D 1269-61, Am. Soc. Testing Mater., Philadelphia 1973, pp. 446/50).

[582] Anonymous (IP Standards for Petroleum and Its Products, Pt. 1, Sect. 2, IP 270/77, Institute of Petroleum, London 1977).

[583] Anonymous (Deutsche Norm, Prüfung von Mineralölerzeugnissen; Bestimmung des Bleigehaltes (Gesamtblei) von Ottokraftstoffen; Komplexometrisches Verfahren, DIN 51769, Pt. 5, Juni 1984).

[584] Rather Jr., J. B. (Mater. Res. Std. 2 [1962] 494/7).

[585] Isizaki, M. (Mem. Defense Acad. Math. Phys. Chem. Eng. [Yokosuka, Japan] 10 [1970] 389/92).

[586] Anonymous (1975 Annual Book of ASTM Standards, Vol. 17, D 526-70, Am. Soc. Testing Mater., Philadelphia 1975, pp. 275/8).

[587] Anonymous (IP Standards for Petroleum and Its Products, Pt. 1, Sect. 1 and 2, IP 96/70, Institute of Petroleum, London 1976).

[588] Ishizaki, M. (Mem. Defense Acad. Math. Phys. Chem. Eng. [Yokosuka, Japan] 11 [1971] 25/8; C.A. 76 [1972] No. 156409).

[589] Anonymous (Deutsche Norm, Bestimmung des Bleigehaltes von Ottokraftstoffen; Volumetrisches Chromat-Verfahren, DIN EN 13, Okt. 1975).

[590] Anonymous (IP Standards for Petroleum and Its Products, Pt. 1, Sect. 1 and 2, IP 248/70, Institute of Petroleum, London 1976).

[591] Anonymous (1985 Annual Book of ASTM Standards, Vol. 5.02, D 2547-82, Am. Soc. Testing Mater., Philadelphia 1985, pp. 464/9).

[592] Bühler, A. E. (Prax. Naturwiss. Chem. 33 [1984] 165/9).

[593] Banerjee, S. (Talanta 33 [1986] 358/9).

[594] Leisey, F. A., Standard Oil Co., Indiana (U.S. 3462244 [1966/69]; C.A. 71 [1969] No. 126973).

[595] Hozman, R. (Sb. Praci Vyzk. Chem. Vyuziti Uhli Dehtu Ropy 17 [1984] 55/69).

[596] Harrison, R. M. (J. Environ. Sci. Health A 11 [1976] 417/23).

[597] Harrison, R. M., Radojević, M. (Environ. Technol. Letters 6 [1985] 129/36).

[598] Pedinelli, M. (Chim. Ind. [Milan] 44 [1962] 651/2).

[599] Barbieri, R., Belluco, U., Tagliavini, G. (Ric. Sci. 30 [1960] 1671/4).

[600] Radojević, M., Allen, A., Rapsomanikis, S., Harrison, R. M. (Anal. Chem. 58 [1986] 658/61).

[601] Chau, Y. K., Wong, P. T. S. (4th Intern. Conf. Organometal. Coord. Chem. Germanium Tin Lead, Montreal 1983, Abstr. 27 (E-4)).

[602] Chau, Y. K., Wong, P. T. S., Bengert, G. A., Dunn, J. L. (Anal. Chem. 56 [1984] 271/4).

[603] Rapsomanikis, S., Donard, O. F. X., Weber, J. H. (Anal. Chem. 58 [1986] 35/8).

[604] Sirota, G. R., Uhte, J. F. (Anal. Chem. 49 [1977] 823/5).

[605] Noden, F. G. (Lead Marine Environ. Proc. Intern. Experts Discuss., Rovinj, Yugoslavia, 1977 [1980], pp. 83/91).

[606] Colombini, M. P., Corbini, G., Fuoco, R., Papoff, P. (Ann. Chim. [Rome] 71 [1981] 609/29).

96

[607] Diehl, K.-H., Rosopulo, A., Kreuzer, W. (Z. Anal. Chem. **317** [1984] 469/70).
[608] Jennen, A., Delafortrie, A., Verdoodt, D., Jacobs, T., Dourte, P. (Rev. Agric. [Brussels] **37** [1984] 1025/7).
[609] Aneva, Z. (Z. Anal. Chem. **321** [1985] 680/1).
[610] Orren, D. K., Caldwell-Kenkel, J. C., Mushak, P. (J. Anal. Toxicol. **9** [1985] 258/61).

[611] Chau, Y. K., Snodgrass, W. J., Wong, P. T. S. (Water Res. **11** [1977] 807/9).
[612] Fleischmann, M., Mengoli, G., Pletcher, D. (Electrochim. Acta **18** [1973] 231/5).
[613] Blaszkewicz, M., Baumhoer, G., Neidhart, B. (Heavy Metals Environ. 4th Intern. Conf., Heidelberg 1983, pp. 99/102).
[614] Blaszkewicz, M., Baumhoer, G., Neidhart, B. (Z. Anal. Chem. **317** [1984] 221/5).
[615] Bond, A. M., McLachlan, N. M. (Anal. Chem. **58** [1986] 756/8).
[616] Tagliavini, G., Belluco, U., Schiavon, G., Riccoboni, L. (Ric. Sci. **28** [1958] 2349/54).
[617] Good, W. D., Scott, D. W. (Pure Appl. Chem. **2** [1961] 77/82).
[618] Cox, J. D., Pilcher, G. (Thermochemistry of Organic and Organometallic Compounds, Academic, London 1970, p. 477).
[619] Tel'noi, V. I., Rabinovich, I. B. (Usp. Khim. **49** [1980] 1137/73; Russ. Chem. Rev. **49** [1980] 603/22).
[620] Crowder, G. A., Gorin, G., Kruse, F. H., Scott, D. W. (J. Mol. Spectrosc. **16** [1965] 115/21).

[621] Good, W. D., Scott, D. W. (in: Skinner, H. A., Experimental Thermochemistry, Vol. 2, Interscience, New York 1962, pp. 57/7).
[622] Skinner, H. A. (Advan. Organometal. Chem. **2** [1964] 49/114).
[623] Tel'noi, V. I., Rabinovich, I. B. (Zh. Fiz. Khim. **39** [1965] 2076/7; Russ. J. Phys. Chem. **39** [1965] 1108/9).
[624] Lippincott, E. R., Tobin, M. C. (J. Am. Chem. Soc. **75** [1953] 4141/7).
[625] Somayajulu, G. R., Zwolinski, B. J. (J. Chem. Soc. Faraday Trans. II **1974** 973/93).
[626] Dewar, M. J. S., Holloway, M. K., Grady, G. L., Stewart, J. J. P. (Organometallics **4** [1985] 1973/80).
[627] Skinner, H. A., Connor, J. A. (Pure Appl. Chem. **57** [1985] 79/88).
[628] Hawker, D. W., Wells, P. R. (Organometallics **4** [1985] 821/5).
[629] Blais, J. S., Marshall, W. D. (Appl. Organometal. Chem. **1** [1987] 251/60).
[630] Siergiejczyk, L., Boleslawski, M., Synoradzki, L. (Polimery [Warsaw] **31** [1986] 397/8).

[631] Chau, Y. K., Wong, P. T. S., Mojesky, C. A., Carty, A. J. (Appl. Organometal. Chem. **1** [1987] 235/9).
[632] Rapsomanikis, S., Donard, O. F. X., Weber, J. H. (Appl. Organometal. Chem. **1** [1987] 115/8).
[633] D'Ulivo, A., Papoff, P. (J. Anal. At. Spectrom. **1** [1986] 479/84).
[634] Ebdon, L., Hill, S., Jones, P. (J. Anal. At. Spectrom. **2** [1987] 205/10).
[635] Nerin, C., Olavide, S., Cacho, J. (Anal. Chem. **59** [1987] 1918/21).
[636] Bye, R. (J. Chem. Educ. **64** [1987] 188).
[637] Røyset, O., Thomassen, Y. (Anal. Chim. Acta **188** [1986] 247/55).
[638] Standard Oil Co., Indiana (Belg. 601371 [1961]).

1.1.1.1.2   The Molecule. Spectra

1.1.1.1.2.1   Structure

From electron diffraction studies, distances and angles in the tetrahedral $Pb(CH_3)_4$ molecule have been determined [12] and collected below together with values from older measurements [1, 7]:

	distances in Å		angles in °
Pb-C	2.238(9) [12], 2.203(10) [7], 2.29(5) [1]	Pb-C-H	104.6(5.4) [12], 109.0(4.0) [7]
C-H	1.08(2) [12]	C-Pb-C	109.5(3.6) [12]
C···C	3.66(7) [12]		
Pb···H	2.72(6) [12], 2.79(5) [7]		

From stretching and bending force constants, a C-Pb-C angle of 109.5° and distance of 3.74 Å between the methyl groups was estimated [4], whereas from $\gamma(CH)$ vibrations a C-H distance of 1.094 Å was predicted [13]; see also [10]. The bond lengths Pb-C = 2.17 Å and C-H = 1.10 Å, and the angles C-Pb-C = 109.5°, H-C-Pb = 109.3°, and H-C-H = 109.8° have been calculated by the MNDO method [18, 19], whereas a value of 2.247 Å has been obtained from Hartree-Fock calculations accounting for relativistic effects [20].

The minimum distance of approach of rotating methyl groups was estimated to be 3.046 Å [2] and it was inferred that, using a Pb-C distance of 2.23 Å, the rotation of the methyl groups is not [2] or probably not hindered [3]. The barrier to methyl rotation is rather low [15].

Inelastic neutron scattering experiments at 3 K showed quantum mechanical tunnel splitting of the vibrational ground state of the methyl group [14, 16]. Rotational potentials of methyl groups are determined in [16]. Tunnel splittings $\hbar\omega_t^0$ in the ground state of 74 µeV [16] and 30.7 µeV [16] or 35 µeV [14] have been observed; see also [15]. Torsional excitation E_{01} from the ground state at 15 K measured by neutron scattering is 0.3 kJ/mol, the activation energy E_A is 0.7 kJ/mol; E_A = 0.75 kJ/mol is derived from NMR measurements [15]. Reorientation of the methyl groups occurs at near liquid helium temperature [14, 16]. In earlier work methyl group reorientation was observed at temperatures as low as 77 K [10]. The energy barrier to methyl reorientation goes roughly as the inverse sixth power of the methyl separation [9, 10]. Molecular reorientation occurs only very slowly (if at all) at 100 K [9] and was observed above 180 K [14]. From self-diffusion measurements it is estimated that the molecular radius of $Pb(CH_3)_4$ is 1.2% smaller than that of $Sn(CH_3)_4$ [8].

According to X-ray diffraction [17] and neutron scattering measurements [16], a threefold axis exists at the molecular site of solid $Pb(CH_3)_4$, the molecule being compressed along this axis; space group Pa3, Z = 8 [16, 17]. $Pb(CH_3)_4$ is isomorphous with $Sn(CH_3)_4$ [16]. Calorimetric studies showed no evidence that $Pb(CH_3)_4$ can crystallize in more than one form [3, 5]. The van der Waals volume of lead derived from $Pb(CH_3)_4$ data is calculated to be 17.8 cm^3/mol [11].

The partial charge on the methyl groups was estimated to be 0.004 [6]. For measured data of the density of states, see [16].

References:

[1] Brockway, L. O., Jenkins, H. O. (J. Am. Chem. Soc. **58** [1936] 2036/44).

[2] French, F. A., Rasmussen, R. S. (J. Chem. Phys. **14** [1946] 389/94).

[3] Staveley, L. A. K., Paget, H. P., Goalby, B. B., Warren, J. B. (Nature **164** [1949] 787/8).

[4] Sheline, R. K. (J. Chem. Phys. **18** [1950] 602/6).

[5] Staveley, L. A. K., Warren, J. B., Paget, H. P., Dowrick, D. J. (J. Chem. Soc. **1954** 1992/2001).

[6] Sanderson, R. T. (J. Am. Chem. Soc. **77** [1955] 4531/2).

[7] Wong, C.-H., Schomaker, V. (J. Chem. Phys. **28** [1958] 1007/9).

[8] Bambynek, W. (Z. Physik. Chem. [N.F.] **25** [1960] 403/14).

[9] Smith, G. W. (Liquids Struct. Prop. Solid Interactions Proc. Symp., Warren, Mich., 1963 [1965], pp. 219/25).

[10] Smith, G. W. (J. Chem. Phys. **42** [1965] 4229/43).

[11] Bondi, A. (J. Phys. Chem. **70** [1966] 3006/7).

[12] Oyamada, T., Iijima, T., Kimura, M. (Bull. Chem. Soc. Japan **44** [1971] 2638/42).

[13] McKean, D. C., Duncan, J. L., Batt, L. (Spectrochim. Acta A **29** [1973] 1037/49).

[14] Kollmar, A., Alefeld, B. (Proc. Conf. Neutron Scattering, Gatlinburg, Tenn., 1976, Vol. 1, pp. 330/6).

[15] Müller-Warmuth, W., Duprée, K.-H., Prager, M. (Z. Naturforsch. **39a** [1984] 66/79).

[16] Prager, M., Müller-Warmuth, W. (Z. Naturforsch. **39a** [1984] 1187/94).

[17] Krebs, B., Henkel, G., Dartmann, M. (in: Prager, M., Müller-Warmuth, W., Z. Naturforsch. **39a** [1984] 1187/94).

[18] Dewar, M. J. S., Holloway, M. K., Grady, G. L., Stewart, J. J. P. (Organometallics **4** [1985] 1973/80).

[19] Dewar, M. J. (AD-A 173791-5-GAR [1985] 1/10; C.A. **106** [1987] No. 202036).

[20] Almlof, J., Faegri Jr., K. (Theor. Chim. Acta **69** [1986] 437/46).

1.1.1.1.2.2 Ionization. Photoelectron Spectrum

The ionization potential of $Pb(CH_3)_4$ obtained from mass spectroscopic data was given as 8.26 ± 0.17 eV [7], 9.32 ± 0.2 eV [6], and 8.0 ± 0.4 eV [3]. A provisional value of 11.5 eV had been obtained earlier using the molecular beam method [1]; see also [2, 19]. A value of 7.8 eV was calculated [3], and a value of 13.2×10^{-12} erg was estimated by extrapolation from data of related compounds [5]. The first ionization energy I_1 was calculated by the MNDO method using Koopman's theorem to be 10.29 eV [22]. A linear relationship between the total ionization cross-section and the polarizability of $M(CH_3)_4$ (M = Ge, Sn, Pb) was found [4].

The Helium(I) photoelectron spectrum of $Pb(CH_3)_4$ is considerably more complex than are the spectra of the tetramethyl compounds of the lighter group 14 elements. Spectra are shown in [8, 10, 14]; that from [10] is depicted in **Fig. 3**. This spectrum exhibits seven regions of photoelectron activity with maxima at 8.91, 9.75, 11.3, 13.1, 13.7, 15.0, and 16.4 eV. Orbital assignments (given in parentheses) are made for the ionization energies at 11.3 ($2a_1$), 13.1 ($1t_1$), 13.7 ($2t_2$), 15.0 ($1e$), and 16.4 ($1a_1$) eV and compared with the data of the other group 14 $M(CH_3)_4$ compounds. These values and assignments [10] were later cited [9, 22]. Other values and assignments [8] are: 8.38, 8.81 ($3t_2$), 9.09 ($3t_2$), 9.86 ($3t_2$), 13.3 (t_1, e?, $2t_2$?), and 15.3 ($2a_1$); see also [11] and [17].

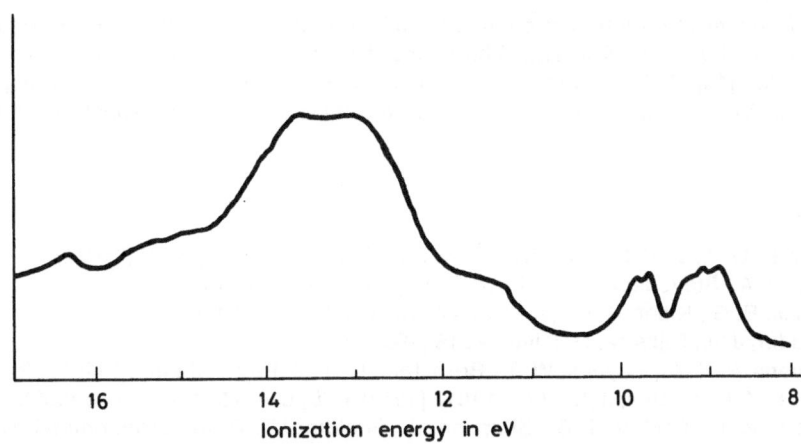

Fig. 3. He(I) photoelectron spectrum of Pb(CH$_3$)$_4$ [10].

Higher ionization potentials have been calculated by the MNDO method 13.87 (1t$_1$), 13.95 (1e), 14.30 (2t$_2$), 20.32 (2a$_1$), 29.88 (1t$_2$), 33.20 (1a$_1$) [22], orbital assignments (given in parentheses) being, however, in substantial disagreement with those given in [10]. The photoelectron spectroscopic onset energy is given as 8.50 eV [10]. Values of the angular parameters β of M(CH$_3$)$_4$ (M = C, Si, Ge, Sn, Pb) show little variation [9, 10]; see also [11]. The adiabatic ionization energy of the lower energy band amounts to 8.38 eV [8, 14].

Splitting of degeneracies is observed throughout all M(CH$_3$)$_4$ spectra (M = C, Si, Ge, Sn, Pb); the mechanism in lighter-M compounds is Jahn-Teller, in heavier-M compounds it is chiefly spin-orbit [10, 11]; see also [21]. Relativistically parametrized Extended Hückel (REX) calculations of orbital energies are reported for Pb(CH$_3$)$_4$, and are compared with experimental values [18]. The first band (A) of the photoelectron spectrum exhibits a relativistic splitting of 0.91 eV [8] or 0.65 eV [10]; a value of 0.46 eV was calculated [18]; see also [16].

From the experimental results and from semiempirical calculations it is inferred that a significant amount of stabilization is imparted by backbonding of molecular orbitals to vacant atomic d orbitals of M in M(CH$_3$)$_4$ (M = C, Si, Ge, Sn, Pb) [10, 11]. However, d-orbital participation has been discounted [12].

Photoelectron spectroscopic data were used to set up a tentative molecular orbital energy scheme based on localized σ-bond orbitals [8]; see also [10].

The electrochemical oxidation potentials [14] as well as the rates of oxidation of (CH$_3$)$_{4-n}$Pb(C$_2$H$_5$)$_n$ (n = 0 to 4) by [IrCl$_6$]$^{2-}$ [13] and the vertical ionization potentials show a linear correlation [13, 14]. The ionization potentials decrease monotonically with increasing substitution of ethyl for methyl groups around lead [14]. The vertical ionization potentials of M(CH$_3$)$_4$ (M = C, Si, Ge, Sn, Pb) decrease monotonically, indicating that ionization is associated with electrons localized close to the central atom [8, 14, 17]. Correlations exist between the Brønsted coefficient [20] or electron transfer [14, 17] and the ionization potential of Pb(CH$_3$)$_4$ and other organometallic compounds.

100

Quasimolecular M radiation of about 8 to 10 keV is emitted in collisions between Xe ions of 4 to 5 MeV and gaseous $Pb(CH_3)_4$. The energy threshold for excitation lies between about 2.0 and 2.5 MeV [15]. With synchroton radiation in the range between 45 and 75 nm, photoionization of the 5 d core levels was observed in the threshold electron spectrum of $Pb(CH_3)_4$ [23].

References:

[1] Fraser, R. G. J., Jewitt, T. N. (Proc. Roy. Soc. [London] A **160** [1937] 563/74).
[2] Hipple, J. A., Stevenson, D. P. (Phys. Rev. [2] **63** [1943] 121/6).
[3] Hobrock, B. G., Kiser, R. W. (J. Phys. Chem. **65** [1961] 2186/9).
[4] de Ridder, J. J., Dijkstra, G. (Nature **216** [1967] 260/1).
[5] Rummens, F. H. A., Raynes, W. T., Bernstein, H. J. (J. Phys. Chem. **72** [1968] 2111/9).
[6] Lappert, M. F., Pedley, J. B. (AD-719817 [1970] 1/78; C.A. **75** [1971] No. 80968).
[7] Lappert, M. F., Pedley, J. B., Simpson, J., Spalding, T. R. (J. Organometal. Chem. **29** [1971] 195/208).
[8] Evans, S., Green, J. C., Joachim, P. J., Orchard, A. F., Turner, D. W., Maier, J. P. (J. Chem. Soc. Faraday Trans. II **68** [1972] 905/11).
[9] Carlson, T. A., McGuire, G. E., Jonas, A. E., Cheng, K. L., Anderson, C. P., Lu, C. C., Pullen, B. P. (Electron Spectrosc. Proc. Intern. Conf., Pacific Grove, Calif., 1971 [1972], pp. 207/31).
[10] Jonas, A. E., Schweitzer, G. K., Grimm, F. A., Carlson, T. A. (J. Electron Spectrosc. Relat. Phenom. **1** [1972/73] 29/66).

[11] Jonas, A. E. (Diss. Univ. Tennessee 1971, Diss. Abstr. Intern. B **32** [1972] 6275).
[12] Boschi, R., Lappert, M. F., Pedley, J. B., Schmidt, W., Wilkins, B. T. (J. Organometal. Chem. **50** [1973] 69/73).
[13] Gardner, H. C., Kochi, J. K. (J. Am. Chem. Soc. **96** [1974] 1982/4).
[14] Gardner, H. C., Kochi, J. K. (J. Am. Chem. Soc. **97** [1975] 1855/65).
[15] Lutz, H. O., McMurray, W. R., Pretorius, R., van Heerden, I. J., van Reenen, R. J., Fricke, B. (J. Phys. B **9** [1976] L157/L160).
[16] Fadini, A., Glozbach, E., Krommes, P., Lorberth, J. (J. Organometal. Chem. **149** [1978] 297/307).
[17] Kochi, J. K. (Pure Appl. Chem. **52** [1980] 571/605).
[18] Lohr Jr., L. L., Hotokka, M., Pyykkö, P. (Intern. J. Quantum Chem. **18** [1980] 347/55).
[19] Liepins, R., Campbell, M., Clements, J. S., Hammond, J., Fries, R. J. (J. Vac. Sci. Technol. **18** [1981] 1218/26).
[20] Klingler, R. J., Kochi, J. K. (J. Am. Chem. Soc. **103** [1981] 5839/48).

[21] Eland, J. H. D. (Photoelectron Spectroscopy, 2nd Ed., Butterworths, London 1984, pp. 167/8).
[22] Dewar, M. J. S., Holloway, M. K., Grady, G. L., Stewart, J. J. P. (Organometallics **4** [1985] 1973/80).
[23] Nagaoka, S., Suzuki, S., Koyano, J. (Phys. Rev. Letters **58** [1987] 1524/7).

1.1.1.1.2.3 Dissociation

The dissociation energy D_1 of the first Pb-C bond in $Pb(CH_3)_4$ is reported to be 48.6 [12, 14], 48.8 [13], and 49 kcal/mol [21]. Values of 35.9 [5] and 41.3 ± 2 kcal/mol (converted units from 173 ± 8 kJ/mol) [16] were derived from kinetic data. D_1 values of 23.5 [1] and 28.2 kcal/mol [2] are deemed to be erroneous [5].

The energy to dissociate all four methyl groups is given as 130.3 [8], 139.6 [4], 143.7 [5], 146.9 (at 298 K) and ca. 142.9 (at 0 K) [15], 148.0 [6], and 150.2 kcal/mol [7].

The mean bond dissociation energy $\overline{D}$ in kcal/mol (for reviews, see [8, 22]) is calculated to be 32.6 [8], 34 [4], 35.7 [15], 35.9 [5], 36.5 ± 1 [9], 37.0 [6], 37.5 [7, 11], 37.6 [14], 38 ± 1 [22], 38.4 [12, 13, 20], and 43.4 ± 2 [3]. A value of 55 to 58 kcal/mol was determined from very-low-pressure-pyrolysis and activation energy data [23]; see also [10]. Values given in kJ/mol are 160.8 [19] and 168 ± 4 [18]. An empirical equation, which considers electronegativity, polarizability, and atomic number, yielded a contributing bond energy for the Pb-C bond in $Pb(CH_3)_4$ of 38.2 kcal/mol [24].

The thermochemical bond energy, E(Pb-C), was calculated to be 32.9 [12, 13] and 34.9 kcal/mol [4]. The heat of atomization (equal to the atomic heat of formation) was reported as $\Delta H^\circ_{298.15} = 1323.0$ kcal/mol [13, 22].

The activation energy for the methyl-lead bond dissociation at 700 K is calculated to be $E_a = 54 \pm 3$ kcal/mol [23]. The C-H dissociation energy in $Pb(CH_3)_4$ is predicted from $\nu(CH)$ to be 102.0 kcal/mol [17].

References:

[1] Romm, F. S. (Zh. Obshch. Khim. **10** [1940] 1784/92; C.A. **1941** 3880).
[2] Eltenton, G. C. (J. Chem. Phys. **15** [1947] 465/74).
[3] Mortimer, C. T. (J. Chem. Educ. **35** [1958] 381/4).
[4] Good, W. D., Scott, D. W., Lacina, J. L., McCullough, J. P. (J. Phys. Chem. **63** [1959] 1139/42).
[5] Long, L. H. (Pure Appl. Chem. **2** [1961] 61/9).
[6] Lautsch, W. F. (Habilitationsschr. Halle 1952, in: Lautsch, W. F., Tröber, A., Körner, H., Wagner, K., Kaden, R., Blase, S., Z. Chem. [Leipzig] **4** [1964] 441/54).
[7] Jehne, S. (Dipl.-Arbeit Leuna-Merseburg T. H. 1963, in: Lautsch, W. F., Tröber, A., Körner, H., Wagner, K., Kaden, R., Blase, S., Z. Chem. [Leipzig] **4** [1964] 441/54).
[8] Lautsch, W. F., Tröber, A., Körner, H., Wagner, K., Kaden, R., Blase, S. (Z. Chem. [Leipzig] **4** [1964] 441/54).
[9] Skinner, H. A. (Advan. Organometal. Chem. **2** [1964] 49/114).
[10] Tel'noi, V. I., Rabinovich, I. B. (Zh. Fiz. Khim. **39** [1965] 2076/7; Russ. J. Phys. Chem. **39** [1965] 1108/9).

[11] Cox, J. D., Pilcher, G. (Thermochemistry of Organic and Organometallic Compounds, Academic, London 1970, pp. 476/7, 517).
[12] Lappert, M. F., Pedley, J. B. (AD-719817 [1970] 1/78; C.A. **75** [1971] No. 80968).
[13] Lappert, M. F., Pedley, J. B., Simpson, J., Spalding, T. R. (J. Organometal. Chem. **29** [1971] 195/208).
[14] Price, S. J. W. (in: Bamford, C. H., Tipper, C. F. H., Comprehensive Chemical Kinetics, Vol. 4, Elsevier, Amsterdam 1972, p. 247).
[15] Gilroy, K. M., Price, S. J., Webster, N. J. (Can. J. Chem. **50** [1972] 2639/41).
[16] Homer, J. B., Hurle, I. R. (Proc. Roy. Soc. [London] A **327** [1972] 61/79).
[17] McKean, D. C., Duncan, J. L., Batt, L. (Spectrochim. Acta A **29** [1973] 1037/49).
[18] Steele, W. V. (Ann. Rept. Progr. Chem. [Chem. Soc. London] A **71** [1974/75] 103/18).
[19] Pilcher, G. (MTP [Med. Tech. Publ. Co.] Intern. Rev. Sci. Phys. Chem. Ser. Two **10** [1975] 45/80).
[20] Ribeiro da Silva, M. A. V., Reis, A. M. M. V. (Rev. Port. Quim. **20** [1978] 47/62).

[21] Jackson, R. A. (J. Organometal. Chem. **166** [1979] 17/9).

[22] Tel'noi, V. I., Rabinovich, I. B. (Usp. Khim. **49** [1980] 1137/73; Russ. Chem. Rev. **49** [1980] 603/22).

[23] Smith, G. P., Patrick, R. (Intern. J. Chem. Kinet. **15** [1983] 167/85).

[24] Lubing Yuan, Aihua Qi (Huaxue Tongbao **1985** No. 8, pp. 17/20).

1.1.1.1.2.4 Nuclear Magnetic Resonance Spectra

The ^{1}H NMR spectrum of $Pb(CH_3)_4$ exhibits one singlet for the methyl protons along with an overlapping doublet, due to coupling with the ^{207}Pb nucleus ($I = ^1/_2$, 22% natural abundance). Chemical shifts and coupling constants measured for the neat liquid or in various solvents are listed in Table 3. The chemical shift in the vapor state at 30°C was estimated to be 10.9 Hz (60 MHz, downfield from ethane) [29] and at 170°C a value of -34.0 Hz (60 MHz, relative to CH_4 gas) was obtained [3]; corrections for bulk susceptibility of the gases were made. A plot of chemical shifts as a function of the concentration in CCl_4 solution with a value of 161.5 ± 0.7 extrapolated to infinite dilution (40 MHz, relative to H_2O) was reported [1]. For a relationship between the chemical shift of the CH_3 protons and the electronic charge of the σ-C-H bonds involving also $Pb(CH_3)_4$, see [31]. Paramagnetic contributions to the shielding of protons in $M(CH_3)_4$ (M = Si, Ge, Sn, Pb) are attributed to the anisotropy of the carbon atom [3, 20, 21]. No isotopic effect on the ^{1}H NMR spectra of $(CH_3)_{4-n}Pb(CD_3)_n$ is observed [60]; see also [86].

For the ^{13}C chemical shift of $Pb(CH_3)_4$ in the liquid state the following values are reported: 196.2 ppm (referred to CS_2) [32], 131.6 ± 3 ppm (at 15.1 MHz; referred to C_6H_6) [3], and 2416 ± 5 Hz (at 15.1 MHz; referred to $(CH_3O)_2{}^{13}CO$) [3]. A ^{13}C chemical shift value $\delta = -4.2$ ppm (referred to TMS) has been measured in toluene solution [66, 67], and $\delta = -4.2 \pm 0.1$ ppm in $CDCl_3$ solution [69, 70].

^{207}Pb NMR spectra have been obtained directly and by INDOR studies. The central frequency of the multiplet ranged from 12.551 to 12.552 MHz [39]. In the directly obtained spectrum it is found at 12.55 MHz at 14.1 kG [5]. The ^{207}Pb resonance frequency in $Pb(CH_3)_4$ has a value of 20.920657 MHz [48], and 20.920597 MHz when corrected for a polarizing field strength such that TMS gives a proton resonance of exactly 100 MHz [64].

The chemical shift δ^{207}Pb of neat $Pb(CH_3)_4$ is nearly -3000 ppm with regard to $Pb(NO_3)_2$ in water (spectrometer frequency 12.55 MHz) [5]. The following chemical shift values δ^{207}Pb have been measured from solutions of $Pb(CH_3)_4$: 2961 ppm (3.6 M, in toluene), 2960.0 ppm (0.36 M, in CCl_4), 2962.7 ppm (0.36 M, in acetone), and 2954.1 ppm (0.36 M; in CS_2), each with respect to 1.0 M aqueous solution of $Pb(NO_3)_2$ [35], ca. -65 ppm (in toluene, with respect to external $Pb(C_4H_9)_4$) [32], -430 ppm (with respect to $(CH_3)_3PbCl$ in CH_2Cl_2) [78], 2.39 ± 0.05 ppm (in a mixture of toluene and pyridine, with respect to $Pb(CH_3)_4$ in toluene) [48], and 3.3 ppm (in $CDCl_3$, with respect to $Pb(CH_3)_4$ in toluene-C_6D_6) [83]. $Pb(CH_3)_4$ is used as a reference for ^{207}Pb NMR spectroscopy [46, 48, 64, 66, 84].

From INDOR spectra of $Pb(CH_3)_4$ measured in various solvents, the influence of the solvents on the chemical shifts δ^{207}Pb, determined as the total effect of the solvent on the nucleus, is analyzed and compared with effects of the medium on δ^{29}Si and $\delta^{117, 119}$Sn values of $M(CH_3)_4$ (M = Si, Sn). From the δM values of $M(CH_3)_4$ (M = Si, Sn, Pb) it is inferred that the solvent has the same sort of influence on the different central atoms; $\delta^{117, 119}$Sn and δ^{207}Pb are linearly correlated [68].

The secondary $^{13}C/^{12}C$ isotope effect $\Delta\delta^i$ for ^{207}Pb shielding for $Pb(CH_3)_4$ dissolved in toluene-C_6D_6 or in $CDCl_3$ is -0.089 ± 0.003 and -0.087 ± 0.003, respectively, indicating no significant influence of the solvent [83].

Estimated van der Waals contributions to the medium shift (chemical shift in the liquid state minus that in the gas phase at zero pressure) of $M(CH_3)_4$ (M = C, Si, Ge, Sn, Pb) and their binary combinations compare satisfactorily with experimental values. The medium shift and its temperature dependence in magnetically isotropic solvents was calculated [29]. The contribution by dispersion interaction on the screening constant was considered [68].

The coupling constants $^2J(^{207}Pb, {}^1H)$, $^1J(^{207}Pb, {}^{13}C)$, and $^1J(^{13}C, {}^1H)$ obtained with neat $Pb(CH_3)_4$ or in various solvents are summarized in Table 3 (columns 3 to 5) on p. 105. Further coupling constants have been measured $^2J(^{13}C, {}^{13}C) = 5.0 \pm 0.2$ [41], $^3J(^{13}C, {}^1H) = 1.05$ [32], 1.0 ± 0.1 [26], and $^4J(^1H, {}^1H) < 0.05$ Hz [26] from neat $Pb(CH_3)_4$. From a highly deuterated sample in $CDCl_3$ ($CHCl_3$) the coupling constant $^2J(^1H, {}^1H) = -10.94 \pm 0.07$ Hz was obtained [42]. The normalized coupling constant $^2J(^{207}Pb, {}^1H)$ has a negative sign, whereas $^1J(^{13}C, {}^1H)$, $^1J(^{207}Pb, {}^{13}C)$, and $^3J(^{13}C, {}^1H)$ are positive [26, 27]. Valence bond eigen functions yield -49 Hz for $^2J(^{207}Pb, {}^1H)$ [6]. For other calculations of $^2J(^{207}Pb, {}^1H)$, see [10, 22]. For $^1J(^{207}Pb, {}^{13}C)$ a value of 276 Hz [32] (also 432 Hz [50, 71]) was calculated; see also [72, 75]. Reduced coupling constants K_{MC} of $M(CH_3)_4$ compounds, corrected to sp^3 hybridization, were determined from calculated values of the spin density at the nuclei of the isolated atoms. In contrast to the data of the other group 14 elements, the calculated value, $K_{PbC} = 810 \times 10^{20}$ cm^{-3} disagrees markedly with the experimental value of 508×10^{20} cm^{-3} [43, 44]. $^2J(^{207}Pb, {}^1H)$ depends on the bulk properties of the solution and is well correlated with the change in the dispersion interaction [35]; see also [68]. The variation in chemical shift of the methyl protons on changing from one solvent to another does not correlate with the solvating power of the solvents; however, a relationship of $^2J(^{207}Pb, {}^1H)$ with the donor activity of the solvent was established. From this correlation a series of solvents with increasing solvating power for $Pb(CH_3)_4$ can be established [34]; see also Section 1.1.1.1.5.

The relationship between $^2J(^nM, {}^1H)$ ($^nM = {}^{119}Sn, {}^{207}Pb$) and hybridization of the valence orbitals of M used in bonding to carbon in $M(CH_3)_4$ was studied [61]; see also [54]. Relations of $^1J(^{207}Pb, {}^{13}C)$ coupling constants of PbR_4 and $(CH_3)_{4-n}PbR_n$ compounds and correlations with the s character of the metal hybrid orbital were studied [76]. The s-electron densities at the Pb and C nuclei, and the hybridization of the Pb and the C bonding orbitals account more for variations of coupling constants of $Pb(CH_3)_4$, other tetraalkyllead compounds, and alkyllead acetates than the mean excitation energy [69]. $^2J(^{207}Pb, {}^1H)$ correlates with the s character α^2 of the hybrid orbital of lead in the Pb-C bond through the relation $^2J(^{207}Pb, {}^1H) = 2.46 \, \alpha^2$. This equation yields an s character of 25% in C_6H_{12} solution, 27.2% in THF solution, and 28.8% in $CH_3S(O)CH_3$ solution [34]; see also [51]. A comparable relationship exists between $^1J(^{207}Pb, {}^{13}C)$ and the s character of the hybrid orbital of carbon [79]. The s character of the carbon orbitals in the C-H bonds, α^2_{CH}, and of the carbon orbitals in the Pb-C bonds, α^2_{PbC} in $Pb(CH_3)_4$, is calculated from $^1J(^{13}C, {}^1H)$ to be 0.2680 and 0.1960, respectively [42]; see also [11, 30].

The coupling constants $^1J(^nM, {}^{13}C)$ ($^nM = {}^1H, {}^{29}Si, {}^{119}Sn, {}^{207}Pb$) are linearly correlated [72, 77, 79], for example, $^1J(^{207}Pb, {}^{13}C) = 0.4980 \, |^1J(^{199}Hg, {}^{13}C)| - 106.7$ [72]. Linear relationships also exist between $^1J(^{207}Pb, {}^{13}C)$ and $^2J(^{207}Pb, {}^1H)$ for $(CH_3)_3PbR$ compounds [55]. $^1J(^{207}Pb, {}^{13}C)$ of $Pb(CH_3)_4$ and lead tetraalkyls are compared with $^1J(^nM, {}^{13}C)$ of analogous tin and mercury compounds and are correlated with the inductive substituent effect [80].

The contribution of the Fermi contact term is assumed to be dominant in the coupling mechanism [6, 9, 10, 16, 26, 32, 43, 54, 55, 69, 72, 73, 76]; see also [27, 36, 41, 51, 77]. A technique for estimating the contribution of the core s electrons to $^1J(^{207}Pb, {}^{13}C)$ was proposed [77, 79]; see also [24].

The nuclear spin-spin coupling tensor components and the full tensor K(PbC) for $Pb(CH_3)_4$ are calculated, using Relativistically parametrized Extended Hückel (REX) wave functions and the relativistic analogue of Ramsey's theory. Relativistic effects are predicted to increase the relative anisotropy R; R(PbC) should be larger than R(SnC) [73]; see also [75].

$^1J(^{207}Pb, ^{13}C)$ in lead tetraalkyls decreases from methyl to ethyl to butyl [67]. Coupling constants of $M(CH_3)_4$ (M = C, Si, Ge, Sn, Pb) are correlated with the atomic number of M [9, 16 to 19, 26, 32, 43, 51]; see also [27]. For an additivity relationship for $^2J(^{207}Pb, ^1H)$ and inductive σ^*-parameters, see [30]. A linear relationship exists between $^1J(^{13}C, ^1H)$ and the symmetrical M-CH$_3$ stretching frequency of $M(CH_3)_4$ (M = C, Si, Ge, Sn, Pb) [25]. For a correlation of $\nu(CH)$ with the coupling constant $^1J(^{13}C, ^1H)$, see [47]. A direct correlation of the coupling constant $^1J(^{13}C, ^1H)$ with the activation energy and the velocity constant of the reaction of methyl radicals with $M(CH_3)_4$ (M = Si, Ge, Sn, Pb) was proposed [37].

Relative electronegativities for group 14 elements in sp^3 valence states are determined from chemical shifts of the tetramethyl compounds in CCl_4 at infinite dilution. This orbital electronegativity scale which shows especially for lead a large discrepancy with established values [1, 4] is rejected [2, 11] and, considering shielding effects, it is concluded that the electronegativities for Si, Ge, Sn, and Pb are essentially constant or slightly decreasing [2, 11]; see also [3, 8, 20, 21]. A linear relationship between coupling constants $^1J(^{13}C, ^1H)$ and ^1H chemical shifts for a series of lead methyl compounds including $Pb(CH_3)_4$ is reported, and it is inferred that electronegativities of the group 14 elements bonding to the methyl groups cannot be evaluated from chemical shift data [20, 21]. Orbital overlap is considered to be the primary factor affecting the trend in $^1J(^{13}C, ^1H)$ for $M(CH_3)_4$ (M = Si, Ge, Sn, Pb) with respect to electronegativity differences [20]. Experimental ^1H and ^{13}C chemical shifts in $M(CH_3)_4$ (M = C, Si, Ge, Sn, Pb) and other $M'(CH_3)_n$-type compounds are correlated with electronegativity and the number of lone pairs present in M' [33]. Relations between the magnitudes of $^1J(^{13}C, ^1H)$ and $^2J(^1H, ^1H)$ and the electronegativities in the series $M(CH_3)_4$ (M = Si, Ge, Sn, Pb) were studied [42].

The longitudinal nuclear relaxation time T_1 of protons in $Pb(CH_3)_4$ as measured by the spin-echo method in the liquid phase between 250 and 325 K correlates with the self-diffusion coefficient [28]. Temperature-dependent ^1H NMR measurements showed the low-frequency reorientation about the molecular center of gravity is observed for solid $Si(CH_3)_4$ and $Ge(CH_3)_4$, but not for solid $Sn(CH_3)_4$ and $Pb(CH_3)_4$. The energy barrier to methyl rotation in these compounds as found from T_1 studies goes as $1/r^n$ ($n \approx 6$) [12, 23]; see also [62]. The spin-lattice relaxation time $T_1 = 0.6$ s is remarkably short, compared to T_1 values for ^{13}C and ^{29}Si [46, 49]. Spin-rotation is the predominant relaxation mechanism in the ^{207}Pb spin-lattice relaxation [52]. Reanalysis of data [52] gave $T_1 = 0.53$ s at 298 K, a spin-rotation constant $C_0(^{207}Pb) = 20$ kHz, an angular correlation time $\tau_\omega = 0.3 \times 10^{-13}$ s, and a molecular reorientation time $\tau_\theta = 4 \times 10^{-12}$ s [65]. ^{207}Pb spin-lattice relaxation studies at magnetic fields of 2.35 and 7.05 T exhibit T_1 values of 0.586 and 0.601 s, respectively, at 306 K which are equal within experimental error. This field independence of T_1 indicates that spin rotation and not chemical shift anisotropy is the dominant spin-lattice relaxation mechanism [74]. The experimentally determined spin-lattice relaxation rate of $Pb(CH_3)_4$ [82] is explained in terms of superposition of two nonequivalent methyl rotators [81]. The relaxation is tunneling induced [82]. Values of the apparent activation energies E'_A, determined between 30 and 65 K, and E''_A, determined at 5 and 10 K, are 0.62 and 0.23 kJ/mol, respectively. Tunneling energies $\hbar\omega_t$ are 2.6 µeV at 29 K and 20 µeV at 100 K [82]. For assignments and other relaxation data τ'_0, τ''_0, C, C_{AE}, and C_{EE}, see [82]; for a plot of the temperature dependence of $1/T_1$ showing three maxima, see [81, 82].

Explanation for Table 3: Chemical shifts given in τ or Δν in the original papers are presented as δ values referred to TMS. Chemical shifts referred to H₂O [8], cyclohexane [34], or CH₂Cl₂ [59] were recalculated using the shift values $\delta = 4.61$ (H₂O), $\delta = 1.44$ (C₆H₁₂), and $\delta = 5.33$ ppm (CH₂Cl₂). The abbreviations THF and TMEDA were used for tetrahydrofuran and $(CH_3)_2NCH_2CH_2N(CH_3)_2$, respectively.

Table 3
^{1}H NMR Shifts δ of Pb(CH₃)₄ with TMS as Standard ($\delta = 0$ ppm) in ppm and Coupling Constants in Hz Obtained in Various Solvents.

solvent	δ	$^2J(^{207}Pb, {}^1H)$	$^1J(^{207}Pb, {}^{13}C)$	$^1J(^{13}C, {}^1H)$	Ref.
neat liquid		61.2 ± 0.2	249 ± 3	133.9 ± 0.2	[26]
		60.9 ± 0.2	250 ± 1	135	[27]
	0.87	60.5		132.5	[7]
		61.0 ± 0.3		134.0 ± 0.2	[10]
		55.3 [6]		133 [3]	[3,6]
		61.0 ± 0.4 [9]	250 [32]	134.2 [32]	[9,32]
		61.2			[5]
CCl₄	0.70	60			[45,63]
	0.71			134 ± 1	[20,21]
	0.73	61		135.3	[13]
	0.74 [56]	61.2 ± 0.2 [26]	249 ± 3 [26]	133.9 ± 0.2 [26]	[26,56]
	0.87	60.5		135.1	[14]
	0.57 [8]	61.4 ± 0.8 [27]	250 ± 1 [27]	134.3 ± 0.2 [42]	[8,27,42]
CDCl₃ (CHCl₃)	0.74 ± 0.05	61 ± 3	251 ± 1		[69,70]
	0.74	61.8 ± 0.1			[30]
	0.87 [38]	60.5 [38]	248.3 [83]		[38,83]
		62.0 ± 0.8 [35,51]	251 [50,71]		[35,50,51,71]
CCl₄ + CDCl₃	0.72	62			[77]
CH₂Cl₂		62.8 ± 0.8			[35]
fluorocarbon		63.4 ± 0.8			[35]
C₆H₆		61.5 ± 0.8			[35]
C₆H₅CH₃	0.73	61.4 ± 0.2		134.2 ± 0.3	[39]
	0.66 [15,48]	62.0 ± 0.2 [15,48,84,85]	250 ± 1 [32,66,67]		[15,32,48,66,67,84,85]
	0.65 [85]				[85]
C₆H₆ + C₆H₅CH₃	0.72	62			[57]
C₆D₆ + C₆H₅CH₃			248.5		[83]
C₆H₅NO₂	0.85 [40]	61.7 ± 0.8 [35]			[35,40]
C₆H₁₂	0.75	61.5			[34]

Table 3 (continued)

solvent	δ	$^2J(^{207}Pb, ^1H)$	$^1J(^{207}Pb, ^{13}C)$	$^1J(^{13}C, ^1H)$	Ref.
CH_3OH	0.73	62.5			[57, 58]
		62.7 ± 0.8			[35]
C_5H_5N	0.79	63.0			[34, 53]
		62.7 ± 0.8			[35]
$C_6H_5CH_3 + C_5H_5N$	0.71 [48]	62 [57]			[48, 57]
CH_3COCH_3	0.71	68.5			[34]
CD_3COCD_3		63.3 ± 0.8			[35]
C_6H_5CHO		61.9 ± 0.8			[35]
THF	0.7				[48]
	0.70	67.0			[34]
$CH_3OCH_2CH_2OCH_3$	0.67				[53]
	0.77	62.0			[34]
dioxane-1,4	0.69 [53]	62.4 ± 0.8 [35]			[35, 53]
	0.75	62.0			[34]
$N(C_2H_5)_3$	0.70	64.5			[34]
TMEDA	0.70	68.5			[34]
$C_6H_5NH_2$		61.2 ± 0.8			[35]
$HCON(CH_3)_2$	0.73	69.5			[34]
		62.9 ± 0.8			[35]
$P(C_6H_5)_3$	0.69	67.5			[34]
$OP(N(CH_3)_2)_3$	0.74	62.0			[34]
		62.1 ± 0.8			[35]
$CH_3S(O)CH_3$	0.74 [53]	62.24 [35]			[35, 53]
	0.70	71.0			[34]
CH_3CN		63.9 ± 0.8			[35]
$CD_3CN + CD_3COOD$	0.73	63.5			[59]
C_6H_5CN		61.8 ± 0.8			[35]
tetrahydrothiophene	0.73	63.5			[34]
CS_2		60.7 ± 0.8			[35]
not specified		61.0			[61]

References:

[1] Allred, A. L., Rochow, E. G. (J. Inorg. Nucl. Chem. **5** [1958] 269/88).
[2] Drago, R. S. (J. Inorg. Nucl. Chem. **15** [1960] 237/41).
[3] Spiesecke, H., Schneider, W. G. (J. Chem. Phys. **35** [1961] 722/30).

[4] Allred, A. L., Rochow, E. G. (J. Inorg. Nucl. Chem. **20** [1961] 167/70).
[5] Schneider, W. G., Buckingham, A. D. (Discussions Faraday Soc. No. 34 [1962] 147/55).
[6] Klose, G. (Ann. Physik [7] **9** [1962] 262/78).
[7] Flitcroft, N., Kaesz, H. D. (J. Am. Chem. Soc. **85** [1963] 1377/80).
[8] McCoy, C. R., Allred, A. L. (J. Inorg. Nucl. Chem. **25** [1963] 1219/23).
[9] Dreeskamp, H. (Z. Physik. Chem. [N.F.] **38** [1963] 121/4).
[10] Smith, G. W. (J. Chem. Phys. **39** [1963] 2031/4).

[11] Matwiyoff, N. A. (Diss. Univ. Illinois 1963; Diss. Abstr. **24** [1963] 502).
[12] Smith, G. W. (Liquids Struct. Properties Solid Interact. Proc. Symp., Warren, Mich., 1963 [1965], pp. 219/25).
[13] Schmidbaur, H., Hussek, H. (J. Organometal. Chem. **1** [1964] 257/67).
[14] Schmidbaur, H. (Chem. Ber. **97** [1964] 270/81).
[15] Fritz, H. P., Schwarzhans, K.-E. (J. Organometal. Chem. **1** [1964] 297/8).
[16] Dreeskamp, H. (Z. Naturforsch. **19a** [1964] 139/42).
[17] Wells, E. J., Reeves, L. W. (J. Chem. Phys. **40** [1964] 2036/7).
[18] Smith, G. W. (J. Chem. Phys. **40** [1964] 2037/8).
[19] Reeves, L. W. (J. Chem. Phys. **40** [1964] 2128/31).
[20] Drago, R. S., Matwiyoff, N. A. (J. Organometal. Chem. **3** [1965] 62/9).

[21] Drago, R. S. (Record Chem. Progr. **26** [1965] 157/67).
[22] Smith, G. W. (J. Chem. Phys. **42** [1965] 435/6).
[23] Smith, G. W. (J. Chem. Phys. **42** [1965] 4229/43).
[24] Dreeskamp, H. (Proc. Colloq. AMPERE **13** [1964] 400).
[25] Jouve, P. (Compt. Rend. B **262** [1966] 815/7).
[26] Dreeskamp, H., Stegmeier, G. (Z. Naturforsch. **22a** [1967] 1458/64).
[27] McFarlane, W. (Mol. Phys. **13** [1967] 587/8).
[28] Kessler, D., Weiss, A., Witte, H. (Ber. Bunsenges. Physik. Chem. **71** [1967] 3/19).
[29] Rummens, F. H., Raynes, W. T., Bernstein, H. J. (J. Phys. Chem. **72** [1968] 2111/9).
[30] Singh, G. (J. Organometal. Chem. **11** [1968] 133/43).

[31] Bykov, G. V. (Izv. Akad. Nauk SSSR Ser. Khim. **1968** 1773/9; Bull. Acad. Sci. USSR Div. Chem. Sci. **1968** 1677/82).
[32] Weigert, F. J., Winokur, M., Roberts, J. D. (J. Am. Chem. Soc. **90** [1968] 1566/9).
[33] Bucci, P. (J. Am. Chem. Soc. **90** [1968] 252/3).
[34] Petrosyan, V. S., Voyakin, A. S., Reutov, O. A. (Zh. Org. Khim. **6** [1970] 889/93; J. Org. Chem. [USSR] **6** [1970] 895/8).
[35] Laszlo, P., Speert, A. (J. Magn. Resonance **1** [1969] 291/7).
[36] Jameson, C. J., Gutowsky, H. S. (J. Chem. Phys. **51** [1969] 2790/803).
[37] Chaudhry, A. U., Gowenlock, B. G. (J. Organometal. Chem. **16** [1969] 221/6).
[38] Fong, C. W., Kitching, W. (J. Organometal. Chem. **21** [1970] 365/75).
[39] Banney, P. J., McWilliam, D. C., Wells, P. R. (J. Magn. Resonance **2** [1970] 235/42).
[40] Bade, V., Huber, F. (J. Organometal. Chem. **24** [1970] 691/5).

[41] Schumann, C., Dreeskamp, H. (J. Magn. Resonance **3** [1970] 204/17).
[42] Lacey, M. J., Macdonald, C. G., Pross, A., Shannon, J. S. (Australian J. Chem. **23** [1970] 1421/9).
[43] Dalling, D. K., Gutowsky, H. S. (U.S. Clearinghouse Fed. Sci. Tech. Inform. AD-721717 [1971] 1/32).
[44] Dalling, D. K., Gutowsky, H. S. (J. Chem. Phys. **55** [1971] 4959/66).
[45] Haupt, H.-J., Schubert, W., Huber, F. (J. Organometal. Chem. **54** [1973] 231/8).

[46] Maciel, G. E., Dallas, J. L. (J. Am. Chem. Soc. **95** [1973] 3039).
[47] McKean, D. C., Duncan, J. L., Batt, L. (Spectrochim. Acta A **29** [1973] 1037/49).
[48] Cooper, M. J., Holliday, A. K., Makin, P. H., Puddephatt, R. J., Smith, P. J. (J. Organometal. Chem. **65** [1974] 377/82).
[49] Maciel, G. E. (Nuclear Magnetic Resonance Spectroscopy of Nuclei Other Than Protons, Wiley, New York 1974, pp. 347/75).
[50] Mann, B. E. (Advan. Organometal. Chem. **12** [1974] 135/213).

[51] Barbieri, G., Benassi, R. (Atti Soc. Nat. Mat. Modena **105** [1974] 39/46).
[52] Hawk, R. M. (Diss. University Michigan 1973; Diss. Abstr. Intern. B **35** [1974] 156/7).
[53] Fedorov, L. A., Kalinin, V. N., Gasanov, K. G., Zakharkin, L. I. (Zh. Obshch. Khim. **45** [1975] 591/9; J. Gen. Chem. [USSR] **45** [1975] 581/7).
[54] Barbieri, G., Benassi, R., Taddei, F. (Gazz. Chim. Ital. **105** [1975] 807/26).
[55] Singh, G. (J. Organometal. Chem. **99** [1975] 251/62).
[56] Lukevics, E., Erchak, N. P., Popelis, J., Zolotoyabko, R. M. (Khim. Elementoorg. Soedin. [Moscow] **1976** 63/7).
[57] Arnold, D. P., Wells, P. R. (J. Organometal. Chem. **111** [1976] 269/83).
[58] Arnold, D. P., Wells, P. R. (J. Organometal. Chem. **108** [1976] 345/52).
[59] Gardner, H. C., Kochi, J. K. (J. Am. Chem. Soc. **98** [1976] 2460/9).
[60] Arnold, D. P., Wells, P. R. (J. Organometal. Chem. **111** [1976] 285/96).

[61] Reeves, L. W., Suzuki, M., Vanin, J. A. (Anais Acad. Brasil. Cienc. **48** [1976] 739/42).
[62] Kollmar, A., Alefeld, B. (Proc. Conf. Neutron Scattering, Gatlinburg, Tenn., 1976, Vol. 1, pp. 330/6).
[63] Wieber, M., Baudis, U. (J. Organometal. Chem. **125** [1977] 199/207).
[64] Kennedy, J. D., McFarlane, W., Pyne, G. S. (J. Chem. Soc. Dalton Trans. **1977** 2332/6).
[65] Lassigne, C. R., Wells, E. J. (J. Magn. Resonance **26** [1977] 55/69).
[66] Mitchell, T. N., Gmehling, J., Huber, F. (J. Chem. Soc. Dalton Trans. **1978** 960/4).
[67] Cox, R. H. (J. Magn. Resonance **33** [1979] 61/70).
[68] Bakhbukh, M., Grishin, Yu. K., Ustynyuk, Yu. A., Zemlyanskii, N. N. (Vestn. Mosk. Univ. Khim. **34** No. 4 [1979] 366/8; Moscow Univ. Chem. Bull. **34** No. 4 [1979] 69/72).
[69] de Vos, D., van Beelen, D. C., Wolters, J. (Bull. Soc. Chim. Belges **89** [1980] 791/6).
[70] van Beelen, D. C. (Diss. Leiden 1980).

[71] Mann, B. E., Taylor B. F. (^{13}C NMR Data for Organometallic Compounds, Academic, London 1981).
[72] Radeglia, R., Steinborn, D., Taube, R. (Z. Chem. [Leipzig] **21** [1981] 365/6).
[73] Pyykkö, P., Wiesenfeld, L. (Mol. Phys. **43** [1981] 557/80).
[74] Hays, G. R., Gillies, D. G., Blaauw, L. P., Clague, A. D. H. (J. Magn. Resonance **45** [1981] 102/7).
[75] Pyykkö, P. (J. Organometal. Chem. **232** [1982] 21/32).
[76] Steinborn, D., Taube, R., Radeglia, R. (J. Organometal. Chem. **229** [1982] 159/68).
[77] Fedin, E. I., Fedorov, L. A. (Dokl. Akad. Nauk SSSR **267** [1982] 1159/62; Dokl. Phys. Chem. Proc. Acad. Sci. USSR **262/267** [1982] 1007/10).
[78] Kubicki, M. M., Kergoat, R., Guerchais, J.-E., L'Haridon, P. (J. Chem. Soc. Dalton Trans. **1984** 1791/3).
[79] Fedorov, L. A. (Zh. Strukt. Khim. **25** No. 3 [1984] 43/8; J. Struct. Chem. [USSR] **25** [1984] 379/84).
[80] Fedorov, L. A. (Zh. Strukt. Khim. **25** No. 4 [1984] 35/42; J. Struct. Chem. [USSR] **25** [1984] 538/44).

[81] Prager, M., Müller-Warmuth, W. (Z. Naturforsch. **39a** [1984] 1187/94).

[82] Müller-Warmuth, W., Duprée, K.-H., Prager, M. (Z. Naturforsch. **39a** [1984] 66/79).

[83] Kerschl, S., Sebald, A., Wrackmeyer, B. (Magn. Resonance Chem. **23** [1985] 514/20).

[84] Capek, E., Schwarzhans, K. E. (Monatsh. Chem. **118** [1987] 419/26).

[85] Fritz, H. P., Schwarzhans, K. E. (Chem. Ber. **97** [1964] 1390/7).

[86] Wei, Y.-C., Wells, P. R., Lambert, L. K. (Magn. Resonance Chem. **24** [1986] 659/62).

1.1.1.1.2.5 Vibrational Spectra

Calculated fundamental vibrational frequencies of $Pb(CH_3)_4$ and the appropriate assignments are listed in Table 4 [8, 24, 31, 42, 46]. For the assignments, see also [4]. Observed IR absorptions of gaseous and liquid $Pb(CH_3)_4$ and Raman lines of gaseous, liquid, and solid $Pb(CH_3)_4$ have been comprehensively reported; assignments of fundamentals and non-fundamentals are summarized in Table 5 [1, 4, 5, 10, 24, 25, 47, 51] and Table 6, p. 111 [2, 5, 6, 8, 10, 13, 17, 22, 42, 47], respectively. A graphical comparison of the fundamental of $M(CH_3)_4$ (M = C, Si, Ge, Sn, Pb) is given [8, 11]. From the Raman spectra of $Pb(CH_3)_4$ in the vapor phase at 344 K fundamentals ν_1, ν_2, and ν_3 are reported [35]. Calculated and observed fundamental IR and Raman frequencies of gaseous and liquid $Pb(CD_3)_4$ are collected in Table 7, p. 113. The fundamentals of a series of $M(CD_3)_4$ compounds (M = Si, Ge, Sn, Pb) are listed [42, 46, 47].

Table 4
Calculated Fundamental Vibrations of $Pb(CH_3)_4$.
Wave numbers in cm^{-1}.

T_d		frequencies					assignment
		[8,9]	[24]	[31]	[42]	[46]	
ν_1	a_1	2919	2928	2931	2921	2920	$\nu_s(CH_3)$
ν_2	a_1	1158	1172	1167	1166	1172	$\delta_s(CH_3)$
ν_3	a_1	460	460	462*)	466	465	$\nu_s(PbC_4)$
ν_4	a_2			inactive			$\tau(CH_3)$
ν_5	e	3000	3003	2985	3006	3002	$\nu_{as}(CH_3)$
ν_6	e	1439	1452	1431	1405	1438	$\delta_{as}(CH_3)$
ν_7	e	770	765	777	771	768	$\varrho(CH_3)$
ν_8	e	136	117	145	130	129	$\delta_s(PbC_4)$
ν_9	f_1		3003				$\nu_{as}(CH_3)$
ν_{10}	f_1		1452				$\delta_{as}(CH_3)$
ν_{11}	f_1		763				$\varrho(CH_3)$
ν_{12}	f_1			inactive			$\tau(CH_3)$
ν_{13}	f_2	3000	3003		3006	3002	$\nu_{as}(CH_3)$
ν_{14}	f_2	2920	2928		2921	2920	$\nu_s(CH_3)$
ν_{15}	f_2	1439	1452		1416	1438	$\delta_{as}(CH_3)$
ν_{16}	f_2	1158	1172		1153	1172	$\delta_s(CH_3)$
ν_{17}	f_2	772	767		772	769	$\varrho(CH_3)$
ν_{18}	f_2	472	475	477	478	476	$\nu_{as}(PbC_4)$
ν_{19}	f_2	136	120	131	129	121	$\delta_{as}(PbC_4)$

*) T_d model; 459 for the C_{3v} model.

For figures of IR spectra of gaseous $Pb(CH_3)_4$, see [1, 5, 24, 25, 30]; of liquid $Pb(CH_3)_4$, see [5, 24]; and of gaseous and of liquid $Pb(CD_3)_4$, see [42]. A low frequency IR spectrum of $Pb(CH_3)_4$ in cyclopentane is depicted in [24, 25]. The frequencies in the spectrum of $Pb(CH_3)_4$ are compared with those of the tetramethyl derivatives of C, Si, Ge, and Sn [4, 13, 30]; see also [42]. IR absorptions and assignments of gaseous $(CH_3)_3PbCD_3$ are listed, too [47].

Table 5
Fundamental Vibrations [1, 10, 24, 25, 47] and IR Spectra of $Pb(CH_3)_4$ [5, 10] in Various States. Wave numbers in cm^{-1}. For the assignments, see Table 4.

gas [1]	[5]	[24,25], see also [10]	[47]	liquid [5], see also [24]	[10], see also [51]	assignment
	4350 (vw)			4360 (w)		$v_{14} + v_{15}$
				4100 (w)		$v_2 + v_{14}$
				4050 (w)		$v_{14} + v_{16}$
	3760 (vw)			3770 (w)		$v_{13} + v_{17}$
				3380 (vw)		$v_{14} + v_{18}$
2999	2992 2982 } (s)	3003	3005 (s)	3000 (vs)	3000 (vs)	v_{13} [10, 24, 47], v_5, v_{13} [5], v_5, v_{13}, (v_9) [1]
2918	2910 (s)	2928	2927.7 (s)	2905 (s, sh)	2920 (vs)	v_{14} [10, 24, 47], v_{14}, v_1 [1,5]
	2292 (w)			2290 (m)	2295 (m)	$2v_{16}$
				2108 (vw)		$2v_{18} + v_{16}$
	1925 (w)			1922 (m)	1920 (w)	$v_{18} + v_{15}$ [10], $v_{16} + v_{17}$ [5]
	1621 (w)			1618 (m)	1624 (m)	$v_2 + v_3$ [10], $v_{16} + v_{18}$ [5]
1462	1453 (m)	1452	1456 (w)	1440 (s)	1448 (vs)	v_{15} [10, 24, 47], v_6, v_{15} [5], v_{15}, (v_6, v_{10}) [1]
	1392 (m)			1390 (s)	1400 (s)	$2v_3 + v_{18}$ [5]
	1306 (w)			1296 (m)	1295 (w)	$v_2 + v_{19}$ [5], $v_{19} + v_{16}$ [10]
1170		1178				v_2
					1166 (sh)	$v_3 + v_7$
1169	1162 1147 1140 } (s,sh)	1166	1165.8 (w)	1164 1152 } (s,sh)	1148 (s)	v_{16}
				1122 (s)		liquid effect (?)
	1076 (w)			1067 (w)		$v_3 + v_{18} + v_{19}$
				1051 (vw)		$v_{16} - v_8$
	1022 (vw)			1022 (w)	1017 (w)	$v_{16} - v_8$ [10], $v_{16} - v_{19}$ [5]
	933 (vw)			931 (w)	930 (w)	$v_3 + v_{18}$

Table 5 (continued)

gas [1]	[5]	[24,25], see also [10]	[47]	liquid [5], see also [24]	[10], see also [51]	assignment
767	769 (vs)	767	769.3 (vs)	771 (vs)		ν_{17} [10,24,47], ν_7, ν_{17} [5], ν_7, ν_{17}, (ν_{11}) [1]
		599 (sh)				$\nu_{18} + \nu_{19}$
		580 (w)				$\nu_3 + \nu_{19}$
473	476 (s)	483R 476Q $\Big\}$ (s) 468P	477.5 (s)	475 (vs)		ν_{18}
		355 (w)				$\nu_{18} - \nu_{19}$
		243 (vw)				$2\nu_{19}$
130		120 (s)	121 (s)	130 (s)		ν_{19}

Table 6
Fundamental Raman Bands [6, 8, 22, 42, 47] and Raman Spectra of Pb(CH$_3$)$_4$ [2, 5, 10, 22] in Liquid and Solid (at −70 °C) State.
Wave numbers in cm^{-1} and bracketed bands not clearly resolved. For the assignments, see Table 4.

liquid [2,5]	[6,8]	[10]	[22]	[42]	[47]	solid [13,22]	assignment
3755 (vw)							$\nu_{13} + \nu_{17}$
3679 (vw)							$\nu_{14} + \nu_{17}$
2999 (s)	3001	2996 (vs)	2998 (m)	2998	2995 (w)	2998 (w)	ν_5 [47], $\nu_5 + \nu_{13}$ [5,6,10,22,42]
2918 (s)	2920	2924 (vs)	2918 (s)	2918	2915 (m)	2918 (m)	ν_1 [47], $\nu_1 + \nu_{14}$ [5,6,10,22,42]
			2318 (vvw)				$2\nu_2$
2292 (w)		2292 (vw)	2293 (w)				$2\nu_{16}$
			1644 (vw)				$\nu_2 + \nu_{18}$
		1622 (w)	1622 (vw)				$\nu_2 + \nu_3$ [10], $\nu_{16} + \nu_{18}$ [22]
		1544 (vw)	1531 (vw)				$2\nu_{17}$
(1453)	1439	1450 (w)	1455 (w)		1442 (vw)		ν_{15} [10,22,47], ν_6, ν_{15} [5,6]
		1400 (w)	1412 (w)	1410	1410 (vw)		ν_6 [10,22,47], ν_6, ν_{15} [42]
			~1360 (vw)				$\nu_{16} + \nu_{12}$ (?)
		1300 (vw)	1286 (vw)				$\nu_{16} + \nu_{19}$
1170 (m)	1167	1170 (vs)	1169 (vs)	1169	1169 (s)	1157 (vs)	ν_2

112

Table 6 (continued)

liquid [2,5]	[6,8]	[10]	[22]	[42]	[47]	solid [13,22]	assignment
1155 (m)	1151	1154 (vs)	1157 (vs)	1156	1155 (m)	1150 (w)	ν_{16}
			1033 (vvw)			1141 (w)	?
		1019 (w)					$\nu_{16} - \nu_{18}$
930 (vw)		930 (w)	920 (vvw)				$\nu_3 + \nu_{18}$ [5,10], $2\nu_3$ [22]
			~800 } (w)				ν_7
767 (w)	771	767 (m)	769	767	769 (w)		ν_7 [47], ν_{17} [10,22], ν_7, ν_{17} [5,6,42]
		700 (vw)					ν_7
		634 (vw)					$\nu_{17} - \nu_8$
577 (vw)		590 (vw)	598 (vw)				$\nu_3 + \nu_8$ [10], $\nu_8 + \nu_{18}$ [5,22]
473 (m)	472	478 (vs)	475 (vvs)	474	473 (m)	474 / 471 } (s)	ν_{18}
460 (vs)	460	459 (vs)	462 (vvs)	462	461 (vs)	463 (vs)	ν_3
			342 (vw)				$\nu_8 + \nu_{12}$ (?)
		145 (vs)	~145 } (vs)				ν_{19} [10], ν_8 [22]
130 (s)	136	130 (s)	130	130	129 (s)		ν_8 [10,47], ν_{19} [5,22], ν_8, ν_{19} [6,42]

Raman lines of liquid $Pb(CH_3)_4$ have been reported for the first time and have been partly assigned [3]. For figures of vapor-phase Raman spectra of $Pb(CH_3)_4$ at 344 K, see [35]. The totally symmetrical Raman bands ν_1, ν_2, and ν_3 of gaseous $Pb(CH_3)_4$ at about 1-atm pressure were obtained at 2920, 1167, and 460 cm^{-1}, respectively, and compared with those of the other group 14 methyl compounds. Intensity measurements under various excitation conditions were undertaken [17]; see also polarizability below. In addition to the Raman bands in Table 6 further extremely weak and not assigned bands at 692, 1004, 1833, 1854, 1928, 2096, 2567, and 3706 cm^{-1} were given in [22]. Raman lines of liquid $(CH_3)_3PbCD_3$ and assignments are reported in [47]. For Raman spectra of liquid $M(CD_3)_4$ (M = Si, Ge, Sn, Pb), see [42, 47]. Raman spectra of crystalline and of liquid $Pb(CH_3)_4$ at various conditions are reproduced [22].

Previous assignments are critically considered in [10, 13, 22, 24, 30, 42]; see also [45]. Reassignments for C-H stretching and deformation vibrations proposed in [30] are rejected [33]. The previous assignments are approved on the basis of extensive studies of spectra of $Pb(CH_3)_4$, $(CH_3)_3PbCD_3$, $Pb(CD_3)_4$, and other tetramethyl compounds of group 14 elements [33]. Positions of unperturbed $\nu_s(CH_3)$ and $\nu(CD_3)$ in $Pb(CX_3)_4$ (X = H, D) are estimated from $\nu_{CH}(CHD_2)$ and $\nu_{as}(CH_3)$ and the extent of perturbance by Fermi resonance is discussed [36]. Analogies between vibrational fine structure in photoelectron spectra of $Pb(CH_3)_4$ and stretching vibrations ν_2 and ν_{16}, and ν_3 and ν_{16} are discussed [34].

Table 7

Calculated and Observed Fundamental IR and Raman Frequencies of $Pb(CD_3)_4$.
Wave numbers in cm^{-1}. For the assignments, see Table 4.

IR (gas)				Raman (liquid)		assignment
calculated		observed				
[42]	[46]	[42]	[47]	[42]	[47]	
2120	2128			2120 (w)	2119 (vs, p)	ν_1
892	909			898 (m)	899 (m, p)	ν_2
420	424			424 (s)	425 (vs, p)	ν_3
2238	2252			2249 (w)	2247	ν_5
1016	1049			1010 (vw)		ν_6
572	583			575 (vw)		ν_7
115	114			114 (m)	114 (s, b)	ν_8
2238	2252	2249 (m)	2253.0 (m)	2249 (w)		ν_{13}
2120	2127	2122 (m)	2127.5 (m)	2120 (w)		ν_{14}
1024	1049		1036.5 (vw)	1032 (vw)		ν_{15}
895	907	908 (w)	909.2 (w)	900 (sh)		ν_{16}
575	583	586 (s)	586.0 (s)	583 (vw)		ν_{17}
428	438	438 (s)	435.9 (vs)	433 (sh)	434 (w)	ν_{18}
116	108	105 (s)	107 (m)	114 (m)		ν_{19}
			1015 (vw)			$\nu_{17} + \nu_{18}$

Force constants have been obtained following various procedures: from wave equations derived from pure valence force potentials using effective methyl masses [7], from a refined valence force model: $f_{CPb} = 1.90$ [8, 9], 1.776 [29] mdyn/Å, $f_{CH} = 4.83$ mdyn/Å [8, 9]; from symmetry force constants: $f_{CPb} = 1.88$ [42], 1.94 [5], 1.95 [17] mdyn/Å, $f_{CH} = 4.80$ [42], 4.870 [5, 17] mdyn/Å; from kinetic constants [49]; by using the Urey-Bradley force field: $f_{CPb} = 1.783$ [14], 1.847 [24], 1.853 [39] mdyn/Å, $f_{CH} = 4.489$ [19], 4.863 mdyn/Å [24]; see also [31]. Valence compliance constants are calculated in [43]. Sets of symmetry force constants which have been calculated recently, are given in [42, 46]; see also [29]. A value of the symmetry constant $f_{CPb} = 1.92$ mdyn/Å is cited in [40]. A force constant $f_{CPb} = 2.14$ aJ/Å^2 has been obtained from Hartree-Fock calculations accounting for relativistic effects by using the Breit-Pauli Hamiltonian and first-order perturbation theory [50]. From the experimental fundamental frequencies of $M(CH_3)_4$, $M(CD_3)_4$, and $(CH_3)_3PbCD_3$ (M = Si, Ge, Sn, Pb), a transferable local symmetry type force field for $(CH_3)_{4-n}M$ groups and appropriate force constants are calculated [46]. Pseudo-exact force constants and symmetrized force constants are calculated and compared with values obtained from other approximation methods [41]. The adequacy of various force fields such as Orbital Valence (OVFF), Urey-Bradley (UBFF), and approximate General Valence force fields (GVFF) in terms of frequency reproduction is compared [39]. Pb-C stretching force constants are compared with those of $Hg(CH_3)_2$ and of tetracyclopropyllead [44]. Normal coordinate calculations are carried out by a symmetry force field treating the isotopic species $Pb(CH_3)_4$ and $Pb(CD_3)_4$ simultaneously without assuming methyl groups as a point mass [42]; see also [46]. The repulsive force constant between two methyl groups $F(CH_3-CH_3)$ is calculated to be $196.5\ r_{CC}^{-6.8}$ mdyn/Å [19]. Rigidity coefficient for the Pb-C bond is $K_{11} \cdot 10^6 = 3.17$ cm^{-2} [23]. The Coriolis coupling constant ξ_3 is calculated to be 0.08 [35, 39, 41, 49]; see also [43]. For an analysis of the Coriolis coupling constants of $M(CH_3)_4$ (M = C, Si, Ge, Sn, Pb) using different force fields, see [39]. The rotational constants for $Pb(CH_3)_4$ and

$Pb(CD_3)_4$ are 726×10^{-4} and $569 \times 10^{-4}\,cm^{-1}$, respectively [47]. For centrifugal distortion constants, see [49]. The rotational branch structure to the skeletal fundamentals is studied and OR, RS branch separations are observed and compared with calculated values [35]. The PR separations are 15.7 and 13.9 cm^{-1}, respectively [47]. The inertia defect, bringing out the influence of vibration-rotation interaction on molecule structure was calculated and compared with that of other tetrahedral molecules [48].

Vibrational mean amplitudes for bonded and nonbonded distances are reported to be 0.0594 Å at 288 K [46], 0.0583 Å [39] for the Pb-C bond, and 0.154 Å at 288 K [46], 0.1591 Å [39] for two methyl groups. The values for mean amplitudes of vibration were estimated for the Pb-C bond at 298 and 500 K as 0.0601 and 0.0666 Å, respectively, and for two methyl groups as 0.1563 and 0.1940 Å, respectively [29]; see also [49]. The Bastiansen-Morino shrinkage is given in [39].

The effective methyl masses of the compounds $M(CH_3)_4$ (M = C, Si, Ge, Sn, Pb) give a linear relationship when plotted against M-C stretching and bending frequencies [7]. From a linear correlation of average $\nu(CH)$ values and C-H distances of various compounds containing CH_3, CH_2, and CH groups, a C-H distance of 1.094 Å for $Pb(CH_3)_4$ was predicted. A similar correlation gave the dissociation energy $D_{298}^o = 102.2$ kcal/mol [37]. A plot of $\nu_s(CH_3)$ [27] or of the average $\nu(CH)$ [37] of $M(CH_3)_4$ (M = C, Si, Ge, Sn, Pb) against the coupling constant $^1J(^{13}C, {}^1H)$ gave a linear relation [32]. $\nu(CH)$ and the substituent resonance constant σ_R^o of $Pb(CH_3)_4$ and of other methyl compounds show a good linear correlation [32]. $\delta_s(CH_3)$ of $M(CH_3)_m$ (M = element of groups 14 to 17, m = valence of M) varies with the electronegativity of M; frequency variations are reflected in changes in the deformation force constants [12]. Frequency shifts of $\varrho(CX_3)$, ascribed to changes in the XCM deformation force constants, and of $\delta_s(CX_3)$ in $(CX_3)_nMX_{m-n}$ compounds (X = H, D; M = group 14 to 17 element, m = valence of M, n = integer from 1 to m), including $Pb(CH_3)_4$ are correlated to electronegativity χ of M and to M-C bond length; $\delta_s(CX_3)$ and $\varrho(CX_3)$ show linear dependence upon log $(\chi_M/r^2(M\text{-}C))$; electronegativities of M (perturbed by the adjacent atoms = effective group electronegativity of the group directly bonded to CX_3) can therefore be derived from $\delta_s(CX_3)$ and $\varrho(CX_3)$ [20, 21].

Molecular and bond polarizabilities are estimated from Raman intensities and force constants. The Wolkenstein bond polarizability is tested using relative Raman intensities of $M(CH_3)_4$ (M = C, Si, Ge, Sn, Pb) [17]; see also [26]. The mean molecular polarizability was calculated to be 15.9×10^{-24} [38], $17.626 \times 10^{-24}\,cm^3$ [28]. For calculation of mean molecular polarizability derivatives using a delta-function potential model, see [26].

IR absorption spectra observed during flash photolysis of $Pb(CH_3)_4$ are proposed to arise from radical species [46]; see also [15, 16].

References:

[1] Kettering, C. F., Sleator, W. W. (Physics [N.Y.] **4** [1933] 39/49).
[2] Duncan, A. B. F., Murray, J. W. (J. Chem. Phys. **2** [1934] 636/43).
[3] Duncan, A. B. F., Murray, J. W. (J. Chem. Phys. **2** [1934] 146).
[4] Young, C. W., Koehler, J. S., McKinney, D. S. (J. Am. Chem. Soc. **69** [1947] 1410/5).
[5] Sheline, R. K., Pitzer, K. S. (J. Chem. Phys. **18** [1950] 595/601).
[6] Siebert, H. (Z. Anorg. Allgem. Chem. **263** [1950] 82/6).
[7] Sheline, R. K. (J. Chem. Phys. **18** [1950] 602/6).
[8] Siebert, H. (Z. Anorg. Allgem. Chem. **268** [1952] 177/90).
[9] Siebert, H. (Z. Anorg. Allgem. Chem. **271** [1952] 75).
[10] Lippincott, E. R., Tobin, M. C. (J. Am. Chem. Soc. **75** [1953] 4141/7).

[11] Edgell, W. F., Ward, C. H. (J. Am. Chem. Soc. **77** [1955] 6486/91).

[12] Sheppard, N. (Trans. Faraday Soc. **51** [1955] 1465/9).

[13] Jackson Jr., J. A. (Diss. Univ. Oklahoma 1955; Diss. Abstr. **16** [1956] 357/8).

[14] Shimizu, K. (Nippon Kagaku Zasshi **77** [1956] 1284/7).

[15] Cook, C. L., Clouston, J. G. (Nature **177** [1956] 1178/9).

[16] Clouston, J. G., Cook, C. L. (Nature **179** [1957] 1240/1).

[17] Waters, D. N., Woodward, L. A. (Proc. Roy. Soc. [London] A **246** [1958] 119/32).

[18] Clouston, J. G., Cook, C. L. (Trans. Faraday Soc. **54** [1958] 1001/7).

[19] Overend, J., Scherer, J. R. (J. Opt. Soc. Am. **50** [1960] 1203/7).

[20] Takenaka, T., Gotoh, R. (Proc. Intern. Symp. Mol. Struct. Spectrosc., Tokyo 1962, A 211-1/-4).

[21] Takenaka, T., Gotoh, R. (Nippon Kagaku Zasshi **83** [1962] 997/1002).

[22] Jackson, J. A., Nielsen, J. R. (J. Mol. Spectrosc. **14** [1964] 320/41).

[23] Gastilovich, E. A., Shigorin, D. N., Komarov, N. V. (Opt. Spektroskopiya **16** [1964] 46/51; Opt. Spectrosc. [USSR] **16** [1964] 24/6).

[24] Crowder, G. A., Gorin, G., Kruse, F. H., Scott, D. W. (J. Mol. Spectrosc. **16** [1965] 115/21).

[25] Crowder, G. A., Scott, D. W. (U.S. Bur. Mines Rept. Invest. No. 6630 [1965] 1/37).

[26] Lippincott, E. R., Nagarajan, G. (Bull. Soc. Chim. Belges **74** [1965] 551/64).

[27] Jouve, P. (Compt. Rend. B **262** [1966] 815/7).

[28] private communication from: The Ethyl Corp., Detroit, Mich. (cited in: Rummens, F. H. A., Raynes, W. T., Bernstein, H. J., J. Phys. Chem. **72** [1968] 2111/9).

[29] Durig, J. R., Nagarajan, G. (Monatsh. Chem. **100** [1969] 1948/59).

[30] Graham, S. C. (Spectrochim. Acta A **26** [1970] 345/63).

[31] Oyamada, T., Iijima, T., Kimura, M. (Bull. Chem. Soc. Japan **44** [1971] 2638/42).

[32] Tupitsyn, I. F., Zatsepina, N. N., Kolodina, N. S., Kirova, A. V. (Reakts. Sposobnost Org. Soedin. **8** [1971] 765/86).

[33] Bürger, H., Biedermann, S. (Spectrochim. Acta A **28** [1972] 2283/6).

[34] Jonas, A. E., Schweitzer, G. K., Grimm, F. A., Carlson, T. A. (J. Electron Spectrosc. Related Phenom. **1** [1972/73] 29/66).

[35] Bosworth, Y. M., Clark, R. J. H., Rippon, D. M. (J. Mol. Spectrosc. **46** [1973] 240/55).

[36] McKean, D. C. (Spectrochim. Acta A **29** [1973] 1559/74).

[37] McKean, D. C., Duncan, J. L., Batt, L. (Spectrochim. Acta A **29** [1973] 1037/49).

[38] Sharma, D. K., Pandey, A. N., Panday, B. N. (Acta Ciencia Indica **1** [1974] 35/8).

[39] Sanyal, N. K., Verma, D. N., Dixit, L. (Spectrosc. Letters **9** [1976] 697/713).

[40] Höfler, F. (Monatsh. Chem. **107** [1976] 705/19).

[41] Dublish, A. K., Srivastava, B. B., Sharma, D. K., Verma, U. P., Pandey, A. N. (Z. Naturforsch. **32a** [1977] 76/8).

[42] Watari, F. (Spectrochim. Acta A **34** [1978] 1239/44).

[43] Kaila, R., Dixit, L., Gupta, P. L. (Bull. Soc. Chim. Belges **87** [1978] 93/103).

[44] Czuchajowski, L., Habdas, J., Kucharski, S. A., Rogosz, K. (J. Organometal. Chem. **155** [1978] 185/93).

[45] Anacona, J. R. (9th Intern. Conf. Organometal. Chem., Dijon 1979, Abstr. P 24 W).

[46] Biedermann, S., Bürger, H., Hassler, K., Höfler, F. (Monatsh. Chem. **111** [1980] 715/25).

[47] Biedermann, S., Bürger, H., Hassler, K., Höfler, F. (Monatsh. Chem. **111** [1980] 703/14).

[48] Lalitha, M., Srinivasamoorthy, R., Savariraj, G. A. (Indian J. Pure Appl. Phys. **19** [1981] 330/4).

[49] Mohan, S., Bhoopathy, T. J. (Acta Phys. Polon. A **69** [1986] 135/8).

[50] Almlof, J., Faegri Jr., K. (Theor. Chim. Acta **69** [1986] 437/46).

[51] Fritz, H. P., Schwarzhans, K.-E. (Chem. Ber. **97** [1964] 1390/7).

116

1.1.1.1.2.6 UV Spectrum

$Pb(CH_3)_4$ is unstable to UV radiation [2]. Gaseous $Pb(CH_3)_4$ shows continuous absorption in the UV region [1, 2]; for spectra between 255 and 275 nm at various pressures, see [2]. Limits of absorption at pressures corresponding to the vapor pressures of liquid $Pb(CH_3)_4$ at -25, 0, and $+25°C$ are 265, 270, and 275 nm, respectively [1]. The long wavelength absorption limit of gaseous $Pb(CH_3)_4$ was observed at about 280 nm and of $Pb(CH_3)_4$ in a 2,2,4-trimethylpentane solution at 310 nm [2]. The absorption at 240 nm obeys Beer's law for gas-phase concentrations of 0.012 to 0.2 vol% $Pb(CH_3)_4$ in dried Ar, O_2, or air. The absorption coefficient, α, was given for 295 K; at about 1000 K it was apparently larger by a factor of about 5 [6]. UV absorption coefficients between 290 and 340 nm were reported graphically [9]. The extinction coefficient, ε, at 206 nm is 5×10^3 L·cm^{-1}·mol^{-1} [8, 10]. The variation of the extinction coefficient of gaseous $Pb(CH_3)_4$ with wavelength in the far UV was given graphically [11]. $\lambda_{max}(\varepsilon) = 206(8900)$ nm is reported for solutions of $Pb(CH_3)_4$ in acetonitrile [7].

$Pb(CH_3)_4$ absorbs weakly in the tropospheric solar UV region, the energy of the absorbed light (>350 kJ/mol) being well in excess of the mean Pb-C bond dissociation energy [9]. The absorption spectra arising during flash photolysis of $Pb(CH_3)_4$ were studied [3, 4, 5].

References:

[1] Duncan, A. B. F., Murray, J. W. (J. Chem. Phys. **2** [1934] 636/43).
[2] Leighton, P. A., Mortensen, R. A. (J. Am. Chem. Soc. **58** [1936] 448/54).
[3] Cook, C. L., Clouston, J. G. (Nature **177** [1956] 1178/9).
[4] Clouston, J. G., Cook, C. L. (Nature **179** [1957] 1240/1).
[5] Clouston, J. G., Cook, C. L. (Trans. Faraday Soc. **54** [1958] 1001/7).
[6] Homer, J. B., Hurle, I. R. (Proc. Roy. Soc. [London] A **327** [1972] 61/79).
[7] Gardner, H. C., Kochi, J. K. (J. Am. Chem. Soc. **98** [1976] 2460/9).
[8] Tarunin, B. I., Aleksandrov, Yu. A., Baklanov, N. V., Perepletchikova, V. P., Petrova, A. A. (Khim. Elementoorg. Soedin. [Gorkii] No. 6 [1978] 48/50).
[9] Harrison, R. M., Laxen, D. P. H. (Environ. Sci. Technol. **12** [1978] 1384/92).
[10] Aleksandrov, Yu. A., Tarunin, B. I., Perepletchikov, M. L., Seliverstov, N. N. (Kinetika Kataliz **22** [1981] 1384/8; Kinet. Catal. [USSR] **22** [1981] 1092/6).
[11] Bell, C. F., Husain, D. (J. Photochem. **29** [1985] 267/83).

1.1.1.1.3 Physical Properties

1.1.1.1.3.1 Melting Point. Boiling Point. Density. Refractive Index. Molar Refraction. Critical Constants

Tetramethyllead is a colorless liquid (mobile at room temperature) having a relatively high vapor pressure. It volatilizes in air about as fast as benzene and is very volatile with ether vapor [6]. Its smell is sweetish and resembles that of raspberries [3, 6]. Careful and cautious handling of the pure compound is necessary.

$Pb(CH_3)_4$ can be distilled at normal pressure without decomposition in the presence of air [3] or advisably in an inert atmosphere [1, 2]. It decomposes explosively when overheated [11, 19, 20]. When distilling $Pb(CH_3)_4$, the bath temperature should be kept below 150 to 160°C [19]. Explosive decompositions have been reported during distillation [19] and shortly before the end of a distillation over sodium [23].

The boiling point was estimated to be 110°C [3, 4, 6 to 9, 12], 108 to 109°C [18] at normal pressure (probably 760 Torr). Boiling points (b.p. in °C) at other pressures (p in Torr) with corrected values from [16, 29] are collected below:

Ref.	[29]	[24]	[23, 25]	[15]	[30]	[16]
b.p./p	110.4/760	110.0/725	109/720	48 to 50/100	26 to 28/40	6/10

Other reported values of ca. 160°C [1, 2], 101°C [22], and 180°C [32] at 760 Torr are obviously wrong. An estimated value of 376.5 K is not considered significant [17]. The boiling point at 10 mm is calculated to be 12°C [16]. The boiling points of the compounds $M(CH_3)_4$ (M = C, Si, Sn, Pb) and the molecular polarizabilities at 20°C show a linear relationship [31].

The melting point of $Pb(CH_3)_4$ has been reported to be −30.2°C (given as 242.92 K) [26] and −30°C (given as 243 ± 2 K) [33]. A higher value of −27.5°C has been measured earlier [4, 5, 12]; see also [9, 21].

Experimental values of the density of liquid $Pb(CH_3)_4$ in the temperature range between 0 and 120°C [38] and 15 to 30°C [29] are represented graphically in **Fig. 4**a and b, respectively. Equations for the temperature dependence of the density are given as d = $2.03777 - (2.369 \times 10^{-3})t$ [34] and according to [38] as d = $2.0397 \pm 0.0001 - (2.44 \pm 0.01 \times 10^{-3})t$ fitting the straight line in Fig. 4a. Other measured values are $d_4^0 = 2.034$ [3], $d_4^{20} = 1.9952$ [4, 6, 7], and $d_4^{25} = 1.979$ [28].

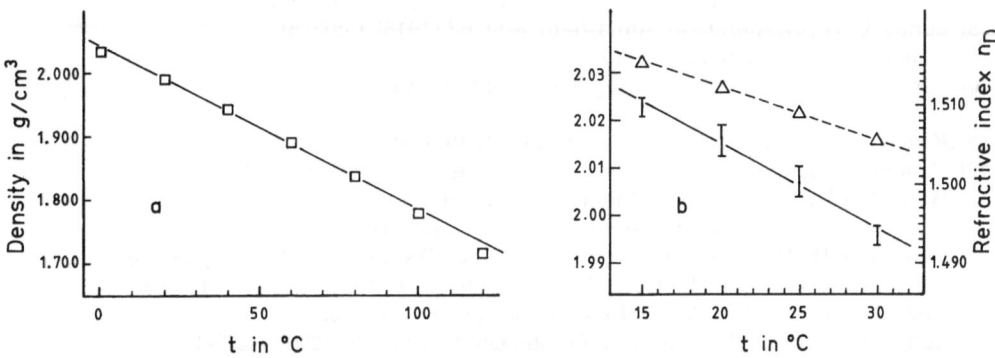

Fig. 4. Density over a large temperature range (a, left scale) [38] of $Pb(CH_3)_4$; density (b, left scale, solid line) and refractive index (b, right scale, broken line) in the 15 to 30°C range [29].

The zero point density was calculated to be 2.514 g/cm³ [13]. An equation to calculate the liquid and the vapor density based on a variant of the lattice theory of liquids was derived. Values of the vapor density d_g (in mg/cm³) are given below [38]:

t in °C	0	20	40	60	80	100	120
d_g	0.11328	0.33076	0.82676	1.82238	3.62641	6.63889	11.3544

The refractive index n_D^t (dashed line) in the 15 to 30°C range is shown graphically in Fig. 4b [29]. Other measured values are $n_D^{20} = 1.5128$ [4], 1.5120 [6, 7], 1.5037 [37], $n_D^{18} = 1.5130$ [18], and $n_{H\alpha}^{20} = 1.5068$ [6, 7]. The zero point refractive index was calculated to be 1.680 [13]. Dispersion for the violet and red and for the blue and red hydrogen line are $n_\gamma - n_\alpha = 0.03040$ and $n_\beta - n_\alpha = 0.01877$ [6, 7], 0.01881 [4], respectively. Specific refractions are $\Sigma R_{H\alpha} = 14.91$, $\Sigma R_D = 15.04$; specific dispersions $\Sigma \Delta_{\gamma-\alpha} = 0.75$, $\Sigma \Delta_{\beta-\alpha} = 0.46$ [6, 7]. Molar refractions are $[R_L]_{H\alpha} = 39.83$, $[R_L]_D = 40.18$ [6, 7, 16], 40.13 [10], 41.04 (calculated: 40.18) [27]. Molar dispersions were estimated as $\Delta_{\gamma-\alpha} = 2.00$, $\Delta_{\beta-\alpha} = 1.24$ [6, 7].

Critical constants calculated from vapor pressure data are: $T_c = 569.0$ K, $P_c = 3.120$ MPa, $V_c = 0.39850$ dm^3/mol, V_0 (apparent zero point molar volume) = 0.1063 [14], 0.10383 dm^3/mol [36]; see also [13]. From the critical constants the Lennard-Jones (12-6) potential parameters are calculated as $\sigma = 6.20 \times 10^{-6}$ cm [35], 6.44×10^{-6} cm [36]; $\varepsilon/k = 453$ [35], 380 K [36]; see also [38].

References:

[1] Cahours, A. (Ann. Chim. Phys. [3] **62** [1861] 257/350).
[2] Cahours, A. (Liebigs Ann. Chem. **122** [1862] 48/71).
[3] Butlerow, A. (Jahresber. Fortschr. Chem. **1863** 474/6).
[4] Grüttner, G., Krause, E. (Ber. Deut. Chem. Ges. **49** [1916] 1415/28).
[5] Grüttner, G., Krause, E. (Ber. Deut. Chem. Ges. **49** [1916] 1125/33).
[6] Krause, E. (Diss. Friedrich-Wilhelms-Universität Berlin 1917).
[7] Grüttner, G., Krause, E. (Liebigs Ann. Chem. **415** [1918] 338/62).
[8] Jones, L. W., Werner, L. (J. Am. Chem. Soc. **40** [1918] 1257/75).
[9] Calingaert, G. (Chem. Rev. **2** [1925] 43/83).
[10] Tanaka, Y., Nagai, Y. (Kogyo Kagaku Zasshi **31** [1928] 20/3).

[11] Krause, E. (Ber. Deut. Chem. Ges. **62** [1929] 1877/8).
[12] Paneth, F., Hofeditz, W. (Ber. Deut. Chem. Ges. **62** [1929] 1335/47).
[13] Herz, W. (Z. Anorg. Allgem. Chem. **182** [1929] 173/6).
[14] Biltz, W., Sapper, A. (Z. Anorg. Allgem. Chem. **186** [1930] 387/91).
[15] Simons, J. H., McNamee, R. W., Hurd, C. D. (J. Phys. Chem. **36** [1932] 939/48).
[16] Jones, W. J., Evans, D. P., Gulwell, T., Griffiths, D. C. (J. Chem. Soc. **1935** 39/47).
[17] Kelley, K. K. (U.S. Bur. Mines Bull. No. 383 [1935] 1/132, 62).
[18] Leighton, P. A., Mortensen, R. A. (J. Am. Chem. Soc. **58** [1936] 448/54).
[19] Zscharn, A. (Chemiker-Ztg. **64** [1940] 498).
[20] Gilman, H., Jones, R. G. (J. Am. Chem. Soc. **68** [1946] 517/20).

[21] Stull, D. R. (Ind. Eng. Chem. **39** [1947] 517/40).
[22] Heap, R., Saunders, B. C. (J. Chem. Soc. **1949** 2983/8).
[23] Siebert, H. (Z. Anorg. Allgem. Chem. **263** [1950] 82/6).
[24] Sheline, R. K., Pitzer, K. S. (J. Chem. Phys. **18** [1950] 595/601).
[25] Lippincott, E. R., Tobin, M. C. (J. Am. Chem. Soc. **75** [1953] 4141/7).
[26] Staveley, L. A. K., Warren, J. B., Paget, H. P., Dowrick, D. J. (J. Chem. Soc. **1954** 1992/2001).
[27] Vogel, A. I., Cresswell, W. T., Leicester, J. (J. Phys. Chem. **58** [1954] 174/7).
[28] Good, W. D., Scott, D. W., Lacina, J. L., McCullough, J. P. (J. Phys. Chem. **63** [1959] 1139/42).
[29] Bambynek, W. (Z. Physik. Chem. [Frankfurt] **25** [1960] 403/14).
[30] Bawn, C. E. H., Whitby, F. J. (J. Chem. Soc. **1960** 3926/31).

[31] Lutskii, A. E., Obukhova, E. M. (Zh. Fiz. Khim. **35** [1961] 1960/5; Russ. J. Phys. Chem. **35** [1961] 962/5).
[32] Prösch, U., Zöpfl, H.-J. (Z. Chem. **3** [1963] 97/100).
[33] Smith, G. W. (J. Chem. Phys. **42** [1965] 4229/43).
[34] The Ethyl Corp., Detroit, Mich. (cited in: Rummens, F. H. A., Raynes, W. T., Bernstein, H. J., J. Phys. Chem. **72** [1968] 2111/9).
[35] Rummens, F. H. A., Raynes, W. T., Bernstein, H. J. (J. Phys. Chem. **72** [1968] 2111/9).
[36] Rummens, F. H. A., Rajan, S. (Can. J. Chem. Eng. **57** [1979] 349/54).
[37] van Beelen, D. C. (Diss. Univ. Leiden 1980).
[38] Zorin, A. D., Kut'in, A. M., Kuznetsova, T. V., Feshchenko, I. A. (Zh. Fiz. Khim. **59** [1985] 154/8; Russ. J. Phys. Chem. **59** [1985] 85/7).

1.1.1.1.3.2 Vapor Pressure

Vapor pressure of $Pb(CH_3)_4$ in relationship to temperature was determined by two authors:

t in °C	−29.0	−6.8	0.00	4.4	15.00	16.6	20.00	25.00	30.00	30.3	35.00
p in Torr [1]	1	5		10		20				40	
p in Torr [4]			7.00		17.09		22.52	29.29	37.72		48.12

t in °C	39.2	40.00	45.00	50.00	50.8	55.00	60.00	68.8	89.0	110.0
p in Torr [1]	60				100			200	400	760
p in Torr [4]		60.78	76.06	94.51		116.30	142.24			

Other vapor pressures p at various temperatures were measured as follows (t in °C in parentheses): 31 (25.0), 51 (35.0) [2], 50 (33.2) [3], 17.5 (15), 100.00 (50) [5], and 29.30 (25) (given as 3906 Pa at 298.15 K) [7].

The temperature dependence of the vapor pressure is given by the equation log p = B − A/t', with A = 1378.7, B = 6.9381, t' = (t + 230) [3], and A = 1335.317, B = 6.93767, t' = (t + 219.084) between t = 0 to 60 °C [4], and A = 1865, B = 7.751, t' = T (in K) [2]; see also [6]. According to [6], vapor pressures given in [4] are considered more reliable than data in [2]; for calculation of saturated vapor pressures, see [8].

References:

[1] Stull, D. R. (Ind. Eng. Chem. **39** [1947] 517/40).
[2] Tanaka, Y., Nagai, Y. (Proc. Imp. Acad. [Tokyo] **5** [1929] 78/9; J. Soc. Chem. Ind. Japan Suppl. **32** [1929] 59B/60B).
[3] Calingaert, G., Beatty, H. A., Neal, H. R. (J. Am. Chem. Soc. **61** [1939] 2755/8).
[4] Good, W. D., Scott, D. W., Lacina, J. L., McCullough, J. P. (J. Phys. Chem. **63** [1959] 1139/42).
[5] Goodacre, G. L., Foord, D. (Riv. Combust. **16** [1962] 340/9).
[6] Rummens, F. H. A., Rajan, S. (Can. J. Chem. Eng. **57** [1979] 349/54).
[7] Abraham, M. H., Irving, R. J. (J. Chem. Thermodyn. **12** [1980] 539/44).
[8] Zorin, A. D., Kut'in, A. M., Kuznetsova, T. V., Feshchenko, I. A. (Zh. Fiz. Khim. **59** [1985] 154/8; Russ. J. Phys. Chem. **59** [1985] 85/7).

1.1.1.1.3.3 Enthalpy and Entropy of Transformation

The heat of fusion was determined as $\Delta H_f = 2.581$ kcal/mol and the entropy of fusion as $\Delta S_f = 10.62$ cal $\cdot$ mol$^{-1} \cdot$ K^{-1} [4].

The heat of vaporization ΔH_v (in kcal/mol) was estimated to be 9.075 at 298.16 K at saturation pressure [5], 8.2 at 25°C [3], 9.36 ± 0.11 between -15 and $+14$°C [7], 9.11 [11]. Recent reported ΔH_v values (originally given in kJ/mol) at temperatures between 0 and 120°C are collected below [12]:

t in °C	0	20	40	60	80	100	120
ΔH_v in kcal/mol	9.1417	8.9919	8.8267	8.6429	8.4379	8.2077	7.9481

A value of 9.1 kcal/mol as calculated from data in [1] was criticized [2]. For a review, see [8]. The energy of vaporization at that temperature at which the ratio of molal volume to van der Waals volume is equal to 1.70 is given as 9.76 kcal/mol [9]. Thermodynamic values for the vaporization of $Pb(CH_3)_4$ to the ideal gas state at 760 Torr (101325 Pa) and 298.15 K are calculated: $\Delta G_v^\circ = 1.93$ kcal/mol (8.08 kJ/mol), $\Delta H_v^\circ = 9.082$ kcal/mol (38.00 kJ/mol), $\Delta S_v^\circ = 24.00$ cal $\cdot$ mol$^{-1} \cdot$ K^{-1} (100.4 J $\cdot$ mol$^{-1} \cdot$ K^{-1}) [10]. Nomograms for estimating the heat of vaporization and the entropy of vaporization at temperatures between 0 and 300°C have been published [6]. The Trouton rule quotient has been determined to be $\Delta H_v/Tb = 22.3$ cal $\cdot$ mol$^{-1} \cdot$ K^{-1} [1].

References:

[1] Tanaka, Y., Nagai, Y. (Proc. Imp. Acad. [Tokyo] **5** [1929] 78/9; J. Soc. Chem. Ind. Japan Suppl. **32** [1929] 59 B/60 B).
[2] Kelley, K. K. (U.S. Bur. Mines Bull. No. 383 [1935] 1/132, 62).
[3] Lautsch, W. F. (Habilitationsschr. Univ. Halle 1952 cited in [8]).
[4] Staveley, L. A. K., Warren, J. B., Paget, H. P., Dowrick, D. J. (J. Chem. Soc. **1954** 1992/2001).
[5] Good, W. D., Scott, D. W., Lacina, J. L., McCullough, J. P. (J. Phys. Chem. **63** [1959] 1139/42).
[6] Othmer, D. F., Zudkevitch, D. (Ind. Eng. Chem. **51** [1959] 791/6).
[7] Jehne, S. (Dipl.-Arbeit TH Leuna-Merseburg 1963 cited in [8]).
[8] Lautsch, W. F., Tröber, A., Körner, H., Wagner, K., Kaden, R., Blase, S. (Z. Chem. [Leipzig] **4** [1964] 441/54).
[9] Bondi, A. (J. Phys. Chem. **70** [1966] 3006/7).
[10] Abraham, M. H., Irving, R. J. (J. Chem. Thermodyn. **12** [1980] 539/44).

[11] Ducros, M., Sannier, H. (Thermochim. Acta **75** [1984] 329/40).
[12] Zorin, A. D., Kut'in, A. M., Kuznetsova, T. V., Feshchenko, I. A. (Zh. Fiz. Khim. **59** [1985] 154/8; Russ. J. Phys. Chem. **59** [1985] 85/7).

1.1.1.1.3.4 Thermodynamic Functions

The molar heat capacity C_p° (in cal $\cdot$ mol$^{-1} \cdot$ K^{-1}), the enthalpy and free enthalpy functions $(H^\circ - H_0^\circ)/T$ and $(G^\circ - H_0^\circ)/T$, and the entropy S° (all in cal $\cdot$ mol$^{-1} \cdot$ K^{-1}) of gaseous $Pb(CH_3)_4$ as a function of temperature T (in K), calculated from spectroscopic data [1, 2], are listed

Table 8
Molar Heat Capacity C_p°, Enthalpy and Free Enthalpy Functions $(H^\circ - H_0^\circ)/T$ and $(G^\circ - H_0^\circ)/T$, and Entropy S° for $Pb(CH_3)_4$ [1, 2].

T in K	C_p°		$(H^\circ - H_0^\circ)/T$		$(G^\circ - H_0^\circ)/T$ in cal$\cdot$mol$^{-1}\cdot$K^{-1}		S°	
	[1]	[2]	[1]	[2]	[1]	[2]	[1]	[2]
273.15	—	32.3	—	22.2	—	75.4	—	97.6
298.15	—	34.1	—	23.1	—	77.4	—	100.5
298.2	34.42	—	22.97	—	76.89	—	99.86	—
300	34.53	34.3	23.03	23.2	77.03	77.5	100.06	100.7
400	41.57	41.4	26.78	26.9	84.14	84.7	110.92	111.6
500	47.73	47.4	30.42	30.4	90.55	91.0	120.97	121.4
600	52.80	52.6	33.72	33.7	96.42	96.9	130.15	130.6
700	57.30	57.1	36.81	36.7	101.83	102.3	138.64	139.0
800	60.17	61.0	39.61	39.5	106.94	107.4	146.55	146.9
900	64.62	64.5	41.20	42.1	111.75	112.2	153.85	154.3
1000	67.65	67.6	44.60	44.5	116.32	116.7	160.92	161.2
1100	70.30	—	46.80	—	120.58	—	167.38	—
1200	72.67	—	48.84	—	124.83	—	173.67	—
1300	74.81	—	50.81	—	128.86	—	179.67	—
1400	76.60	—	52.58	—	132.71	—	185.29	—
1500	78.23	—	54.24	—	136.39	—	190.63	—

in Table 8. An additional value for the entropy of the ideal gas S° is given as 100.48 ± 0.20 cal$\cdot$mol$^{-1}\cdot$K^{-1} [3]. The molar heat capacities C_p° of crystalline and of liquid $Pb(CH_3)_4$ obtained from calorimetric measurements [4] are as follows:

T in K	100	110	120	130	140	150	160	170
C_p° in cal$\cdot$mol$^{-1}\cdot$K^{-1} . .	24.2	25.15	26.05	27.0	27.95	28.9	29.85	30.8

T in K	180	190	200	210	220	230	250*)	260*)
C_p° in cal$\cdot$mol$^{-1}\cdot$K^{-1} . .	31.75	32.75	33.8	34.7	35.45	36.4	46.65	47.7

*) liquid

The specific heat at 25°C was estimated to be 0.181 cal$\cdot$g$^{-1}\cdot$K^{-1}. The entropy of liquid $Pb(CH_3)_4$ at 298.16 K is given as $S_{saturated} = 76.48 \pm 0.20$ cal$\cdot$mol$^{-1}\cdot$K^{-1}; $(\partial U/\partial P)_T = -0.0044$ cal$\cdot$atm$^{-1}\cdot$g^{-1} [3].

References:

[1] Lippincott, E. R., Tobin, M. C. (J. Am. Chem. Soc. **75** [1953] 4141/7).
[2] Crowder, G. A., Gorin, G., Kruse, F. H., Scott, D. W. (J. Mol. Spectrosc. **16** [1965] 115/21).
[3] Good, W. D., Scott, D. W., Lacina, J. L., McCullough, J. P. (J. Phys. Chem. **63** [1959] 1139/42).
[4] Staveley, L. A. K., Warren, J. B., Paget, H. P., Dowrick, D. J. (J. Chem. Soc. **1954** 1992/2001).

1.1.1.1.3.5 Other Physical Properties

The molar volume of $Pb(CH_3)_4$ (in cm^3/mol) was found to be 134.0 (calculated 132.3) [2], 135.9 at 20°C [6, 10], and 133 at 298.15 K [13]. For the zero point volume, see Section 1.1.1.1.3.1. The expansion between 0 and 100°C is appreciable [1].

The viscosity η (in mPa·s) of $Pb(CH_3)_4$ between 20 and 90°C can be calculated from the equation: $\log \eta = 434 \pm 10/T - 0.689 \pm 0.001$ [14]. Measured values for the viscosity are reported as 5.723 mP at 20°C [6], and as 6.00 ± 0.03 mP at 15°C, 5.61 ± 0.03 mP at 20°C, 5.25 ± 0.02 mP at 25°C, and 4.95 ± 0.01 mP at 30°C. The activation energy of the viscosity is 2.23 ± 0.09 kcal/mol [8]. The surface tension σ at various temperatures t is given below [14]:

σ in mN/m . .	27.225	24.797	22.425	20.113	17.866	15.688	13.584
t in °C	0	20	40	60	80	100	120

The surface tensions σ (in mN/m) for temperatures t between 20 and 90°C can be calculated from the equation $\sigma = (26.352 \pm 0.009) - (1.1018 \pm 0.0008) \cdot t$ [14]. The atomic parachor of lead in $Pb(CH_3)_4$ was estimated to be 76.1 [3].

The self-diffusion coefficient (± 0.05) was measured to be 1.74×10^5 cm^2/s at 15.00 ± 0.002°C, 1.83×10^5 cm^2/s at 20.00 ± 0.02°C, 1.95×10^5 cm^2/s at 25.00 ± 0.02°C, 2.05×10^5 cm^2/s at 30.00 ± 0.02°C [8]. A lower value of 1.54×10^5 cm^2/s at 25°C, and other values which are also about 20% lower than the values in [8], are reported for the temperature range of 250 to 325 K [11]. Values of the activation energy of self-diffusion are 2.1 ± 0.4 [8], 1.96 [11] kcal/mol. For the determination of the self-diffusion coefficient of $Pb(CH_3)_4$ in the liquid phase by the spin-echo method and for a comparison of experimental self-diffusion coefficients of $Pb(CH_3)_4$ and other methyl element compounds with theories of diffusion, see [11]. For a correlation of the viscosity with the self-diffusion coefficient, see [8]. The diffusion coefficients in the system $Pb(CH_3)_4$-$Sn(CH_3)_4$ show no linear dependence with concentration [12].

The molar diamagnetic susceptibility χ_{mol} is given as -108×10^{-6} emu [10]. χ_{Pb} in $Pb(CH_3)_4$ is estimated to be -49.6×10^{-6} [4], -49.8×10^{-6} cm^3/mol [3, 5, 7]; see also [9].

References:

[1] Butlerow, A. (Jahresber. Fortschr. Chem. **1863** 474/6).
[2] Jones, W. J., Evans, D. P., Gulwell, T., Griffiths, D. C. (J. Chem. Soc. **1935** 39/47).
[3] Kadomtzeff, I. (Compt. Rend. **226** [1948] 661/3).
[4] Pascal, P. (Propriétés magnétiques et constitution chimique, in: Grignard, V., Traité de chimie organique, Vol. 2, Masson, Paris 1948, p. 571).
[5] Kadomtzeff, I. (Bull. Soc. Chim. France **1949** D394/D396).
[6] Hugel, G. (Kolloid-Z. **131** [1953] 4/10).
[7] Foex, G. (Tables de Constantes et Données Numériques, Vol. 7, Diamagnétisme et Paramagnétisme, Masson, Paris 1957).
[8] Bambynek, W. (Z. Physik. Chem. [Frankfurt] **25** [1960] 403/14).
[9] Pascal, P., Gallais, F., Labarre, J. F. (Compt. Rend. **256** [1963] 335/9).
[10] Rummens, F. H. A., Raynes, W. T., Bernstein, H. J. (J. Phys. Chem. **72** [1968] 2111/9).

[11] Kessler, D., Weiss, A., Witte, H. (Ber. Bunsenges. Physik. Chem. **71** [1967] 3/19).
[12] Freise, V. (Z. Physik. Chem. [Frankfurt] **67** [1969] 132/7).
[13] Abraham, M. H., Irving, R. J. (J. Chem. Thermodyn. **12** [1980] 539/44).
[14] Zorin, A. D., Kut'in, A. M., Kuznetsova, T. V., Feshchenko, I. A. (Zh. Fiz. Khim. **59** [1985] 154/8; Russ. J. Phys. Chem. **59** [1985] 85/7).

1.1.1.1.4 Chemical Reactions

Pb(CH$_3$)$_4$ is a colorless liquid, soluble in the usual organic solvents and insoluble in water.

1.1.1.1.4.1 Thermal Decomposition

Pb(CH$_3$)$_4$ starts to decompose in an inert atmosphere slightly above its boiling point [1], and overheating during distillation at normal pressure has to be strictly avoided [13, 17]. Explosion occurred during sealing a tube for determining vapor density according to Dumas at 130 °C [2, 28]. Pb(CH$_3$)$_4$ should therefore not be sealed in ampules [39]; see also [4]. Pb(CH$_3$)$_4$ can be stored in brown glass-stoppered flasks without decomposition for years [4, 39], and remains practically unchanged when it is heated in benzene solution in an autoclave to 200 °C for 24 h [6, 10]. For measures to inhibit the thermal decomposition of Pb(CH$_3$)$_4$, see Section 1.1.1.1.8, p. 169.

Gaseous Pb(CH$_3$)$_4$ displays a half-life of 320 h in purified dry air in the dark at 295 ± 3 K. The decay in the dark is first-order in Pb(CH$_3$)$_4$. It is enhanced by NO$_2$, but inhibited by the presence of water vapor [40]. Pb(CH$_3$)$_4$ is assumed to decompose on adsorption on activated coal [27].

Major gaseous products of pyrolysis are CH$_4$ and C$_2$H$_6$ in a flow system between 550 and 820 °C and at pressures between 0.5 and 2.0 Torr, as well as in a static system between 265 and 620 °C and at pressures of about 700 to 1900 Torr [7] or between 240 and 371 °C and at pressures below 25 Torr [30]. Appreciable amounts of C$_2$H$_4$ and hydrogen are also observed in the static system above ∼460 °C and in the flow system at 820 °C [7]; see also [21]. During pyrolysis at 330 °C/5 Torr in a static system, C$_2$H$_6$ was observed as the major product, smaller amounts of CH$_4$ and C$_3$H$_8$, and C$_2$H$_4$ as a minor product [32]. After pyrolysis of Pb(CH$_3$)$_4$, using Xe as carrier gas, the reaction products, frozen out at 4 K, were identified as CH$_4$, C$_2$H$_6$, C$_2$H$_4$, and C$_3$H$_8$ [19]. For the variation of the CH$_4$/C$_2$H$_6$ ratio with temperature of decomposition, see [21]; see also [18]. Pyrolysis between 671 and 753 K in a toluene carrier flow system showed, based upon product analysis (CH$_4$, C$_2$H$_6$, ethylbenzene), that approximately four methyl radicals are released for each molecule of Pb(CH$_3$)$_4$ that enters the decomposition reaction [34]. Addition of toluene markedly increases the initial amount of CH$_4$ formed per mole of Pb(CH$_3$)$_4$ consumed during decomposition at temperatures between about 270 and 371 °C [30]. For thermal decomposition of Pb(CH$_3$)$_4$ in the presence of di-tert-butyl peroxide, see [30]. From investigations of the decomposition kinetics of Pb(CH$_3$)$_4$, autocatalysis is inferred [42].

Controlled thermal decomposition of streaming Pb(CH$_3$)$_4$ vapor in a quartz tube is a method for producing methyl radicals [5, 8, 9, 18], e.g., employing a temperature of 550 °C [33]. Existence of free methyl radicals in decomposition of Pb(CH$_3$)$_4$ in a flow system on tungsten, below 1500 K at 10^{-5} Torr, was inferred from mass spectroscopic measurements [20, 22]. Decomposition was shown to be independent of the composition of the hot wire (Pt, Ni, or W) up to 1500 K due to carbonization of the surface. Poor reproducibility of results was referred to wall reactions of radicals with deposited lead [22]. For the dependence of methyl radical concentration and temperature in a mixture of Pb(CH$_3$)$_4$ and different gases, see [18]. The production of methyl radicals by thermal decomposition of Pb(CH$_3$)$_4$ and their reaction with C$_2$H$_4$, C$_2$H$_6$, and C$_3$H$_6$ were studied by mass spectrometry [15].

Metallic lead is deposited during decomposition of Pb(CH$_3$)$_4$ as a sputtered film [7] or as a mirror [3, 5, 7, 8, 9]. Mirrors produced at low temperature contain carbon [7]; see also [5, 30]. At first, lead exists as a vapor, which then condenses at a temperature of about 900 K on a millisecond time scale into particles, provided that the lead vapor exceeds a critical saturation ratio [35]. In the presence of air, a smoke of PbO is formed by oxidation of lead particles [35, 37]. PbO particles, once formed, do not catalyze further decomposition of $Pb(CH_3)_4$ [37].

The decomposition of $Pb(CH_3)_4$ in a static system at 550°C was complete in less than 10 min, whereas a bulb held at 265°C for 20 h still contained $Pb(CH_3)_4$ [7]. The decomposition of $Pb(CH_3)_4$, present at concentrations of 0.1% in different carrier gases, was found to be complete at low pressures and with a contact time of 0.001 s at 730°C [15]. $Pb(CH_3)_4$ was found to be decomposed in an engine at a rate of 4.7% per millisecond at 505°C (940°F), while a rate of about 40% per millisecond is assumed at 725°C (1340°F) [23].

The rate of decomposition of $Pb(CH_3)_4$ in dilute mixtures (<0.2 vol%) with argon, argon and oxygen, and with air behind incident shock waves between 890 and 1060 K is first-order with a rate constant k (in s^{-1}) represented by the equation $\log k = (13.25 \pm 0.4) - (9.0 \pm 0.4) \cdot 10^3/T$ [35]. From studies of pyrolysis of $Pb(CH_3)_4$ in a toluene carrier flow system over the temperature range 671 to 753 K, the rate equation $k_1 = 5.0 \times 10^{14} \exp(-49400/RT)$ s^{-1} was derived [34]; see also [36]. An earlier equation, $k_1 = 1.5 \times 10^{10} \exp(-28200/RT)$ s^{-1}, for the decomposition of $Pb(CH_3)_4$ in an inert gas was given [18]. Assuming a first-order reaction, an activation energy for the decomposition of $Pb(CH_3)_4$ of 41.35 kcal/mol (173 ± 8 kJ/mol) was obtained [35]; see also earlier values of 23.500 kcal/mol [14] and 39.3 kcal/mol [24]. The dissociation of $Pb(CH_3)_4$ was shown to be unimolecular [14, 18, 34, 35]. For an estimation of the unimolecular frequency factor for the decomposition of $Pb(CH_3)_4$ in a gas stream, see [18]. The Arrhenius parameter for the first-order decomposition of $Pb(CH_3)_4$ was estimated to be $\log A = 12.3$ [24]. From the fact that the log A values for methyl rupture from C, Si, and Ge correspond with the vibrational-rotational entropies of $M(CH_3)_4$, an Arrhenius A factor of $\log [A(s^{-1})/methyl] = 14.3$ was extrapolated for $Pb(CH_3)_4$ [38].

At low temperatures the decomposition of $Pb(CH_3)_4$ is concluded to be slow and takes place as a wall reaction producing carbon, tarry substances, and some isobutene [7]. In contrast to this interpretation, it was later suggested that the rate-controlling step of the pyrolysis at low temperature occurs only in the gas phase [30]. At high temperatures a fast homogeneous decomposition in the vapor phase is assumed to be favored [7]. The rate-controlling step of the decomposition of $Pb(CH_3)_4$ is probably the loss of the first methyl group [18, 34, 35, 36]. It is concluded that the pyrolysis of $Pb(CH_3)_4$, between 240 and 371°C and at pressures below 25 Torr, is not a chain reaction [30]; see also [32]. From comparison of thermal decomposition of $Pb(CH_3)_4$, and of other metal alkyls, a mechanism for decomposition below 340°C is proposed assuming CH_4 to be formed through reactions of methyl radicals among themselves and not from abstraction reaction of methyl radical on $Pb(CH_3)_4$ [21]. Later, however, this latter reaction was postulated to be responsible for CH_4 formation [30, 32].

The relationship of rate constants of decomposition of $Pb(CH_3)_4$, other lead tetraalkyls, and various other antiknock compounds as determined in engines and temperature is depicted graphically. The decomposition rate is unaffected by pressure, and is also independent of fuel type and the presence of other compounds in the fuel when $Pb(CH_3)_4$ is employed as the antiknock agent [23].

Thermal decomposition of $Pb(CH_3)_4$ was employed to measure the ionization potential of methyl radicals by the molecular beam method [11, 12].

Concentrations of methyl and ethyl radicals have been determined during pyrolysis of mixtures of $Pb(CH_3)_4$ and $Pb(C_2H_5)_4$. The concentration of the pyrolysis products, C_2H_6, C_4H_{10} and (mainly at higher temperatures) C_2H_4, are given graphically. Pyrolysis of a mixture of $Pb(CH_3)_4$ and $Pb(C_2H_5)_4$ in the presence of 1,2-dibromoethane results in the formation of CH_3Br and C_2H_5Br [29]; see also [16].

Single pulse shock tube experiments at temperatures between 731 and 931 K show $Pb(CH_3)_4$ to be thermally more stable than $(CH_3)_{4-n}Pb(C_2H_5)_n$ (n = 1 to 4) [24]. After 5.6 ms at 744 K, about 90% of the original $Pb(CH_3)_4$, but only 37% of $Pb(C_2H_5)_4$, remained [26]. However, $Pb(CH_3)_4$ was reported to be more explosive and sensitive to shock than $Pb(C_2H_5)_4$ and the explosive characteristics of $Pb(CH_3)_4$ to be comparable to those of picric acid. The sensitivity of $Pb(CH_3)_4$ is reduced by addition of toluene [25].

UV irradiation of thermal degradation products of $Pb(CH_3)_4$ together with water and ammonia at 77 K for 3.5 h yields serine, glycine, and alanine [31].

Suspensions of $Pb(CH_3)_4$ in water decompose slowly in the dark. After 22 d, 16% of $Pb(CH_3)_4$, present in a concentration of 12 to 15×10^{-5} mol/L, had reacted to give the cation $[Pb(CH_3)_3]^+$. Hydrolysis of $Pb(CH_3)_4$ adsorbed on silica gel takes place quantitatively [41]. For decomposition reactions of $Pb(CH_3)_4$ in water, see also Section 1.1.1.1.4.6, p. 138.

References:

[1] Cahours, A. (Ann. Chim. Phys. [3] **62** [1861] 257/350).
[2] Butlerow, A. (Jahresber. Fortschr. Chem. **1863** 474/6).
[3] Paneth, F., Matthies, M., Schmidt-Hebbel, E. (Ber. Deut. Chem. Ges. **55** [1922] 775/89).
[4] Krause, E. (Ber. Deut. Chem. Ges. **62** [1929] 1877/8).
[5] Paneth, F., Hofeditz, W. (Ber. Deut. Chem. Ges. **62** [1929] 1335/47).
[6] Ipatiew, W. N., Rasuwajew, G. A., Bogdanow, I. F. (J. Russ. Phys. Chem. Soc. **61** [1929] 1791/9; Ber. Deut. Chem. Ges. **63** [1930] 335/42).
[7] Simons, J. H., McNamee, R. W., Hurd, C. D. (J. Phys. Chem. **36** [1932] 939/48).
[8] Rice, F. O., Johnston, W. R., Evering, B. L. (J. Am. Chem. Soc. **54** [1932] 3529/43).
[9] Paneth, F. A., Hofeditz, W., Wunsch, A. (J. Chem. Soc. **1935** 372/9).
[10] Ipatieff, V. N. (Catalytic Reactions at High Pressures and Temperatures, Macmillan, New York 1936, pp. 350/8).

[11] Fraser, R. G. J., Jewitt, T. N. (Phys. Rev. [2] **50** [1936] 1091).
[12] Fraser, R. G. J., Jewitt, T. N. (Proc. Roy. Soc. [London] A **160** [1937] 563/74).
[13] Zscharn, A. (Chemiker-Ztg. **64** [1940] 498).
[14] Romm, F. S. (J. Gen. Chem. [USSR] **10** [1940] 1784/92; C.A. **1941** 3880).
[15] Eltenton, G. C. (J. Chem. Phys. **10** [1942] 403).
[16] Hipple, J. A., Stevenson, D. P. (Phys. Rev. [2] **63** [1943] 121/6).
[17] Gilman, H., Jones, R. G. (J. Am. Chem. Soc. **68** [1946] 517/20).
[18] Eltenton, G. C. (J. Chem. Phys. **15** [1947] 465/74).
[19] Mador, I. L. (J. Chem. Phys. **22** [1954] 1617).
[20] Le Goff, P., Letort, M. (Compt. Rend. **239** [1954] 970/2).

[21] Long, L. H. (J. Chem. Soc. **1956** 3410/6).
[22] Le Goff, P., Letort, M. (J. Chim. Phys. **53** [1956] 480/92).
[23] Rifkin, E. B. (Preprint, Proc. Am. Petrol. Inst. III **38** [1958] 60/7).
[24] Ryason, P. R. (Combust. Flame **7** [1963] 235/43).
[25] California Research Corp. (Brit. 941742 [1960/63]; C.A. **60** [1964] 3932).
[26] Richardson, W. L., Ryason, P. R., Kautsky, G. J., Barusch, M. R. (9th Symp. Intern. Combust. Proc., Ithaca, N.Y., 1962 [1963], pp. 1023/33).
[27] Snyder, L. J. (Anal. Chem. **39** [1967] 591/5).
[28] Gilman, H. (Advan. Organometal. Chem. **7** [1968] 1/52).
[29] Butzert, H., Dockey, H. D. (Z. Physik. Chem. [Frankfurt] **62** [1968] 83/102).
[30] Hoare, D. E., Li, T.-M., Walsh, A. D. (12th Intern. Symp. Combust. Proc., Poitiers 1968 [1969], pp. 357/64).

[31] Czuchajowski, L., Francik, H., Górecka, W., Kostuch, A. (Bull. Acad. Polon. Sci. Ser. Sci. Chim. **18** [1970] 241/5).

[32] Hoare, D. E., Walsh, A. D., Li, T.-M. (13th Intern. Symp. Combust. Proc., Salt Lake City, Utah, 1970 [1971], pp. 461/9).

[33] Hoffmann, P., Bächmann, K., Bögl, W., Klenk, H., Lieser, K. H. (Radiochim. Acta **16** [1971] 172/9).

[34] Gilroy, K. M., Price, S. J., Webster, N. J. (Can. J. Chem. **50** [1972] 2639/41).

[35] Homer, J. B., Hurle, I. R. (Proc. Roy. Soc. [London] A **327** [1972] 61/79).

[36] Price, S. J. W. (Compr. Chem. Kinet. **4** [1972] 246/57).

[37] Homer, J. B., Prothero, A. (J. Chem. Soc. Faraday Trans. I **69** [1973] 673/84).

[38] Scott, R. L., Richardson, A. E., Simons, J. W., Hase, W. L. (Intern. J. Chem. Kinet. **7** [1975] 547/55).

[39] Baudler, M. (Brauer, G., Handbuch der Präparativen Anorganischen Chemie, 3rd Ed., Vol. 2, Enke, Stuttgart 1978, pp. 781/3).

[40] Harrison, R. M., Laxen, D. P. H. (Environ. Sci. Technol. **12** [1978] 1384/92).

[41] Jarvie, A. W. P., Markall, R. N., Potter, H. R. (Environ. Res. **25** [1981] 241/9).

[42] Egsgaard, H., Bo, P., Carlsen, L. (J. Anal. Appl. Pyrolysis **8** [1985] 3/14).

1.1.1.1.4.2 With Radiation

1.1.1.1.4.2.1 Decomposition

$Pb(CH_3)_4$ gradually decomposes in sunlight. The pure compound, however, can be stored according to [1, 26] in brown, glass-stoppered flasks for years. On exposure of commercial $Pb(CH_3)_4$ to sunlight for 6 months in sealed flasks, slow decomposition was observed to give lead, CH_4, C_2H_6, C_3H_8, and traces of C_3H_6 [28, 30, 32]; see also [29]. The decomposition in the presence of sea water under otherwise equal conditions was less fast and CH_4, C_2H_6, C_3H_8, and traces of C_2H_4 were identified as gaseous products [28, 30, 32]. In aqueous suspensions of $Pb(CH_3)_4$ after 22 h of standing in sunlight, 59% had reacted to give $[Pb(CH_3)_3]^+$. Photolysis in the presence of water is proposed to follow a radical mechanism; whereas for the dark reaction, hydrolysis is suggested [31]. The photolytic decay of $Pb(CH_3)_4$ conforms to a first-order reaction, and the decay rate constant for $Pb(CH_3)_4$ in purified dry air and in natural sunlight at atmospheric pressure and 295 ± 3 K is approximately 1.4×10^{-3} min^{-1} at a solar zenith angle of about 40°. The decay rate determined during irradiation by UV lamps (equivalent solar zenith angle about 75°) is $3.38 \pm 0.34 \times 10^{-4}$ min^{-1}. The rate of photolytic reaction is not altered by admixed SO_2 or $BrCH_2CH_2Br$, but significantly enhanced by NO_2. Photolytic experiments may be affected by nitric acid adsorbed in vessels in which experiments with NO_2 had been performed before. From rate measurements in a hydrocarbon-NO_x reaction mixture, in which OH is generated by photolysis, a rate constant of 13.3 ± 2.3 ppm$^{-1} \cdot$ min^{-1} is estimated for the reaction of OH with $Pb(CH_3)_4$ at 293 ± 3 K and 760 Torr. H abstraction is assumed as the first reaction step [27]. $Pb(CH_3)_4$ decomposes in sunlight more slowly than $Pb(C_2H_5)_4$ [27, 28, 30, 31, 32].

$Pb(CH_3)_4$ is decomposed into lead and hydrocarbons (mainly C_2H_6) [4] on irradiation with UV light in the vapor state [2, 4], as well as in the pure liquid and in octane solution [4]. The quantum yield at a vapor pressure of 22 to 31 Torr and at 253.7 nm was determined to be 1.11. The presence of oxygen reduces the quantum yield. The limiting values of the quantum yield at zero exposure are 0.37 and 0.42 for 0.67 and 2.78×10^{-3} M solutions of $Pb(CH_3)_4$ in 2,2,4-

trimethylpentane, respectively [4]. Photodissociation of $Pb(CH_3)_4$ in solution by near UV radiation takes place at a triplet level and leads to the appearance of spin adducts in aromatic solvents [33]. UV photolytic decomposition of $Pb(CH_3)_4$ in the presence of water and ammonia at 77 K for 3.5 h yields serine, glycine, and alanine [17].

Raman measurements of gaseous $Pb(CH_3)_4$ employing mercury lines, give poor results since photochemical decomposition generates lead dust causing scattering of primary light [10]. Slight decomposition of liquid $Pb(CH_3)_4$ occurred on excitation with the 4047 cm^{-1} line [3]. $Pb(CH_3)_4$ is less sensitive to Raman excitation than $Pb(C_2H_5)_4$ and other lead alkyls [15].

Methyl radicals are formed during photolysis of $Pb(CH_3)_4$ vapor [4, 30]. Yet after irradiation of a 2% solution of $Pb(CH_3)_4$ in a frozen glass with light of 280 to 340 nm, no methyl radicals were detected [11]. No photolysis of $Pb(CH_3)_4$ was observed after irradiation of a 0.01 M solution in EPA glass for 1 h with a hydrogen lamp at $-183°C$ [8, 12]. However, $Pb(CH_3)_4$ is apparently photolyzed when subjected to a high intensity light flash [12]. Photodissociation of gaseous $Pb(CH_3)_4$ diluted in argon with flashlamps ($\lambda \geqq 176$ nm) produces exited lead atoms [23].

Decomposition of $Pb(CH_3)_4$ in the vapor phase by flash photolysis at pressures up to 12 Torr and at a flash energy of 2500 J is largely complete in 1.2 ms after the initial light flash [8]. Lead is formed and coats the walls of the reaction vessel; the gaseous products contain mainly C_2H_6, and small amounts of CH_4, C_2H_4, and C_2H_2 [8, 12]; see also [9]. $Pb(CD_3)_4$, on photolysis, behaves in an identical manner [8]. Flash photolysis of $Pb(CH_3)_4$ was examined at pressures of 0.2 to 1 Torr in the presence of Ar, H_2, C_6H_{14}, C_6H_{12}, acetic acid, or methylacetate [14], and in the presence of N_2 to which was admixed O_2, CH_3ONO, CH_3CHO, CH_2O, CD_2O, $(CHO)_2$, or H_2S, respectively. Bands observed in the spectrum at 3096 and 3196 Å were assigned to $PbCH_3$ [19]. Flash photolysis of $Pb(CH_3)_4$, mixed with Ar and Ne in the pressure range of 130 to 350 Pa with a flash energy of 1000 J, gives C_2H_6 as the only product when the initial adiabatic temperature rise was low. However, if it was of about 1200 K, C_2H_4, CH_4, and C_2H_2 were also produced [21]; see also [18]. Products of flash photodecomposition of $Pb(CH_3)_4$ in an inert gas at different concentrations and conditions were determined by time-resolved mass spectrometry. At 50 µs after the flash, C_2H_6 was the main product. For products at 400 µs after the flash, see original [18]. The initial step in the photolytic process involves the production of methyl radicals [18]. For flash photolysis of $Pb(CH_3)_4$ as a method to obtain Pb $6p^2(^3P_2)$ and $(^3P_1)$, see [24].

Essentially complete decomposition of $Pb(CH_3)_4$ occurs on X-ray irradiation (50 keV) as revealed from measurements of relative abundance and recoil energy spectra of fragment ions [16]. Irradiation with X-rays at 4 K produced CH_4. On subsequent warming to about 40 K the irradiated compound turned opaque, presumably because of the formation of elemental lead [7].

Studies on bond rupture by β-decay in ^{210}Pb-labelled $Pb(CH_3)_4$, in solution and in the vapor phase, indicate that rupture results from secondary processes involving neighboring atoms following the decay process [6]; see also [2, 5, 13]. Very short-lived transients are observed after radiolysis of $Pb(CH_3)_4$ doped with naphthalene and toluene by 200 ns single pulses of 40 MeV electrons [25]; see also [34, 36]. After irradiation of $Pb(CH_3)_4$ with ^{60}Co γ-rays at 77 K no methyl radicals have been detected [11], but the radicals $(CH_3)_3PbCH_2^\bullet$ and $(CH_3)_3Pb^\bullet$ have been identified [22]. γ-Irradiation of a solution containing 3 mol% $Pb(CH_3)_4$ in CCl_3F at 85 K generates a species which was inferred to be $[Pb(CH_3)_4]^+$ from ESR spectra [35]. No radiochemical decomposition of $Pb(CH_3)_4$ was caused by the radiation of a ^{224}Ra source [20].

For lifetime, stability, and degradation reactions of $Pb(CH_3)_4$ in the atmosphere, see Sections 1.1.1.1.4.4 and 1.1.1.1.9, pp. 132 and 179.

128

References:

[1] Krause, E. (Ber. Deut. Chem. Ges. **62** [1929] 1877/8).
[2] Mortensen, R. A., Leighton, P. A. (J. Am. Chem. Soc. **56** [1934] 2397/8).
[3] Duncan, A. B. F., Murray, J. W. (J. Chem. Phys. **2** [1934] 636/43).
[4] Leighton, P. A., Mortensen, R. A. (J. Am. Chem. Soc. **58** [1936] 448/54).
[5] Edwards, R. R., Coryell, C. D. (AECU-50 [BNL-C-7] [1948] 63/73; C.A. **1951** 4147).
[6] Edwards, R. R., Day, J. M., Overman, R. F. (J. Chem. Phys. **21** [1953] 1555/8).
[7] Mador, I. L. (J. Chem. Phys. **22** [1954] 1617).
[8] Cook, C. L., Clouston, J. G. (Nature **177** [1956] 1178/9).
[9] Clouston, J. G., Cook, C. L. (Nature **179** [1957] 1240/1).
[10] Waters, D. N., Woodward, L. A. (Proc. Roy. Soc. [London] A **246** [1958] 119/32).

[11] Smaller, B., Matheson, M. S. (J. Chem. Phys. **28** [1958] 1169/78).
[12] Clouston, J. G., Cook, C. L. (Trans. Faraday Soc. **54** [1958] 1001/7).
[13] Kay, J., Rowland, F. S. (J. Am. Chem. Soc. **80** [1958] 3165).
[14] Erhard, K. (Naturwissenschaften **49** [1962] 417/8).
[15] Jackson, J. A., Nielsen, J. R. (J. Mol. Spectrosc. **14** [1964] 320/41).
[16] Carlson, T. A., White, R. M. (J. Chem. Phys. **48** [1968] 5191/4).
[17] Czuchajowski, L., Francik, H., Górecka, W., Kostuch, A. (Bull. Acad. Polon. Sci. Ser. Sci. Chim. **18** [1970] 241/5).
[18] Appleby, S. E., Howarth, S. B., Jones, A. T., Lippiatt, J. H., Orville-Thomas, W. J., Price, D., Heald, P. (Dyn. Mass Spectrom. **1** [1970] 37/57).
[19] Cook, C. L., Napier, I. M. (Australian J. Chem. **24** [1971] 179/82).
[20] Hoffmann, P., Bächmann, K., Bögl, W., Klenk, H., Lieser, K. H. (Radiochim. Acta **16** [1971] 172/9).

[21] Howarth, S. B., Lippiatt, J. H., Price, D., Ward, G. B., Myers, P. (Intern. J. Mass Spectrom. Ion Phys. **9** [1972] 95/105).
[22] Lyons, A. R., Neilson, G. W., Symons, M. C. R. (J. Chem. Soc. Faraday Trans. II **68** [1972] 807/13).
[23] Ewing, J. J., Trainor, D. W., Yatsiv, S. (J. Chem. Phys. **61** [1974] 4433/9).
[24] Trainor, D. W., Ewing, J. J. (J. Chem. Phys. **64** [1976] 222/7).
[25] Hosszu, J. L. (AD-A033409 [1976] 1/12).
[26] Baudler, M. (Brauer, G., Handbuch der Präparativen Anorganischen Chemie, 3rd Ed., Vol. 2, Enke, Stuttgart 1978, pp. 781/3).
[27] Harrison, R. M., Laxen, D. P. H. (Environ. Sci. Technol. **12** [1978] 1384/92).
[28] Charlou, J. L. (CNEXO-COB-17 [1979] 1/15; Contrat CNEXO-ENSCR No. 78-5678).
[29] Radziuk, G., Thomassen, Y., Van Loon, J. C., Chau, Y. K. (Anal. Chim. Acta **105** [1979] 255/62).
[30] Charlou, J. L. (CNEXO-COB-199 [1980] 1/30; Contrat CNEXO-ENSCR No. 79-5943).

[31] Jarvie, A. W. P., Markall, R. N., Potter, H. R. (Environ. Res. **25** [1981] 241/9).
[32] Charlou, J. L., Caprais, M. P., Blanchard, G., Martin, G. (Environ. Technol. Letters **3** [1982] 415/24).
[33] Smirnov, S. G., Konoplev, G. G., Rodionov, A. N., Shigorin, D. N. (Zh. Fiz. Khim. **56** [1982] 964/8; Russ. J. Phys. Chem. **56** [1982] 584/6).
[34] Nielsen, O. J., Nielsen, T., Pagsberg, P. (RISØ-R-463 [1982] 1/17).
[35] Walther, B. W., Williams, F., Lau, W., Kochi, J. K. (Organometallics **2** [1983] 688/90).
[36] Nielsen, O. J. (RISØ-R-480 [1984] 1/126).

1.1.1.1.4.2.2 Mass Spectrum

The mass spectrum of $Pb(CH_3)_4$ was subject of a series of investigations. The lead-containing fragments are collected in Table 9 [6, 9, 14, 18, 27]; see also [5, 36].

Table 9

Monoisotopic Mass Spectrum (^{208}Pb) of $Pb(CH_3)_4$.
Source temperature 203 to 250°C, electron impact energy 70 eV [6, 9, 14, 18] and 50 eV [27].

fragment	intensity in % of total ionization		relative intensities in %		
	[27]	[18]	[6]	[9]	[14]
Pb^+	12.7	18.2	64.0	71.3	69.7
$[PbH]^+$	1.2		5.74	6.65	3.24
$[PbC]^+$			0.87	0.80	
$[PbCH]^+$	0.8		2.78	3.46	
$[PbCH_2]^+$	1.9		6.03	6.80	6.35
$[PbCH_3]^+$	24.1	29.3	85.3	90.9	87.8
$[PbCH_4]^+$			0.37	0.90	
$[PbC_2H_5]^+$	0.4		0.55	0.55	0.53
$[Pb(CH_3)_2]^+$	9.6	8.5	24.8	25.3	25.0
$[PbC_2H_7]^+$	9.9 (?)		0.25	0.45	
$[PbC_3H_7]^+$	0.4				0.27
$[PbC_3H_8]^+$	1.4		1.07	1.00	0.81
$[Pb(CH_3)_3]^+$	44.8	44.4	100.0	100.0	100.0
$[PbC_3H_{10}]^+$			0.33	0.95	
$[Pb(CH_3)_4]^+$	0.3	0.4	0.2	0.22	

For relative abundances of lead isotopes, see [6, 9, 30, 33]. Line diagrams of mass spectra of $Pb(CH_3)_4$ are shown in [18, 26, 40, 41, 44] and of $Pb(CD_3)_4$ in [41]; see also [19]. According to [7], the ion $[(CH_3)_2PbCH_2]^+$ was found to be the only one to be related to $[Pb(CH_3)_3]^+$. The dependence of the mass spectrum of $Pb(CH_3)_4$ on temperature of the ion source was investigated. Above 850°C the compound is completely dissociated, Pb^+ being found as the only ion. The parent ion was not found between 300 and 850°C [13]. In other experiments with ion sources working at lower temperatures, the parent ion was observed, however, with very small abundance [5, 6, 18, 21, 27, 36]. The relative abundance of both the parent molecule ion and the first fragment increases with decreasing ionizing voltage [36]. Hydrocarbon fragment ions are nearly absent [27, 30, 33]; see also [37]. It was concluded that primary ionization results in an electron deficiency on lead [27]. Methyl radicals, CH_4, and C_2H_6 have been identified as low mass products [13]; see also [5, 15]. Formation of C_2H_6 is assumed to occur at the wall [13]. Fragmentation patterns at different ionizing voltages are given in [21, 36]. Probabilities of fragmentation processes have been derived in [6, 9, 18].

Metastable peaks have been observed for the fragmentation reactions: $[PbC_3H_9]^+ \rightarrow [PbC_2H_6]^+ + CH_3$ ($m^* = 113.89$), $[PbC_2H_6]^+ \rightarrow [PbCH_3]^+ + CH_3$ ($m^* = 208.95$), $[PbCH_3]^+ \rightarrow [PbH]^+ + CH_2$ ($m^* = 195.88$), $[PbCH]^+ \rightarrow Pb^+ + CH$ ($m^* = 194.0$?), and $[PbC_3H_9]^+ \rightarrow [PbCH_3]^+ + C_2H_6$ ($m^* = 196.6$) [6, 27].

For a correlation of ion yield and electronegativity, see [37]. Relative abundances and recoil energies of fragment ions formed in the irradiation of $Pb(CH_3)_4$ with 50-keV X-rays were determined with a specially designed mass spectrometer [28]. The appearance potential of $[Pb(CH_3)_3]^+$ was estimated to be 8.77 ± 0.16 eV [30, 33]. Previously found values are 8.9 ± 0.1 [18] and 10.1 ± 0.3 [27]. Appearance potentials of other fragment ions are given in [18, 27]. For calculations of ΔH_f^0 values of fragment ions and radicals, see [30, 33]. The molecular ionization potential is 8.0 eV [23]. The heat of formation, ΔH_f, of the parent molecule ion from $Pb(CH_3)_4$ is determined to be 217 kcal/mol [18].

Neutral-fragment mass spectra of $Pb(CH_3)_4$ have been obtained with a double-ionization-chamber ion source. The unimolecular decomposition of excited $Pb(CH_3)_4$ was studied and dissociative ionization and appearance energies of the neutral fragments $Pb(CH_3)_n$ (n = 1, 2, 3) have been determined [42].

The mass spectrum of an equimolar mixture of $Pb(CH_3)_4$ and $Pb(C_2H_5)_4$ at different ionizing voltages was studied [36]. Concentrations of methyl and ethyl radicals and of their reaction products produced in the pyrolysis of a mixture of $Pb(CH_3)_4$ and $Pb(C_2H_5)_4$, at 10^{-4} Torr and between 0 and 500°C, were measured in a field ion mass spectrometer [29].

The mass spectrum of $Pb(CH_3)_4$ has been discussed in relation to mass spectra of other organolead compounds [6, 14], of methylmercury [6], and other methylmetal compounds [9, 18, 27]. The mass spectra of the group 14 tetramethyl compounds $M(CH_3)_4$ (M = C, Si, Ge, Sn, Pb) are qualitatively similar [9, 27, 34, 37], though distinct group dependences exist; see also [6, 30, 33]. The amounts of rearrangement ions (e.g., $[HM(CH_3)_2]^+$, $[H_2MCH_3]^+$, $[MH_3]^+$, $[MH]^+$) decrease with increasing mass of M [27]. Decreasing stability of $[M(CH_3)_3]^+$ and increasing intensity of $[MCH_3]^+$ with higher atomic number is explained by increasing stability of the lower oxidation state of M [37]. For fragment ion intensities and yield of ions $[M(CH_3)_n]^+$ (n = 0 to 4) as function of electron impact energy, see [34, 37, 38]. From these results data for tetramethylekalead were extrapolated [34, 38].

Aspects of applicability of $Pb(CH_3)_4$ for isotopic abundance measurements were discussed in [6, 9, 16, 21]. The mass spectrum of $Pb(CH_3)_4$ has been used to analyze the isotopic composition of lead of different origins [1, 4, 8, 10, 16, 19]; see also [2, 3]. From such analyses of $Pb(CH_3)_4$, obtained from lead in ores, the geological age of the ore was estimated [7, 8, 10, 12, 22, 24]; see also [2, 4, 11].

Mass spectrometry can be used for the quantitative determination of $Pb(CH_3)_4$ and of mixtures of $Pb(CH_3)_4$ and $Pb(C_2H_5)_4$ in gasoline [17, 39]. Mass spectrometric procedures have been employed for detection of $Pb(CH_3)_4$ in gas chromatography [25, 31, 41]. Development of a time-of-flight mass spectrometer to monitor flash photodecomposition of $Pb(CH_3)_4$ is described [35]; see also [32]. Detection of $Pb(CH_3)_4$ with a small quadrupole mass spectrometer is described in [45].

Interference by silicone polymers is a source of error in mass spectrometry of $Pb(CH_3)_4$ [20, 26]. It also should be considered that lead, e.g., deposited from previous experiments, can be volatilized by CH_3I leading to a spectrum similar to that of genuine $Pb(CH_3)_4$ [43].

References:

[1] Aston, F. W. (Nature **120** [1927] 224).
[2] Piggot, C. S. (J. Wash. Acad. Sci. **18** [1928] 269/73).
[3] Aston, F. W. (Nature **123** [1929] 313).
[4] Aston, F. W. (Proc. Roy. Soc. [London] A **140** [1933] 535/43).
[5] Hipple, J. A., Stevenson, D. P. (Phys. Rev. [2] **63** [1943] 121/6).
[6] Dibeler, V. H., Mohler, F. L. (J. Res. Natl. Bur. Std. **47** [1951] 337/42).

[7] Collins, C. B., Freeman, J. R., Wilson, J. T. (Phys. Rev. [2] **82** [1951] 966/7).
[8] Collins, C. B., Farquhar, R. M., Russell, R. D. (Phys. Rev. [2] **88** [1952] 1275/6).
[9] Dibeler, V. H. (J. Res. Natl. Bur. Std. **49** [1952] 235/9).
[10] Collins, C. B., Russell, R. D., Farquhar, R. M. (Can. J. Phys. **31** [1953] 402/18).

[11] Ducheylard, G., Lazard, B., Roth, E. (J. Chim. Phys. **50** [1953] 497/500).
[12] Collins, C. B., Farquhar, R. M., Russell, R. D. (Bull. Geol. Soc. Am. **65** [1954] 1/22).
[13] Osberghaus, O., Taubert, R. (Z. Physik. Chem. [Frankfurt] **4** [1955] 264/85).
[14] Quinn, E. I., Dibeler, V. H., Mohler, F. L. (J. Res. Natl. Bur. Std. **57** [1956] 41/3).
[15] Le Goff, P., Letort, M. (J. Chim. Phys. **53** [1956] 480/92).
[16] Bate, G. L., Miller, D. S., Kulp, J. L. (Anal. Chem. **29** [1957] 84/8).
[17] Howard, H. E., Ferguson, W. C., Snyder, L. R. (Anal. Chem. **32** [1960] 1814/5).
[18] Hobrock, B. G., Kiser, R. W. (J. Phys. Chem. **65** [1961] 2186/9).
[19] Richards, J. R. (J. Geophys. Res. **67** [1962] 869/84).
[20] Slawson, W. F., Russell, R. D. (Mikrochim. Ichnoanal. Acta **1963** 165/8).

[21] Ghate, M. R., Bhide, K. N. (Indian J. Chem. **2** [1964] 243/4).
[22] Ulrych, T. J., Russell, R. D. (Geochim. Cosmochim. Acta **28** [1964] 455/69).
[23] Kiser, R. W. (Introduction to Mass Spectrometry and Its Applications, Prentice-Hall, Englewood Cliffs, N. J., 1965, p. 314).
[24] Whittles, A. B. L., Slawson, W. F. (Geochim. Cosmochim. Acta **29** [1965] 142/3).
[25] Henneberg, D., Schomburg, G. (Z. Anal. Chem. **215** [1965] 424/30).
[26] Richards, J. R. (Vacuum **16** [1966] 310/1).
[27] de Ridder, J. J., Dijkstra, G. (Rec. Trav. Chem. **86** [1967] 737/45).
[28] Carlson, T. A., White, R. M. (J. Chem. Phys. **48** [1968] 5191/4).
[29] Butzert, H., Beckey, H. D. (Z. Physik. Chem. [Frankfurt] **62** [1968] 83/102).
[30] Lappert, M. F., Pedley, J. B. (AD-71-9817 [1970] 1/78).

[31] Laveskog, A. (Proc. 2nd Intern. Clean Air Congr., Washington, D. C., 1970 [1971], pp. 549/57).
[32] Appleby, S. E., Howarth, S. B., Jones, A. T., Lippiat, J. H., Orville-Thomas, W. J., Price, D., Heald, P. (Dyn. Mass Spectrom. **1** [1970] 37/57).
[33] Lappert, M. F., Pedley, J. B., Simpson, J., Spalding, T. R. (J. Organometal. Chem. **29** [1971] 195/208).
[34] Heumann, K. G., Bächmann, K., Kubassek, E., Lieser, K. H. (Proc. 2nd Intern. Conf. Ion Sources, Vienna 1972 [1973], pp. 24/30).
[35] Howarth, S. B., Lippiat, J. H., Price, D., Ward, G. B., Myers, P. (Intern. J. Mass Spectrom. Ion Phys. **9** [1972] 95/105).
[36] Clinton, N. A., Kochi, J. K. (J. Organometal. Chem. **56** [1973] 243/54).
[37] Heumann, K. G., Bächmann, K., Kubassek, E., Lieser, K. H. (Z. Naturforsch. **28b** [1973] 107/12).
[38] Heumann, K. G., Bächmann, K., Hoffmann, P., Kubassek, E., Lieser, K. H. (Radiochim. Acta **20** [1973] 110/4).
[39] Knof, H., Ewers, H., Albers, G. (Compend. Deut. Ges. Mineralölwiss. Kohlechem. **1974/75** Vol. 2, pp. 798/810).
[40] Charlou, J. L. (CNEXO-COB-199 [1980] 1/30; Contrat CNEXO-ENSCR No. 79-5943).

[41] Nielsen, T., Egsgaard, H., Larsen, E., Schroll, G. (Anal. Chim. Acta **124** [1981] 1/13).
[42] Flesch, G. D., Svec, H. J. (Intern. J. Mass Spectrom. Ion Phys. **38** [1981] 361/70).
[43] Jarvie, A. W. P., Whitmore, A. P. (Environ. Technol. Letters **2** [1981] 197/204).
[44] Charlou, J. L., Caprais, M. P., Blanchard, G., Martin, G. (Environ. Technol. Letters **3** [1982] 415/24].
[45] Batey, J. H. (Intern. J. Mass Spectrom. Ion Processes **60** [1984] 117/26).

1.1.1.1.4.2.3 ESR Spectrum

After exposure to ^{60}Co γ-rays at 77 K, pure $Pb(CH_3)_4$ or a solution in toluene gives rise to an ESR spectrum showing features that are assigned to the radical $Pb(CH_3)_3^\cdot$, with $g_\perp \sim 2.1$ and $g_\parallel \sim 1.9$ and abnormally large anisotropic ^{207}Pb hyperfine coupling constants; to $CH_3^\cdot$ and $(CH_3)_3PbCH_2^\cdot$; and tentatively to the radical anion $[Pb(CH_3)_4]^{-}$. The spectrum is depicted in a figure [1]; see also [2]. On X-ray irradiation of neat $Pb(CH_3)_4$ at 77 K, the ESR spectrum exhibits the parameters of the radical $(CH_3)_3PbCH_2^\cdot$ with the parameters $g_\parallel = 2.0029 \pm 0.0004$ and $g_\perp = 1.9938 \pm 0.0002$ along with those of other unidentified radicals. The contact hyperfine splitting a (^{207}Pb) was found to be 158 oe. The radicals $(CH_3)_3MCH_2^\cdot$ (M = C, Si, Ge, Sn, Pb) were compared and discussed with respect to (d-p)π bonding [3]. The analysis of the ESR spectrum of $[Pb(CH_3)_4]^{+}$, generated from $Pb(CH_3)_4$ by electron detachment, is in accordance with a trigonal-pyramidal structure of C_{3v} symmetry [4]. CIDNP spectra with $Pb(CH_3)_4$ were not obtained; although, by spin trapping, methyl radicals could be detected in the reaction of $Pb(CH_3)_4$ with $AgNO_3$ [5].

References:

[1] Booth, R. J., Fieldhouse, S. A., Starkie, H. C., Symons, M. C. R. (J. Chem. Soc. Dalton Trans. **1976** 1506/15).
[2] Smaller, B., Matheson, M. S. (J. Chem. Phys. **28** [1958] 1169/78).
[3] Mackey, J. H., Wood, D. E. (Mol. Phys. **18** [1970] 783/92).
[4] Walther, B. W., Williams, F., Lau, W., Kochi, J. K. (Organometallics **2** [1983] 688/90).
[5] Janzen, E. G. (Accounts Chem. Res. **4** [1971] 31/40).

1.1.1.1.4.3 With Hydrogen

Hydrogen under 60 atm initial pressure reacts with $Pb(CH_3)_4$ in benzene solution on heating to give lead and methane. The separation of lead commences at 125°C and is quantitative at about 250°C after 24 h [1, 2, 3]; see also [4].

References:

[1] Ipat'ev, V. N., Razuvaev, G. A., Bogdanov, I. F. (J. Russ. Phys. Chem. Soc. **61** [1929] 1791/9).
[2] Ipatiew, W. N., Rasuwajew, G. A., Bogdanow, I. F. (Ber. Deut. Chem. Ges. **63** [1930] 335/42).
[3] Ipatieff, V. N. (Catalytic Reactions at High Pressures and Temperatures, Macmillan, New York 1936, pp. 350/8).
[4] Krause, E. (Ber. Deut. Chem. Ges. **63** [1930] 999/1000).

1.1.1.1.4.4 With Oxygen, Ozone, and the Hydroxy Radical OH

$Pb(CH_3)_4$ shows no spontaneous reactivity with oxygen at room temperature [5, 6]. This is correlated to the nonavailability of low-energy orbitals and a low net negative charge on the methyl groups [6]. When ignited in air, $Pb(CH_3)_4$ burns at first steadily; but as the liquid becomes heated, combustion becomes violent, and a grey-brown product, containing lead,

lead oxide, but no carbon, is left behind [5]. A mixture of $Pb(CH_3)_4$ and air may explode on heating [1]. Even when studying established diffusion flames of $Pb(CH_3)_4$, with oxygen or with air, violent explosions have been observed [5]. Combustion in a hot-air oven by heating to 300°C with an efficient draft was recommended to destroy $Pb(CH_3)_4$ [2]; see also Section 1.1.1.1.4.1, p. 123.

The combustion behavior was studied in diffusion flames of $Pb(CH_3)_4$ with air, oxygen, methane-oxygen, and hydrogen-oxygen as well as CH_3CHO-air. Formation of intermediates by decomposition of $Pb(CH_3)_4$, like lead particles on the fuel side of the flame, and complex oxidation products, some of which may be peroxidic, are described [5]. The solid residues of the combustion of $Pb(CH_3)_4$ in air or in argon-oxygen mixtures in the presence of hydrocarbons are rhombic PbO, $PbCO_3$, and small amounts of Pb_3O_4, and also $Pb(NO_2)_2$ when the combustion was performed in air [10]. Gas-phase oxidation of $Pb(CH_3)_4$, as studied in a flow system using $Pb(CH_3)_4$ and benzene mixtures, starts at about 410°C and gives CH_2O and CO as major products, whereas CO_2 was a less important product at temperatures lower than 550°C. Above 450°C, $Pb(CH_3)_4$ oxidizes more readily than $Pb(C_2H_5)_4$, although the oxidation of the latter commenced already at about 340°C [11]. Oxidation in a static system at 330°C proceeds at a moderate rate and gives PbO, CO_2, H_2, some CH_3OH, and CH_4 as a minor product in the early stages; CO was not observed (H_2O was not determined). The overall reaction was found to be nonchain and the main reaction was assumed to occur via methyl radicals. CH_2O was not detected, but considered to possibly be an intermediate in the oxidation [12].

The inflammability limit and the lower explosive limit of $Pb(CH_3)_4$ in air is 1.8% by volume [14, 19]. Concentrations of $Pb(CH_3)_4$ well below the lower limit of inflammability can detonate in shocked mixtures of oxygen and $Pb(CH_3)_4$ [15]. The flash point is 38°C (100°F) [19].

The heat of combustion of liquid $Pb(CH_3)_4$, corrected to constant pressure, is given to be 837.6 ± 3 [3], 870.2 [4], 871.2 ± 0.8 kcal/mol [9]. $\Delta U^0_{298.16}$ and $\Delta H^0_{298.16}$ for combustion of liquid $Pb(CH_3)_4$, and subsequent formation of $Pb(NO_3)_2$, was estimated to be 884.8 ± 0.3 and 886.9 ± 0.3 kcal/mol, respectively [8]. Oxidation of $Pb(CH_3)_4$ by oxygen molecules at ca. 900 to ca. 1040 K is much slower than its pyrolysis. The rate constant at 900 K for the reaction of $Pb(CH_3)_4$ and O_2 was calculated to be less than $k = 5 \times 10^{-16}$ cm$^3 \cdot$ mol$^{-1} \cdot$ s^{-1} [13]. For a mass spectrometric study of oxidation of $Pb(CH_3)_4$ under steady-state conditions, see [26].

The main homogeneous reactions of $Pb(CH_3)_4$ in the atmosphere are photolytic breakdown and reactions with O_3, $O(^3P)$, and OH. The reactions are principally those of photochemical oxidation, with attack by OH representing the main reaction pathway [17]. For photolytic rate constants in purified dry air, see Section 1.1.1.1.4.2, p. 126. For studies of the lifetime of $Pb(CH_3)_4$ in air, see Section 1.1.1.1.9, p. 179.

The second-order rate constant for the reaction of $O(^3P)$ with $Pb(CH_3)_4$ in nitrogen at 295 ± 3 K and 1 atm was estimated to be $13.7 \pm 3.7 \times 10^2$ ppm$^{-1} \cdot$ min^{-1}, the relative rates for the $O(^3P)$ reactions with toluene, $Pb(CH_3)_4$, and $Pb(C_2H_5)_4$ are $1:5.52 \pm 0.38:15.0 \pm 3.3$. A discussion of the mechanism is made in [17].

The rate constant for the reaction of OH with $Pb(CH_3)_4$ at room temperature was estimated to be $13.3 \pm 2.3 \times 10^3$ ppm$^{-1} \cdot$ min^{-1} [17]. Independently, a value of $3.8 \pm 0.8 \times 10^9$ mol$^{-1} \cdot$ s^{-1} (9.2×10^3 ppm$^{-1} \cdot$ min^{-1}) was determined [21, 25]. H atom abstraction, yielding $(CH_3)_3PbCH_2^\bullet$, was considered to be the first reaction step [17], but also other initial steps have been discussed [21, 25]. $Pb(CH_3)_4$ is nine times less reactive than $Pb(C_2H_5)_4$ toward OH [17], according to another study only two times less reactive [21, 25].

The second-order rate constant for the dark reaction of ozone with $Pb(CH_3)_4$ in air at 295 ± 3 K and 1 atm in a static system at O_3 concentrations of about 0.3 to 0.8 ppm was determined to be $1.88 \pm 0.59 \times 10^{-3}$ $ppm^{-1} \cdot min^{-1}$ [17]. The value of $0.71 \pm 0.07 \times 10^{-3}$ $ppm^{-1} \cdot min^{-1}$ was obtained at reactant concentrations two to three orders of magnitude higher [16]. $Pb(CH_3)_4$ is eight times less reactive toward O_3 than $Pb(C_2H_5)_4$ [17]; see also [18].

The ozonolysis of $Pb(CH_3)_4$ in the gas phase at atmospheric pressure, under static conditions, was found to be a bimolecular reaction at 20 to 65°C, that is, first-order with respect to $Pb(CH_3)_4$ and to O_3. The rate is not influenced by reaction products. According to the Arrhenius equation $k = A \cdot e^{E_a/RT}$, the temperature dependence of the rate constant k (in $L \cdot mol^{-1} \cdot s^{-1}$) is expressed by $\log A = 2.8$ and $E_a = 3.20$ kcal/mol [20]; see also [22]. The gas-phase reaction at 15 to 40°C under O_3 excess is pseudo-monomolecular with $A = 6.9 \times 10^2$ and $E_a = 3.2 \pm 0.8$ kcal/mol [18].

Ozone is more reactive toward $Pb(CH_3)_4$ in solution than in the gas phase [18, 20, 22] and reactivity increases with increasing dielectric constant of the solvent [18]. In solution the reaction was established to be bimolecular, second-order, and first-order with respect to both $Pb(CH_3)_4$ and O_3 [18, 20, 23] except under bubbling conditions [23]. The rate constants k (in $L \cdot mol^{-1} \cdot s^{-1}$) in CCl_4 between 0 and 20°C give the values $A = 1.33 \times 10^8$ and $E_a = 9.0 \pm 0.7$ kcal/mol [18]. A value of $k = 62$ $L \cdot mol^{-1} \cdot s^{-1}$ ($\log A = 8.2$ and $E_a = 8.6$ kcal/mol) under static conditions at 25°C is reported; under bubbling conditions, a value of $k = 44$ $L \cdot mol^{-1} \cdot s^{-1}$ ($\log A = 8.1$ and $E_a = 9.01$ kcal/mol) is obtained [23]. For values of k and of Arrhenius parameters of ozonolysis of $Pb(CH_3)_4$ in CCl_4, $CHCl_2CHCl_2$, CH_3COOH, C_2H_5COOH, and CH_3COOH-H_2O mixtures, see [24]. For comparative studies of the ozonolysis of $Pb(CH_3)_4$ and of other organoelement group 14 compounds, see [20, 23, 24]. An overall isokinetic relationship for the ozonolysis of these compounds [20] and of $Pb(CH_3)_4$ and $Hg(CH_3)_2$ [24], respectively, has been established. From the isokinetic temperatures of the ozonolysis of $Pb(CH_3)_4$ and $Hg(CH_3)_2$, a lifetime of the activated complex of 10^{-13} to 10^{-12} s was calculated [24]. Enthalpies and entropies of activation for the reaction of ozone with $Pb(CH_3)_4$, and a series of organotin and organosilicon compounds in the gas and in the liquid phase, are linearly related by $\Delta H^{\ddagger} = -0.240 \times \Delta S^{\ddagger} + 56.743$ [20].

Ozonolysis of $Pb(CH_3)_4$ is assumed to proceed via insertion of O_3 into the Pb-C bond with subsequent fragmentation into $(CH_3)_3PbOOH$ and CH_2O [17]; see also [22].

References:

[1] Krause, E. (Ber. Deut. Chem. Ges. **62** [1929] 1877/8).

[2] Fraser, R. G. J., Jewitt, T. N. (Proc. Roy. Soc. [London] A **160** [1937] 563/74).

[3] Lippincott, E. R., Tobin, M. C. (J. Am. Chem. Soc. **75** [1953] 4141/7).

[4] Chlupacek, W. (Chem. Tech. [Leipzig] **5** [1953] 460).

[5] Egerton, A., Rudkrakanachana, S. (Proc. Roy. Soc. [London] A **225** [1954] 427/43).

[6] Sanderson, R. T. (J. Am. Chem. Soc. **77** [1955] 4531/2).

[7] Lautsch, W. F. (Chem. Tech. [Leipzig] **10** [1958] 419).

[8] Good, W. D., Scott, D. W., Lacina, J. L., McCullough, J. P. (J. Phys. Chem. **63** [1959] 1139/42).

[9] Lautsch, W. F., Tröber, A., Zimmer, W., Mehner, L., Linck, W., Lehmann, H.-M., Branden-burger, H., Körner, H., Metzschker, H.-J., Wagner, K., Kaden, R. (Z. Chem. [Leipzig] **3** [1963] 415/21).

[10] Gelius, R., Franke, W. (Brennstoff-Chem. **47** [1966] 280/9).

[11] Salooja, K. C. (J. Inst. Petrol. **53** [1967] 186/93).

[12] Hoare, D. E., Walsh, A. D., Ting-Man Li (13th Symp. Intern. Combust. Proc., Salt Lake City, Utah, 1970 [1971], pp. 461/9).

[13] Homer, J. B., Hurle, I. R. (Proc. Roy. Soc. [London] A **327** [1972] 61/79).

[14] Grove, J. (from [15]).

[15] Nettleton, M. A., Stirling, R. (Combust. Flame **22** [1974] 407/14).

[16] Patel, J. C., Strawley, D., Webster, P. (in: Harrison, R. M., Laxen, D. P. H., Environ. Sci. Technol. **12** [1978] 1384/92).

[17] Harrison, R. M., Laxen, D. P. H. (Environ. Sci. Technol. **12** [1978] 1384/92).

[18] Tarunin, B. I., Aleksandrov, Yu. A., Baklanov, N. V., Perepletchikova, V. P., Petrova, A. A. (Khim. Elementoorg. Soedin. [Gorkiy] **1978** 48/50).

[19] Irving Sax, N. (Dangerous Properties of Industrial Materials, 5th Ed., Van Nostrand Reinhold, New York 1979, p. 771).

[20] Aleksandrov, Yu. A., Tarunin, B. I., Perepletchikov, M. L., Seliverstov, N. N. (Kinetika Kataliz **22** [1981] 1384/8; Kinet. Catal. [USSR] **22** [1981] 1092/6).

[21] Nielsen, O. J., Nielsen, T., Pagsberg, P. (RISØ-R-463 [1982] 1/17).

[22] Aleksandrov, Yu. A., Tarunin, B. I. (J. Organometal. Chem. **238** [1982] 125/57).

[23] Aleksandrov, Yu. A., Tarunin, B. I., Seliverstov, N. N. (Kinetika Kataliz **24** [1983] 1496/9; Kinet. Catal. [USSR] **24** [1983] 1271/3).

[24] Seliverstov, N. N., Tarunin, B. I., Aleksandrov, Yu. A. (Khim. Elementoorg. Soedin. **1983** 33/7).

[25] Nielsen, O. J. (RISØ-R-480 [1984] 1/126; INIS Atomindex **15** [1984] No. 069 458).

[26] Faerman, V. I., Agafonov, I. L., Emel'yanov, A. A., Kosyak, A. M., Vantsova, T. F. (Khim. Elementoorg. Soedin. [Gorkiy] **1984** 57/9; C.A. **104** [1986] No. 12 620).

1.1.1.1.4.5 With Halogens and Interhalogen Compounds

Cl_2 reacts with $Pb(CH_3)_4$ by substitution of methyl groups [1]. In solvents like diethyl ether, ethyl acetate, or CCl_4 at -70 to $-75°C$, $(CH_3)_3PbCl$ is produced with immediate discoloration of the solution [2, 3]. Continuing chlorination at temperatures above -60 to $-40°C$ [3] or reaction in $CHCl_3$ at $-20°C$ [49] gives $(CH_3)_2PbCl_2$. Chlorinolysis of $Pb(CH_3)_4$ is slightly faster than that of $Pb(C_2H_5)_4$ [35].

In the reaction of thermalized ^{38}Cl atoms at $20°C$ and 500 to 5000 Torr with $Pb(CH_3)_4$, approximately 18% of the Cl atoms react by a direct bimolecular substitution mechanism to give $CH_3^{38}Cl$ and the radical $(CH_3)_3Pb^{\cdot}$. More than 75% of the Cl atoms react by hydrogen abstraction to give $H^{38}Cl$ and the radical $(CH_3)_3PbCH_2^{\cdot}$. The rate constants (in $cm^3 \cdot mol^{-1} \cdot s^{-1}$) are $k = (3.0 \pm 0.5) \times 10^{-11}$ [47], after reevaluation $(8 \pm 1) \times 10^{-11}$ [50] for the substitution and $k = (1.3 \pm 0.3) \times 10^{-10}$ for the abstraction reaction [47]. For influences of added vinyl bromide, CH_4, $HC\equiv CH$, H_2O_2, and O_2, see [47]. When $Pb(CH_3)_4$ vapors and Cl_2 were passed simultaneously into pure silicone granules at $375°C$, methylchlorosilane formed [8].

$Pb(CH_3)_4$ reacts with Br_2 in diethyl ether, ethyl acetate, CCl_4, $CHCl_3$, or another inert solvent or in inert solvent mixtures at -75 to $-65°C$ to form $(CH_3)_3PbBr$, in high yield, and CH_3Br [3, 22, 30]. A second Pb-C bond is cleaved by excess Br_2 when the reaction mixture is allowed to warm to room temperature [3]. A solution of $Pb(CH_3)_4$, to which Br_2 was added at low temperature as long as discoloration was observed, was treated with wet silver oxide to give $(CH_3)_3PbOH$ [4]. The rate constant for the reaction of $Pb(CH_3)_4$ with Br_2 in methanol at $25°C$ in the presence of NaBr at an ionic strength of 0.2 is $k_2 = 1.1 \times 10^5 \ L \cdot mol^{-1} \cdot s^{-1}$ [19]. Oxidation is assumed to occur via outer-sphere electron transfer and to involve a rate-limiting, one-

136

electron process. In bromination a charge-transfer interaction was inferred to give $[Pb(CH_3)_4]^+[Br_2]^-$ as an intermediate [51]; see also [44]. The substitution is electrophilic [19] and an S_Ei mechanism was proposed [26]. Cleavage of $Pb(CH_3)_4$ by Br_2 in methanol is faster than that of $Pb(C_2H_5)_4$ [19]. Reaction of $Pb(CH_3)_4$ and Br_2 in an inert solvent is used for analytical purposes [5, 7, 9, 14, 20, 27]; see also [48].

Vapor-phase reaction of $Pb(CH_3)_4$ and excess iodine yields predominantly $(CH_3)_2PbI_2$ [11]. According to [1], I_2 reacts with $Pb(CH_3)_4$ in the absence of a solvent at room temperature to give CH_3I and $(CH_3)_3PbI$ as well as some PbI_2. This reaction is used to collect $Pb(CH_3)_4$ from air for analysis [11, 15, 21, 45]. For the same purpose, methanolic [11, 12, 38] and also, though less efficiently, aqueous iodine solutions are used [11, 21]. $Pb(CH_3)_4$ is analogously determined in organic solvents, gasoline [13, 39, 40, 54], and in extracts of aqueous phases [56]. Reaction with I_2 in alcoholic solution was also applied for microdetermination of $Pb(CH_3)_4$ [10] and in a reversed-phase HPLC detector [52, 53]. Reaction of I_2 and $Pb(CH_3)_4$ below $-60\,°C$ in ether solution is used for the preparation of $(CH_3)_3PbI$ [6].

Monoiodination of $Pb(CH_3)_4$ in alkali iodide solutions in different solvents, according to kinetic investigations, is a second-order reaction [16, 17, 24, 28, 31]. The rate constants in the presence of 0.057 M NaI at 25°C (in $L \cdot mol^{-1} \cdot s^{-1}$) are $k = 1.68 \times 10^{-6}$ (in CH_3CN), 1.08×10^{-2} (in CH_3OH), 1.48×10^{-3} (in C_2H_5OH), and 1.14×10^{-3} (in C_3H_7OH-n), and a positive salt effect has been observed [24]. For other kinetic data of the reaction in polar solvents, see [28, 31]; see also [17]. In $(CH_3)_2SO$ the reaction is anomalously slow, probably due to complex formation between iodine and the solvent [28]; see also [33]. Breaking of the lead-carbon bond is assumed to be the rate-determining factor [28]. As activation energy and frequency factor for the iodination of $Pb(CH_3)_4$, values of 12.25 kcal/mol and 4.4×10^9, respectively, are given in [17]. In nonpolar solvents, the rate constants $k = 3.55 \times 10^{-2}$ in CCl_4 and $k = 1.9\ L \cdot mol^{-1} \cdot s^{-1}$ in C_6H_6 have been measured [31]. From these data an activation energy of 10 kcal/mol (in CCl_4 and C_6H_6) has been assumed [33]. The rate of iodination increases when polar solvents are added to CCl_4 [31]. For solvent effects in the iodination of $Pb(CH_3)_4$, and for a listing of calculated second-order rate constants in different solvents, see [33]. From the kinetic data an S_E2 type electrophilic substitution at the saturated carbon atom was assumed for the reaction in polar solvents [24], but also nucleophilic assistance has been proposed [28]. An S_F2-mechanism was proposed to work in nonpolar solvents [31]; see also [18, 26, 44]. Pre-equilibrium formation of charge transfer complexes followed by the rate-limiting iodinolysis was assumed in [44]. The free energies of transfer from CH_3OH and from C_6H_6 to various other solvents of transition states in the iododemetallation of $Pb(CH_3)_4$ were calculated [33, 36]. A dissection of solvent influences into initial-state and transition-state influences has been accomplished in [33].

Cleavage of $Pb(CH_3)_4$ by I_2 in polar solvents is faster than that of $Pb(C_2H_5)_4$ [19, 24, 28], and that of $Pb(n-C_3H_7)_4$ [24]. In nonpolar solvents such as CCl_4 and C_6H_6, a reverse order applies: $k(Pb(CH_3)_4) < k(Pb(C_2H_5)_4)$ [31]; see also [33]. The relative reactivities of $Pb(CH_3)_4$ and $Pb(C_2H_5)_4$ are due to transition state solvent effects [24, 26, 33]; see also [18]. A comparison with appropriate iodination reactions of organotin compounds shows close analogy [19, 28]. Reaction of aqueous iodine solution with $Pb(CH_3)_4$ was reported to be considerably slower than with $Pb(C_2H_5)_4$ [11].

Reaction of $Pb(CH_3)_4$ with ICl in aqueous HCl solution gives dimethyllead compounds [23]. This reaction is employed for sampling $Pb(CH_3)_4$ from air and subsequent determination [23, 25, 37, 46], for analysis of $Pb(CH_3)_4$ in gasoline [29, 34, 41, 42, 43, 55], and for removal of $Pb(CH_3)_4$ from gasoline and other petroleum products [32].

References:

[1] Cahours, A. (Ann. Chim. Phys. [3] **62** [1861] 257/350).
[2] Grüttner, G., Krause, E. (Ber. Deut. Chem. Ges. **49** [1916] 1125/33).
[3] Grüttner, G., Krause, E. (Ber. Deut. Chem. Ges. **49** [1916] 1415/28).
[4] Krause, E., Pohland, E. (Ber. Deut. Chem. Ges. **55** [1922] 1282/9).
[5] Ipatiew, W. N., Rasuwajew, G. A., Bogdanow, I. F. (J. Russ. Phys. Chem. Soc. **61** [1929] 7791/9; Ber. Deut. Chem. Ges. **63** [1930] 335/42).
[6] Calingaert, G., Soroos, H. (J. Org. Chem. **2** [1938] 535/9).
[7] Feldman, M. H., Ricci, J. E., Burton, M. (J. Chem. Phys. **10** [1942] 618/23).
[8] Hurd, D. T., Rochow, E. G. (J. Am. Chem. Soc. **67** [1945] 1057/9).
[9] Gilman, H., Jones, R. G. (J. Am. Chem. Soc. **72** [1950] 1760/1).
[10] Barbieri, R., Belluco, U., Tagliavini, G. (Ric. Sci. **30** [1960] 1671/4).

[11] Snyder, L. J., Henderson, S. R. (Anal. Chem. **33** [1961] 1175/80).
[12] Parker, W. W., Smith, G. Z., Hudson, R. L. (Anal. Chem. **33** [1961] 1170/1).
[13] Henderson, S. R., Snyder, L. J. (Anal. Chem. **33** [1961] 1172/5).
[14] Pedinelli, M. (Chim. Ind. [Milan] **44** [1962] 651/2).
[15] Snyder, L. J., Henderson, S. R., Ethyl Corp. (U.S. 3071446 [1962/63]; C.A. **58** [1963] 6200).
[16] Riccoboni, L., Oleari, L. (Ric. Sci. Rend. A [2] **3** [1963] 1031/2).
[17] Riccoboni, L., Pilloni, G., Plazzogna, G., Bernardin, C. (Ric. Sci. Rend. A [2] **3** [1963] 1231/2).
[18] Gielen, M., Nasielski, J. (J. Organometal. Chem. **1** [1963] 173/90).
[19] Gielen, M., Nasielski, J., Dubois, J. E., Fresnet, P. (Bull. Soc. Chim. Belges **73** [1964] 293/6).
[20] Pilloni, G., Plazzogna, G. (Ric. Sci. Rend. A [2] **4** [1964] 27/32).

[21] Linch, A. L., Davis, R. B., Stalzer, R. F., Anzilotti, W. F. (Am. Ind. Hyg. Assoc. J. **25** [1964] 81/93).
[22] Willemsens, L. C., van der Kerk, G. J. M. (Investigations in the Field of Organolead Chemistry, International Lead Zinc Research Organization, Inc., New York 1965, p. 106).
[23] Moss, R., Browett, E. V. (Autom. Anal. Chem. Technicon Symp., London 1965 [1966], pp. 285/90).
[24] Riccoboni, L., Pilloni, G., Plazzogna, G., Tagliavini, G. (J. Electroanal. Chem. **11** [1966] 340/9).
[25] Moss, R., Browett, E. V. (Analyst **91** [1966] 428/38).
[26] Abraham, M. H., Hill, J. A. (J. Organometal. Chem. **7** [1967] 11/21).
[27] Pilloni, G. (Farmaco [Pavia] Ed. Prat. **22** [1967] 666/76).
[28] Gielen, M., Nasielski, J. (J. Organometal. Chem. **7** [1967] 273/80).
[29] Moss, R., Campbell, K. (J. Inst. Petrol. **53** [1967] 89/93).
[30] Singh, G. (J. Organometal. Chem. **11** [1968] 133/43).

[31] Pilloni, G., Tagliavini, G. (J. Organometal. Chem. **11** [1968] 557/62).
[32] Moss, R., Campbell, K., Griffiths, S. T., Associated Octel Co., Ltd. (Brit. 1126630 [1967/68]; C.A. **69** [1968] No. 98222).
[33] Abraham, M. H. (J. Chem. Soc. Perkin Trans. II **1972** 1343/57).
[34] Campbell, K., Palmer, J. M. (J. Inst. Petrol. **58** [1972] 193/200).
[35] Clinton, N. A., Kochi, J. K. (J. Organometal. Chem. **56** [1973] 243/54).
[36] Abraham, M. H., Grellier, P. L. (J. Chem. Soc. Perkin Trans. II **1975** 1856/63).
[37] Hancock, S., Slater, A. (Analyst **100** [1975] 422/9).
[38] Harrison, R. M. (J. Environ. Sci. Health A **11** [1976] 417/23).

138

[39] Zelaskowski, C. A., Mobil Oil Corp. (U.S. 3934976 [1974/76]; C.A. **84** [1976] No. 167189).

[40] Zelaskowski, C. A., Carlisi, J. J., Mobil Oil Corp. (U.S. 3955927 [1973/76]; C.A. **85** [1976] No. 145572).

[41] Snyder, L. J., Ethyl Corp. (U.S. 3912454 [1974/75]; C.A. **84** [1976] No. 62223).

[42] Anonymous (1979 Annual Book of ASTM Standards, Vol. 25, American Society for Testing and Materials, Philadelphia 1979, ANSI/ASTM D 3116-72, pp. 31/5).

[43] Anonymous (1979 Annual Book of ASTM Standards, Vol. 25, American Society for Testing and Materials, Philadelphia 1979, ANSI/ASTM D 3341-79, pp. 236/9).

[44] Fukuzumi, S., Kochi, J. K. (J. Am. Chem. Soc. **102** [1980] 2141/52).

[45] Birnie, S. E., Noden, F. G. (Analyst **105** [1980] 110/8).

[46] Birch, J., Harrison, R. M., Laxen, D. P. H. (Sci. Total Environ. **14** [1980] 31/42).

[47] Kikuchi, M., Lee, F. S. C., Rowland, F. S. (J. Phys. Chem. **85** [1981] 84/8).

[48] Brondi, M., Dall'Aglio, M., Ghiara, E., Mignuzzi, C., Tiravanti, G. (Sci. Total Environ. **19** [1981] 21/31).

[49] Forsyth, D. S., Marshall, W. D. (Anal. Chem. **55** [1983] 2132/7).

[50] Iyer, R. S., Rogers, P. J., Rowland, F. S. (J. Phys. Chem. **87** [1983] 3799/3801).

[51] Walther, B. W., Williams, F., Lau, W., Kochi, J. K. (Organometallics **2** [1983] 688/90).

[52] Blaszkiewicz, M., Baumhoer, G., Neidhart, B. (Heavy Metals Environ. 4th Intern. Conf., Heidelberg 1983, Vol. 1, pp. 99/102; C.A. **101** [1984] No. 96726).

[53] Blaszkiewicz, M., Neidhart, B. (Intern. J. Environ. Anal. Chem. **14** [1983] 11/21).

[54] Aneva, Z., Iancheva, M. (Anal. Chim. Acta **167** [1985] 371/4).

[55] Ivovic, B., Milosevic, Z. (Nafta [Zagreb] **36** [1985] 55/9; C.A. **103** [1985] No. 73505).

[56] Aneva, Z. (Z. Anal. Chem. **321** [1985] 680/1).

1.1.1.1.4.6 With Nonmetal Compounds

$Pb(CH_3)_4$ is hydrolyzed when heated with H_2O [5, 6], whereas at room temperature no spontaneous reactivity with water was established [12]. Later, however, it was found that liquid $Pb(CH_3)_4$ in contact with water accumulates breakdown products. Trimethyllead ions, dimethyllead ions and Pb^{2+}, and water-soluble carbonates and hydroxides are leached and accumulate in the water phase [92]. In sea water, $Pb(CH_3)_4$ undergoes progressive dealkylation [92, 95]; see also [108, 114]. Breakdown of $Pb(CH_3)_4$ in fresh water and in sea water solution takes place at rates which give half-lives measurable in days. The decomposition rate of $Pb(CH_3)_4$ is less than that of $Pb(C_2H_5)_4$ [92]. It is also reported that $Pb(CH_3)_4$ is hydrolyzed to give $(CH_3)_3PbOH$ or $(CH_3)_2Pb(OH)_2$ in the presence of active carbon, coke, or pulverized coal [72].

Alcoholysis of $Pb(CH_3)_4$ by methanol and ethanol has been reported to occur at about 100°C and to give $(CH_3)_3PbOR$ compounds [32]. The reaction, however, could not be reproduced [36] with methanol at 90°C or with butanol at 125 to 130°C, neither with nor in the absence of catalysts [37]. Methanolysis of $Pb(CH_3)_4$ is also undetectable according to rate studies [85].

$Pb(CH_3)_4$ and hydrochloric acid react on slight warming to give $(CH_3)_3PbCl$ and CH_4 [1, 2]. The same reaction occurs on bubbling HCl through a solution of $Pb(CH_3)_4$ in hexane at room temperature [8] or when starting with an HCl-saturated solution at 0°C [67]. Prolonged action of HCl gives $(CH_3)_2PbCl_2$ [8]. When anhydrous HCl is bubbled for 1 h into a boiling solution of $Pb(CH_3)_4$ in THF, a 95% yield of $(CH_3)_2PbCl_2$ is obtained [38]. Reaction of $Pb(CH_3)_4$ with a saturated solution of HCl in ether gives pure $(CH_3)_3PbCl$ [11]. This procedure is used to synthesize ^{14}C-labelled $(CH_3)_3PbCl$ from ^{14}C-labelled $Pb(CH_3)_4$ [115]. The reaction with a solution of HCl in $HOCH_2CH_2OC_2H_5$ is employed to collect $Pb(CH_3)_4$ and other tetraalkyllead compounds from air [33]. Reaction with concentrated hydrochloric acid is used to extract $Pb(CH_3)_4$ from organic solvents and gasoline into the aqueous phase from which lead is determined quantitatively [73, 78, 86, 87, 88, 93, 97, 103, 104, 105].

The reaction of $Pb(CH_3)_4$ and of other tetraorganolead compounds with HCl proceeds in two steps:

$$Pb(CH_3)_4 + HCl \xrightarrow{k_1} (CH_3)_3PbCl + CH_3Cl \qquad (1)$$

$$(CH_3)_3PbCl + HCl \xrightarrow{k_2} (CH_3)_2PbCl_2 + CH_3Cl \qquad (2)$$

Both steps are of second-order; in CH_3OH and in a CH_3OH-C_6H_6 mixture, a competitive consecutive reaction was inferred [39, 57]. From rate measurements in CH_3OH solution with PbR_4 compounds for the first reaction step, the reactivity series C_4H_9-n $\cong$ C_3H_7-n $<$ C_2H_5 $<$ $CH_3 \ll C_6H_5 < OC_6H_4CH_3$-4 was derived [39, 50, 57]. This order was correlated with solvation effects. For the rate of cleavage of the second Pb-C bond the following series was established: $CH_3 \leq C_2H_5 < C_3H_7$-n $> C_4H_9$-n. Rate constants of the reaction steps (1) and (2) were calculated using the time-ratio-method. The measured reaction rates (in $L \cdot mol^{-1} \cdot min^{-1}$) in methanol at 50°C are $k_1 = 1.13$, $k_2 = 0.19 \times 10^{-2}$ [50, 57], and in dioxane at 30°C, $k_1 = 2.66$, $k_2 = 1.94 \times 10^{-3}$. The corresponding activation energies in methanol have been estimated as $E_a(1) = 14.6$ kcal/mol, and in dioxane as $E_a(1) = 12.9$ kcal/mol, and $E_a(2) = 16.7$ kcal/mol [50]. Reaction (1) is assumed to follow an S_E2 mechanism in polar solvents and an S_Ei mechanism in nonpolar solvents [39, 50]. The monosubstituted product is more strongly solvated causing a higher tendency to an S_E2 mechanism in reaction (2) [50]; even an S_N2 initiation had been discussed [39]. The reaction of $Pb(CH_3)_4$ with HCl in methanol is faster than the analogous reaction of $Sn(CH_3)_4$ [50, 57].

$Pb(CH_3)_4$ and hydrobromic acid react on slight warming to give $(CH_3)_3PbBr$ and CH_4 [1, 2]. Cleavage of $Pb(CH_3)_4$ by HBr in methanol and in dioxane is faster than by HCl. For mechanisms, rate constants, and activation energies, see [50].

Perchloric acid cleaves $Pb(CH_3)_4$ in acetic acid to give $(CH_3)_3PbClO_4$. From kinetic data an electrophilic substitution at the carbon atom is proposed as the reaction mechanism. The reaction rate at 25°C of equimolar solutions (0.01 molar) in acetic acid is 0.22 ± 0.01 $L \cdot mol^{-1} \cdot s^{-1}$. It is appreciably higher than that of cleavage of other tetraalkyllead compounds by $HClO_4$ [21].

$Pb(CH_3)_4$ reacts with nitric acid to give $(CH_3)_3PbNO_3$ [113]. This reaction is employed for analytical purposes [110, 113]. Similarly, mixtures of nitric acid with H_2O_2 [106], with bromine [107], and with HCl and $HClO_4$ [96] are used for degradation of $Pb(CH_3)_4$ prior to analysis.

The reaction of $Pb(CH_3)_4$ with HCOOH, CH_3COOH [25], $ClCH_2COOH$, or iso-valerianic acid with silica gel as a catalyst gives the appropriate trimethyllead carboxylate [11]. Refluxing a mixture of $Pb(CH_3)_4$ and excess glacial acetic acid for 5 min produces $(CH_3)_3PbOOCCH_3$ in a yield of 84% [8]. In ether solution, with $ClCH_2COOH$ (in the presence of a piece of porous tile) and with Cl_3CCOOH on slight warming, $(CH_3)_3PbOOCCH_2Cl$ and $(CH_3)_3PbOOCCCl_3$, respectively, are obtained [11]. Heating $Pb(CH_3)_4$ and glacial acetic acid for 16 h to 250 to 260°C gives $Pb(OOCCH_3)_2$, CH_4, and CH_3COOCH_3 [4]. Excess acetic acid cleaves a second Pb-C bond [25]. The rate of the first step of acetolysis (in s^{-1}) measured at 24.9, 49.8, and 60.0°C is

$1.16 \pm 0.03 \times 10^{-5}$, $16.4 \pm 0.2 \times 10^{-5}$, and $41.2 \pm 1.3 \times 10^{-5}$, respectively. ΔH^* and ΔS^* were determined as 20.8 kcal/mol and -12 e.u., respectively [21]. The mechanism of the acetolysis of $Pb(CH_3)_4$ was concluded to be an electrophilic substitution at the carbon atom [21], and an S_Ei [40] or an S_E2i mechanism with a cyclic transition state was proposed [74, 76]. From rates of acetolysis of R_4Pb, the reactivity series: $CH_3 > C_2H_5 \gg C_3H_7\text{-n} < C_4H_9\text{-n} \cong C_5H_{11}\text{-i}$ was found [21]; see also [50, 74]. The rate constants $k_1 = 2.45 \times 10^{-2}$ L·mol^{-1}·min^{-1} and $k_2 = 0.563 \times 10^{-2}$ L·mol^{-1}·min^{-1} of the two steps of the competitive consecutive reactions of $Pb(CH_3)_4$ and CD_3COOD have been evaluated from reactant concentrations determined by ^{1}H NMR spectroscopy [58]. For rate constants of the competitive acetolysis of mixtures of $Pb(CH_3)_4$ and $Pb(C_2H_5)_4$, see [74, 76] and also [89]. The rate of cleavage of Pb-C bonds in $Pb(CH_3)_4$, and in other tetraorganolead compounds $(CH_3)_nPb(C_2H_5)_{4-n}$ (n = 0 to 4), correlates with the ionization potential [74]. Cu^I compounds catalyze the acetolysis of $Pb(CH_3)_4$ [74, 75]. Reaction of $Pb(CH_3)_4$ with neat $(CH_3)_3CCOOH$ or in a solvent is slow even on warming and essentially $Pb(OOCC(CH_3)_3)_2$ is produced [41]. Heating $Pb(CH_3)_4$ and cyanoacetic acid in butanol to 100 to 110°C gives $(CH_3)_3PbOOCCH_2CN$, but only in a yield of 37%, whereas at higher temperatures mainly elemental lead is obtained [51]. $Pb(CH_3)_4$ and trifluoroacetic acid react in toluene in the 1:1 mole ratio at room temperature to give $(CH_3)_3PbOOCCF_3$; with excess acid and heating in the presence of silica gel, $(CH_3)_2Pb(OOCCF_3)_2$ is produced [34, 41]. From kinetic measurements of the acidolysis by CF_3COOH and CCl_3COOH a second-order competitive consecutive reaction was inferred. Reaction steps (1) and (2) have been measured. The reaction rate increases in the series of solvents: dioxane, acetone, $CHCl_3$, and C_6H_6. With CCl_3COOH, little influence on the dependency of rate from solvent is observed [50]. For both reactions an S_Ei mechanism was proposed. For reaction rates and activation energies, see [50].

Reaction of $Pb(CH_3)_4$ and sulfuric acid was reported to occur on slight warming [1, 2]. p-Toluene sulfonic acid and $Pb(CH_3)_4$ react in the presence of silica gel as a catalyst to give the monosulfonate [11]. CF_3SO_3H cleaves $Pb(CH_3)_4$ in acetic acid with a half-life < 10 s to give $(CH_3)_3PbO_3SCF_3$ and CH_4 [74]. For a study of the competitive protonolysis of an excess of $Pb(CH_3)_4$ and $Pb(C_2H_5)_4$ with CF_3SO_3H in acetic acid solution, see [74]. Reaction of a mixture of $Pb(CH_3)_4$ and $Pb(C_2H_5)_4$ with a strong acidic cation resin for nonaqueous solutions of the RSO_3H-type gives a trialkyllead salt of the resin [22]. SO_2 and $Pb(CH_3)_4$ react in the presence of water to give $(CH_3)_2PbSO_3$ [26, 42]; however, no hydrocarbons have been found during the reaction of sulfurous acid with PbR_4 compounds [68]. $Pb(CH_3)_4$ and SO_3 react in wet 1,2-dichlorethane to give $(CH_3)_2PbSO_4$ in a yield of 55% [43]. Reaction of SO_3 and $Pb(CH_3)_4$ in anhydrous CH_2Cl_2 or 1,2-dichloroethane leads to $(CH_3)_3PbO_3SCH_3$ or $(CH_3)_2Pb(O_3SCH_3)_2$, depending on the mole ratio; traces of water are responsible for the formation of some $((CH_3)_3Pb)_2SO_4$ [43, 52]. SO_2 is inserted into Pb-C bonds of $Pb(CH_3)_4$ to give under anhydrous conditions $(CH_3)_3PbO_2SCH_3$ [44, 52, 59] or $(CH_3)_2Pb(O_2SCH_3)_2$, depending on the reaction conditions [27, 42, 52]. $Pb(CH_3)_4$ and sulfur react in the presence of O_2 in toluene solution at 60°C, ultimately leading to $((CH_3)_3Pb)_2SO_4$ [26]. Reaction of $Pb(CH_3)_4$ with anhydrous SeO_2 is slow and does not yield a defined product [77]. Liquid $Pb(CH_3)_4$ does not react with CO_2 [12]. It reacts with N_2O_4 in $CHCl_3$ solution to yield $[(CH_3)_2Pb((CH_3NO)_2\text{-cis})](NO_3)_2$, while in pentane solution $(CH_3)_3PbNO_3$ and the trans dimer of nitrosomethane are produced [69].

Diborane reacts slowly at room temperature in 1,2-dimethoxyethane with $Pb(CH_3)_4$ to produce methyldiboranes, some $B(CH_3)_3$, and $(CH_3)_3PbH$; the latter decomposes to give elemental Pb, H_2, and $Pb(CH_3)_4$. With excess $Pb(CH_3)_4$, the only methylated product isolated was $B(CH_3)_3$ [45]. No reaction occurs in the absence of a solvent [13]. $(CH_3)_2NBH_2$ and $((CH_3)_2N)_2BH$ do not react with $Pb(CH_3)_4$ in 1,2-dimethoxyethane [45]; see also [70]. $NaBH_4$ does not react with $Pb(CH_3)_4$ in dimethylformamide, formamide, or methyl sulfoxide [116].

Pb(CH$_3$)$_4$ methylates SOCl$_2$ to give (CH$_3$)$_2$SO and (CH$_3$)$_3$PbCl. Incomplete reaction was observed in hexane, whereas yields in boiling C$_6$H$_6$ are nearly quantitative; only little (CH$_3$)$_2$PbCl$_2$ forms [23, 52]. Pb(CH$_3$)$_4$ reacts with SO$_2$Cl$_2$ in anhydrous C$_6$H$_6$ in a 1:1 mole ratio in 5 h at 80°C or in 1 d at room temperature to give CH$_3$SO$_2$Cl and (CH$_3$)$_3$PbCl. If a 1:2 mole ratio is used, then some (CH$_3$)$_2$PbCl$_2$ and appreciable amounts of PbCl$_2$ are also produced [52, 60]. Use of SO$_2$Cl$_2$ is a disadvantage for the decontamination of the apparatus in which Pb(CH$_3$)$_4$ had been handled [60]. Only slow reaction was observed between Pb(CH$_3$)$_4$ and CH$_3$SO$_2$Cl [52]. Reaction of Pb(CH$_3$)$_4$ and Se(SeCN)$_2$ in toluene at room temperature yields (CH$_3$)$_3$PbSeCN [35]. Reaction with TeCl$_4$ in toluene in a 1:1 mole ratio at room temperature gives a 80% yield of (CH$_3$)$_2$TeCl$_2$ and (CH$_3$)$_2$PbCl$_2$. Similarly, (CH$_3$)(C$_6$H$_4$OC$_2$H$_5$-p)TeCl$_2$ and (CH$_3$)$_3$PbCl are obtained from (C$_6$H$_4$OC$_2$H$_5$-p)TeCl$_3$ and Pb(CH$_3$)$_4$ [98]. When Pb(CH$_3$)$_4$ was refluxed for 1 h with excess COCl$_2$ in THF, only low yields of (CH$_3$)$_3$PbCl were obtained [53].

Treatment of Pb(CH$_3$)$_4$ with BF$_3$ in the presence of H$_2$O gives an aqueous solution of (CH$_3$)$_3$PbBF$_4$, which on stirring with a 1:1 mixture of Pb(CH$_3$)$_4$ and Pb(C$_2$H$_5$)$_4$ at room temperature, yields mixed methylethyl lead products [64]. Boron-attached halogen is usually replaceable by methyl on reaction of a boron halogen compound with Pb(CH$_3$)$_4$ in the absence of a solvent. Reaction of BCl$_3$ with an excess of Pb(CH$_3$)$_4$ begins at −78°C and is complete after 1 h at room temperature; the yield of B(CH$_3$)$_3$ is 98%, with (CH$_3$)$_3$PbCl as the only methyllead product [46]. The analogous reaction on warming to 50°C under N$_2$ led to a 90% yield of B(CH$_3$)$_3$ [16]; see also [90]. Reaction of BCl$_3$ · N(CH$_3$)$_3$ and Pb(CH$_3$)$_4$ is, however, negligible. B$_2$Cl$_4$, B$_4$Cl$_4$, Cl$_2$BCH=CHBCl$_2$, and (BCl$_2$)$_2$CHCH(BCl$_2$)$_2$ react at 110°C with Pb(CH$_3$)$_4$ to produce also B(CH$_3$)$_3$ and − except in the reaction with B$_2$Cl$_4$ − other methylated borane derivatives [46]. In later work it was reported that B$_2$Cl$_4$ is methylated to form CH$_3$B$_2$Cl$_3$ [65]. The degree of methylation of BCl$_3$ and of BBr$_3$ by Pb(CH$_3$)$_4$ depends on the molar ratio of the reactants; with ratios lower than 1:3, mixtures of (CH$_3$)$_n$BX$_{3-n}$ (X = Cl, Br; n = 1, 2, 3) are obtained; but under proper conditions, one of the methylboranes can be produced as the main product [90]. In the adduct B$_2$Cl$_4$ · 2 N(CH$_3$)$_3$, with excess Pb(CH$_3$)$_4$, only one Cl is substituted by a methyl group. Reaction of BF$_3$ and excess Pb(CH$_3$)$_4$ results in a mixture of CH$_3$BF$_2$ and (CH$_3$)$_2$BF, besides (CH$_3$)$_3$PbF [46]. Pb(CH$_3$)$_4$ methylates B$_9$Br$_9$ in BBr$_3$ solution at room temperature in 26 h to give CH$_3$B$_9$Br$_8$, whereas reaction of Pb(CH$_3$)$_4$ and B$_9$Br$_9$ in a 4:1 mole ratio gives immediately, upon warming the solution from −196°C, a dianionic B$_9$ species [100].

Pb(CH$_3$)$_4$ methylates phosphorus(III) and phosphorus(V) halogen compounds [20, 47, 79, 80, 94, 99] as shown in Table 10. AlCl$_3$ may be used as a catalyst [20]. The ease of methylation of phosphorus halogen compounds by Pb(CH$_3$)$_4$ corresponds to the following series of reactivities: PBr$_3$ ≅ CH$_3$PBr$_2$ > PCl$_3$ > CH$_3$PCl$_2$ ≅ C$_6$H$_5$PCl$_2$ > (CH$_3$)$_2$PCl ≅ (C$_6$H$_5$)$_2$PCl > PSCl$_3$ ≅ C$_6$H$_5$PSCl$_2$ ≅ CH$_3$PSCl$_2$ > ClCH$_2$PSCl$_2$ > (CH$_3$)$_2$PSCl > POCl$_3$ ≅ C$_6$H$_5$POCl$_2$ > ClCH$_2$POCl$_2$. The reactivity for transferring organic groups to these phosphorus compounds is represented by the series: Pb(CH$_3$)$_4$ ≅ Pb(C$_2$H$_5$)$_4$ > Pb(C$_4$H$_9$)$_4$ > Pb(C$_6$H$_5$)$_4$. CH$_3$AsBr$_2$ is methylated in a strongly exothermic reaction to give a 90% yield of (CH$_3$)$_2$AsBr [20]. When Pb(CH$_3$)$_4$ was passed over finely divided silicon at temperatures from 250 to 400°C, no reaction was observed; but with a mixture of Pb(CH$_3$)$_4$ and Cl$_2$, silicon reacted at 375°C to form methylchlorosilanes [9].

Methyl exchange occurs between Pb(CH$_3$)$_4$ and CH$_3$I enriched with ^{14}C in the vapor phase; but in the studied temperature range of 90 to 260°C, formation of PbI$_2$ was also observed [14]. Pb(CH$_3$)$_4$ reacts under the influence of UV light, or on direct heating with CF$_3$I or C$_2$F$_5$I, to yield (CH$_3$)$_3$PbCF$_3$ or (CH$_3$)$_3$PbC$_2$F$_5$, respectively [15, 17, 18]; see also [24]. CF$_3$ radicals produced in a radio-frequency discharge of C$_2$F$_6$ with Pb(CH$_3$)$_4$ give (CH$_3$)$_3$PbCF$_3$, (CH$_3$)$_2$Pb(CF$_3$)$_2$, and CH$_3$Pb(CF$_3$)$_3$ in yields of 1.9, 0.4, and 0.2%, respectively [111]. Slow reaction was reported to occur between Pb(CH$_3$)$_4$ and CCl$_4$ at room temperature [48]; whereas according to [53], no

Table 10
Reaction of $Pb(CH_3)_4$ with Phosphorus(III) and Phosphorus(V) Compounds in Toluene Solution in a Sealed Tube (Nos. 1, 3, 5, 6, 9 to 17) or in an Open Vessel (Nos. 2, 4, 7, 8).

No.	reactant	conditions	product (yield in %) and remarks	Ref.
1	PCl_3	1:2.1 mole ratio, 122°C/4 h and 125°C/20 h	CH_3PCl_2 (86 and 100)	[20, 47]
2	PBr_3	1:5 mole ratio, reflux/2 h	CH_3PBr_2 (73)	[20]
3	CH_3PCl_2	1:2.2 mole ratio, 125°C/4 h and 125°C/20 h	$(CH_3)_2PCl$ (12 and less)	[20]
4	$C_6H_5PCl_2$	1:3 mole ratio, 100°C/2 h, $AlCl_3$ as catalyst	$(C_6H_5)(CH_3)PCl$ (61)	[20]
5	CH_3PBr_2	1:2.5 mole ratio, 22°C/1 h	$(CH_3)_2PBr$ (13)	[20]
6	$C_6H_5P(O)Cl_2$	1:2.3 mole ratio, 125°C/200 h	$(C_6H_5)(CH_3)P(O)Cl$ (19)	[20]
7	$P(S)Cl_3$	1:3 mole ratio, reflux/37 h	$CH_3P(S)Cl_2$ (18)	[20]
8	like No. 7	1:3 mole ratio, reflux/15 h, $AlCl_3$ as catalyst	$CH_3P(S)Cl_2$ (79), $(CH_3)_2P(S)Cl$ (21)	[20]
9	$CH_3P(S)Cl_2$	1:2.5 mole ratio, 125°C/200 h	$(CH_3)_2P(S)Cl$ (81)	[20]
10	$(CH_3)_2P(S)Cl$	like No. 9	$(CH_3)_3PS$ (12)	[20]
11	$C_6H_5P(S)Cl_2$	1:2.3 mole ratio, 125°C/120 h	$(C_6H_5)(CH_3)P(S)Cl$ (90)	[20]
12	$(CF_3)_3PF_2$	reaction on warming to room temperature/2 d	$(CH_3)(CF_3)_3PF$ (82)	[79, 94]
13	$(CF_3)_3PCl_2$	reaction on warming to room temperature/1.5 h	$(CH_3)(CF_3)_3PCl$ (69)	[79, 94]
14	like No. 13	room temperature	$P(CH_3)_2(CF_3)_3$ (75)	[80]
15	$(CF_3)_2PCl_3$	room temperature	$P(CH_3)_3(CF_3)_2$ (44)	[80]
16	$(CH_3)(CF_3)_3PF$	room temperature	$P(CH_3)_2(CF_3)_3$	[80]
17	CF_3PCl_4	room temperature/several days	$(CH_3)(CF_3)PCl_3$ (30), potentially explosive reaction	[99]

reaction is observed during refluxing a solution of $Pb(CH_3)_4$ in CCl_4 for 72 h. $Pb(CH_3)_4$ and CBr_4 react in CCl_4 in the presence of air to give $(CH_3)_2PbBr_2$ and $COBr_2$ [7]. Later it was found that the same reactants produce, on refluxing in toluene for 8 h, a 60% yield of $(CH_3)_3PbBr$, and on standing at room temperature a 40% yield of $(CH_3)_3PbBr$ [53]. Heating a mixture of $Pb(CH_3)_4$ and 1,2-dibromoethane in the presence of $AlCl_3$ for 3 h at 80°C in an autoclave is described as a method for preparing $(CH_3)_3PbCH_2CH_2Pb(CH_3)_2(CH_2CH_2Br)$. Heating $Pb(CH_3)_4$, $Pb(C_2H_5)_4$, $C_2H_4Br_2$, and some $AlCl_3$ gives a complex mixture of lead alkyls and alkylene-dilead alkyls [10]. $Pb(CH_3)_4$ was reported to be only incompletely decomposed when heated with $CH_3OC(O)Cl$ to 200°C [3], and to give only traces of $(CH_3)_3PbCl$ when reacted with trichloropicrate during 8 h at 50°C in toluene [53].

$Pb(CH_3)_4$ reacts with equimolar amounts of trichloroisocyanuric acid at room temperature in acetonitrile-toluene solution to give $(CH_3)_3Pb(N_3C_3O_3Cl_2)$; whereas in acetonitrile at 80°C using a 1:2 mole ratio, the disubstituted product $((CH_3)_3Pb)_2(N_3C_3O_3Cl)$ is formed. Similarly, substitution of Cl by CH_3 is achieved on reaction with $Pb(CH_3)_4$ in 1,3-dichloroparabanic acid at 25°C in acetonitrile, in 1,3-dichlorohydantoin at 40°C in C_2Cl_4, and in N-chloroacetamide at 25°C in CH_2Cl_2 to give $((CH_3)_3Pb)_2(N_2C_3O_3)$, $(CH_3)_3Pb(N_2C_3H_2O_2Cl)$, and $(CH_3)_3Pb(NC_2H_4O)$, respectively [81]. With N-chlorosuccinimide, no analogous reaction occurred [81]; whereas on stirring $Pb(CH_3)_4$ with N-bromosuccinimide in toluene at room temperature, N-(tri-methyllead)succinimide and CH_3Br are produced in an exothermic reaction in a 98% yield [53, 54, 61]. No methylation of 1-chloro-1,2,3-benzotriazole by $Pb(CH_3)_4$ in refluxing C_6H_6 was observed [71], and no reaction occurred with dithizone [19], or with N-acetylglycine in toluene-methanol solution on refluxing [112]. Heating $Pb(CH_3)_4$ with phenol gives $(CH_3)_2Pb(OC_6H_5)_2$ [28]. Interaction of $Pb(CH_3)_4$ with excess of CS_2 in dimethylformamide at 120°C produced only PbS after 7 h in a yield of < 6%; no other products were identified [52, 62].

$Pb(CH_3)_4$ proved to be a methylating agent for obtaining methyl crotonate from methyl acrylate, and for preparing trans-1-phenyl-1-propene from styrene in the presence of Li_2PdCl_4 in methanol [29, 30, 31, 49]. Charge-transfer bands are observed in the visible region imme-diately upon mixing $Pb(CH_3)_4$ and tetracyanoethene (TCNE) in $CHCl_3$, CH_2Cl_2, or 1,2-di-chloropropane solution [82, 91], and formation of a 1:1 electron donor-acceptor complex $(Pb(CH_3)_4^+ \cdot TCNE^-)$ is inferred [82, 91, 109]. Subsequent to the rate-limiting electron transfer step, TCNE inserts into one Pb-C bond to give $(CH_3)_3Pb(NC)_2CC(CN)_2CH_3$. The rate-limiting electron transfer mechanism was derived from rate and spectroscopic studies and from selectivities observed in analogous reactions of mixed methyl-ethyllead compounds [83, 91, 101]; see also [91, 102]. The charge transfer energies of $Pb(CH_3)_4$ and of other tetraalkyllead compounds of TCNE and the rates of TCNE insertion were correlated with vertical ionization potentials [83, 109]. The insertion of TCNE into $Pb(CH_3)_4$ and into other alkylmetals was quantitatively compared with the Diels-Alder cycloaddition of anthracene to TCNE. For both systems, the activation free energies were found to be equal to the energetics of ion-pair formation, which were evaluated from the charge transfer energies. Differences in the rates of insertion and cycloaddition arise from differences in the ion-pair solvation ΔG^s. For $Pb(CH_3)_4$ a value of $\Delta G^s = -3.28$ eV was calculated. Further support for the charge-transfer formulation of the activation process comes from the fact that the same differences in ΔG^s apply quantitatively to the free ions, e.g., $[Pb(CH_3)_4]^+$, independently derived from electrochemical and chemical oxidations (using Fe^{3+}) by outer-sphere electron transfer [109]. For fluorescence quenching of anthracene by $Pb(CH_3)_4$, see [63, 66].

The Arrhenius parameters for the reaction of methyl radicals with $M(CH_3)_4$ (M = Si, Ge, Sn, Pb) according to the equation:

$$CH_3^\bullet + (CH_3)_3MCH_3 \xrightarrow{k} CH_4 + (CH_3)_3MCH_2^\bullet$$

have been determined. For $Pb(CH_3)_4$ the values log A = 10.2 ± 0.48 mL · mol^{-1} · s^{-1}, E_a = 7.4 ± 0.88 kcal/mol, and log k = 6.25 mL · mol^{-1} · s^{-1} (at 140°C) were given. A direct correla-tion of E_a and k with the coupling constants $J(^{13}C, H)$ were found to exist, but no correlation of these parameters with any of the electronegativity values for group 14 elements was noted [55].

No reaction was observed when NH_4Cl and $Pb(CH_3)_4$ in a 1:1 mole ratio were refluxed in C_6H_6 [53].

144

Solvation of $Pb(CH_3)_4$ was inferred from NMR studies in various solvents. The power of the solvent to solvate $Pb(CH_3)_4$ was indicated to increase along the series: C_6H_{12} < 1,2-dimethoxyethane $\cong$ dioxane $\cong$ hexamethylphosphortriamide < pyridine < tetrahydrothiophene < triethylamine < THF < triethylphosphine < N,N,N',N'-tetramethylethylenediamine $\cong$ acetone < dimethylformamide < dimethyl sulfoxide. Formation of complexes with trigonal bipyramidal geometry was assumed [56].

References:

[1] Cahours, A. (Ann. Chim. [Paris] [3] 62 [1861] 257/350).
[2] Cahours, A. (Liebigs Ann. Chem. 122 [1862] 48/71).
[3] Butlerow, A. (Jahresber. Fortschr. Chem. 1863 474/6).
[4] Jones, L. W., Werner, L. (J. Am. Chem. Soc. 40 [1918] 1257/75).
[5] Ipatiew, W. N., Rasuwajew, G. A., Bogdanow, I. F. (J. Russ. Phys. Chem. Soc. 61 [1929] 1791/9; Ber. Deut. Chem. Ges. 63 [1930] 335/42).
[6] Ipatieff, V. N. (Catalytic Reactions at High Pressures and Temperatures, Macmillan, New York 1936, p. 358).
[7] Hein, F., Nebe, E., Reimann, W. (Z. Anorg. Allgem. Chem. 251 [1943] 125/60).
[8] Calingaert, G., Dykstra, F. J., Shapiro, H. (J. Am. Chem. Soc. 67 [1945] 190/2).
[9] Hurd, D. T., Rochow, E. G. (J. Am. Chem. Soc. 67 [1945] 1057/9).
[10] Wiczer, S. B. (U.S. 2447926 [1948]; C.A. 1948 7975/6).

[11] Heap, R., Saunders, B. C. (J. Chem. Soc. 1949 2983/8).
[12] Sanderson, R. T. (J. Am. Chem. Soc. 77 [1955] 4531/2).
[13] Holliday, A. K., Jeffers, W. (J. Inorg. Nucl. Chem. 6 [1958] 134/7).
[14] Tagliavini, G., Belluco, U., Schiavon, G., Riccoboni, L. (Ric. Sci. 28 [1958] 2349/54).
[15] Kaesz, H. D., Phillips, J. R., Stone, F. G. A. (Chem. Ind. [London] 1959 1409/10).
[16] Stone, L. S., Callery Chemical Co. (U.S. 2923740 [1960]; C.A. 1960 No. 15247).
[17] Kaesz, H. D., Phillips, J. R., Stone, F. G. A. (Abstr. Papers 137th Meeting Am. Chem. Soc., Cleveland, Ohio, 1960, No. 93, pp. 37M/38M).
[18] Kaesz, H. D., Phillips, J. R., Stone, F. G. A. (J. Am. Chem. Soc. 82 [1960] 6228/32).
[19] Van Rysselberge, J., Leysen, R. (Nature 189 [1961] 478).
[20] Maier, L. (J. Inorg. Nucl. Chem. 24 [1962] 1073/81).

[21] Robinson, G. C. (J. Org. Chem. 28 [1963] 843/5).
[22] Imura, S., Tamai, Y., Echiru Kagaku Kogyo K. K. (U.S. 3400142 [1964/68]; C.A. 70 [1969] No. 20226).
[23] Gelius, R. (Z. Anorg. Allgem. Chem. 334 [1964/65] 72/80).
[24] Treichel, P. M., Stone, F. G. A. (Advan. Organometal. Chem. 1 [1964] 143/220, 174).
[25] Willemsens, L. C., van der Kerk, G. J. M. (Investigations in the Field of Organolead Chemistry, Schotanus en Jens, Utrecht 1965, pp. 66/8, 76/7, 107).
[26] Padberg, F.-J. (Diss. Aachen T. H. 1965).
[27] Huber, F. (Angew. Chem. 77 [1965] 1084/5).
[28] Matwiyoff, N. A., Drago, R. S. (J. Organometal. Chem. 3 [1965] 393/9).
[29] Heck, R. F., Hercules Powder Co. (U.S. 3527794 [1965/70]; C.A. 73 [1970] No. 109514).
[30] Heck, R. F., Hercules Inc. (U.S. 3700727 [1965/72]; C.A. 78 [1973] No. 71699).

[31] Heck, R. F., Hercules Inc. (U.S. 3783140 [1965/74]; C.A. 80 [1974] No. 145810).
[32] Pedinelli, M., Magri, R., Randi, M. (Chim. Ind. [Milan] 48 [1966] 144).
[33] Zuliani, G., Perin, G., Rausa, G. (Med. Lavoro 57 [1966] 771/80).

[34] Horn, H. (Diss. Aachen T. H. 1966).

[35] Aynsley, E. E., Greenwood, N. N., Hunter, G., Sprague, M. J. (J. Chem. Soc. A **1966** 1344/7).

[36] Anonymous (International Lead Zinc Research Organization, Inc., ILZRO Research Digest **20** [1967] Pt. V, 3).

[37] Willemsens, L. C. (International Lead Zinc Research Organization, Ltd., Project LC-18, Progress Report No. 32 [1967] 21).

[38] Honeycutt Jr., J. B., Ethyl Corp. (U.S. 3439012 [1967/69]; C.A. **71** [1969] No. 13213).

[39] Huber, F., Horn, H., Bade, V. (Angew. Chem. **79** [1967] 996).

[40] Abraham, M. H., Hill, J. A. (J. Organometal. Chem. **7** [1967] 11/21).

[41] Huber, F., Horn, H., Haupt, H. J. (Z. Naturforsch. **22b** [1967] 918/21).

[42] Huber, F., Padberg, F.-J. (Z. Anorg. Allgem. Chem. **351** [1967] 1/8).

[43] Gelius, R., Müller, R. (Z. Anorg. Allgem. Chem. **351** [1967] 42/7).

[44] Gelius, R. (Z. Anorg. Allgem. Chem. **349** [1967] 22/32).

[45] Holliday, A. K., Jessop, G. N. (J. Organometal. Chem. **10** [1967] 291/3).

[46] Holliday, A. K., Jessop, G. N. (J. Chem. Soc. A **1967** 889/91).

[47] Burg, A. (Studies on Boron Hydrides, Univ. of Southern California, 9th Rept. to U.S. Office of Navel Research, Nov. 1st, 1955 in: Shapiro, H., Frey, F. W., The Organic Compounds of Lead, Interscience/Wiley, New York 1968, p. 83, Ref. 86).

[48] Ethyl Corporation, unpublished data (in: Shapiro, H., Frey, F. W., The Organic Compounds of Lead, Interscience/Wiley, New York 1968, p. 100, Ref. 220).

[49] Heck, R. F. (J. Am. Chem. Soc. **90** [1968] 5518/26).

[50] Bade, V. (Diss. Aachen T. H. 1969).

[51] Willemsens, L. C. (International Lead Zinc Research Organization, Ltd., Project LC-18, Progress Report No. 38 [1969] 16).

[52] Gelius, R. (Habilitationsschr. Univ. Greifswald 1969).

[53] Davidsohn, W. E., Pant, B. C. (International Lead Zinc Research Organization, Ltd., Project LC-28, Report No. 31 [1969] 1/5).

[54] Pant, B. C., Davidsohn, W. E. (J. Organometal. Chem. **19** [1969] P3/P4).

[55] Chaudhry, A. U., Gowenlock, B. G. (J. Organometal. Chem. **16** [1969] 221/6).

[56] Petrosyan, V. S., Voyakin, A. S., Reutov, O. A. (Zh. Org. Khim. **6** [1970] 889/93; J. Org. Chem. [USSR] **6** [1970] 895/8).

[57] Bade, V., Huber, F. (J. Organometal. Chem. **24** [1970] 387/97).

[58] Bade, V., Huber, F. (J. Organometal. Chem. **24** [1970] 691/5).

[59] Fong, C. W., Kitching, W. (J. Organometal. Chem. **21** [1970] 365/75).

[60] Gelius, R. (Z. Anorg. Allgem. Chem. **374** [1970] 297/305).

[61] Pant, B. C. (J. Organometal. Chem. **24** [1970] 697/701).

[62] Gelius, R., Kirbach, E. (Z. Chem. [Leipzig] **10** [1970] 117).

[63] Vander Donckt, E., van Bellinghen, J. P. (Chem. Phys. Letters **7** [1970] 630/2).

[64] Shepard Jr., J. C., Nalco Chemical Co. (U.S. 3725447 [1971/73]; C.A. **78** [1973] No. 159870).

[65] Ritter, J. J. (Diss. Univ. Maryland 1971, cited in [70]).

[66] Vander Donckt, E., van Bellinghen, J.-P. (J. Chim. Phys. **68** [1971] 948/53).

[67] Di Bianca, F., Rivarola, E., Stocco, G. C., Barbieri, R. (Z. Anorg. Allgem. Chem. **387** [1972] 126/33).

[68] Kunze, U., Lindner, E., Koola, J. (J. Organometal. Chem. **38** [1972] 51/68).

[69] Williams, K. C., Imhoff, D. W. (J. Organometal. Chem. **42** [1972] 107/15).

[70] Coyle, T. D., Ritter, J. J. (Advan. Organometal. Chem. **10** [1972] 237/72).

146

[71] Pant, B. C., Noltes, J. G. (J. Organometal. Chem. **36** [1972] 293/6).

[72] Ariga, J., Imura, S., Aoyama, H., Kawamura, H., Tabata, T., Toyo Ethyl Co., Ltd. (Japan. 72-45729 [1969/72]; C.A. **78** [1973] No. 58625).

[73] Anonymous (1973 Annual Book of ASTM Standards, Vol. 17, D 1269-61, American Society for Testing and Materials, Philadelphia 1973, pp. 446/50).

[74] Clinton, N. A., Gardner, H. C., Kochi, J. K. (J. Organometal. Chem. **56** [1973] 227/42).

[75] Clinton, N. A., Kochi, J. K. (J. Organometal. Chem. **56** [1973] 243/54).

[76] Kochi, J. K. (Accounts Chem. Res. **7** [1974] 351/60).

[77] Menge, R. (Diss. Univ. Dortmund 1974).

[78] Anonymous (1975 Annual Book of ASTM Standards, Vol. 17, D 526-70, American Society for Testing and Materials, Philadelphia 1975, pp. 275/8).

[79] The, K. I., Cavell, R. G. (J. Chem. Soc. Chem. Commun. **1975** 279/80).

[80] The, K. I., Cavell, R. G. (J. Chem. Soc. Chem. Commun. **1975** 716/7).

[81] Leimeister, H., Dehnicke, K. (Z. Anorg. Allgem. Chem. **415** [1975] 115/24).

[82] Gardner, H. C., Kochi, J. K. (J. Am. Chem. Soc. **97** [1975] 1855/65).

[83] Gardner, H. C., Kochi, J. K. (J. Am. Chem. Soc. **97** [1975] 5026/7).

[84] Gardner, H. C. (Diss. Univ. Indiana, Bloomington 1975; Diss. Abstr. Intern. B **36** [1976] 3959).

[85] Arnold, D. P., Wells, P. R. (J. Organometal. Chem. **111** [1976] 269/83).

[86] Madec, M., La Villa, F. (Rev. Inst. Franc. Petrole Ann. Combust. Liquides **31** [1976] 687/701).

[87] Anonymous (IP Standards for Petroleum and Its Products, Pt. 1, Sect. 1 and 2, IP 96/70, Institute of Petroleum, London 1976).

[88] Anonymous (IP Standards for Petroleum and Its Products, Pt. 1, Sect. 1 and 2, IP 248/70, Institute of Petroleum, London 1976).

[89] Nugent, W. A., Kochi, J. K. (J. Am. Chem. Soc. **98** [1976] 5979/88).

[90] Haubold, W., Weidlein, J. (Z. Anorg. Allgem. Chem. **420** [1976] 251/60).

[91] Gardner, H. C., Kochi, J. K. (J. Am. Chem. Soc. **98** [1976] 2460/9).

[92] Grove, J. R. (Lead Marine Environ. Proc. Intern. Experts Discuss., Rovinj, Yugoslavia, 1977 [1980], pp. 45/52).

[93] Anonymous (IP Standards for Petroleum and Its Products, Pt. 1, Sect. 2, IP 270/77, Institute of Petroleum, London 1977).

[94] The, K. I., Cavell, R. G. (Inorg. Chem. **16** [1977] 2887/94).

[95] Anonymous (International Lead Zinc Research Organization, Inc., Lead Research Digest **36** [1978] 46/7).

[96] Beccaria, A. M., Mor, E. D., Poggi, G. (Ann. Chim. [Rome] **68** [1978] 607/17).

[97] Anonymous (1979 Annual Book of ASTM Standards, Vol. 24, D 2547-70, American Society for Testing and Materials, Philadelphia 1979, pp. 453/6).

[98] Kumar, V., Bird, P. H., Pant, B. C. (Syn. React. Inorg. Metal-Org. Chem. **9** [1979] 203/12).

[99] Yap, N. T., Cavell, R. G. (Inorg. Chem. **18** [1979] 1301/5).

[100] Kutz, N. A., Morrison, J. A. (Inorg. Chem. **19** [1980] 3295/9).

[101] Fukuzumi, S., Kochi, J. K. (J. Phys. Chem. **84** [1980] 2246/54).

[102] Kochi, J. K. (Pure Appl. Chem. **52** [1980] 571/605).

[103] Anonymous (Deutsche Norm, Prüfung von Mineralölerzeugnissen; Bestimmung des Bleigehaltes (Gesamtblei) von Ottokraftstoffen mit einem Bleigehalt über 25 mg/l; Direkte Bestimmung durch Atomabsorptionsspektroskopie, DIN 51769 Teil 7, Okt. 1981).

[104] Anonymous (Deutsche Norm, Prüfung von Mineralölerzeugnissen; Bestimmung des Bleigehaltes (Gesamtblei) von Ottokraftstoffen mit einem Bleigehalt von 5 bis 25 mg/l; Direkte Bestimmung durch Atomabsorptionsspektroskopie, DIN 51769 Teil 8, Okt. 1981).

[105] Anonymous (Deutsche Norm, Prüfung von Mineralölerzeugnissen; Bestimmung des Bleigehaltes (Gesamtblei) von Schmierstoffen; Direkte Bestimmung durch Atomabsorptionsspektroskopie, DIN 51769 Teil 10, Okt. 1981).

[106] Jiang, S. G., Chakraborti, D., De Jonghe, W., Adams, F. (Z. Anal. Chem. **305** [1981] 177/80).

[107] Brondi, M., Dall'Aglio, M., Ghiara, E., Mignuzzi, C., Tiravanti, G. (Sci. Total Environ. **19** [1981] 21/31).

[108] Chakraborti, D., Jiang, S. G., Surkijn, P., De Jonghe, W., Adams, F. (Anal. Proc. [London] **18** [1981] 347/50).

[109] Fukuzumi, S., Kochi, J. K. (Tetrahedron **38** [1982] 1035/49).

[110] Diehl, K.-H., Rosopulo, A., Kreuzer, W. (Z. Anal. Chem. **314** [1983] 755/7).

[111] Guerra, M. A., Armstrong, R. L., Bailey Jr., W. I., Lagow, R. J. (J. Organometal. Chem. **254** [1983] 53/8).

[112] Roge, G., Huber, F., Preut, H., Silvestri, A., Barbieri, R. (J. Chem. Soc. Dalton Trans. **1983** 595/600).

[113] Jennen, A., Delafortrie, A., Verdoodt, D., Jacobs, T., Dourte, P. (Rev. Agric. [Brussels] **37** [1984] 1025/7).

[114] Harrison, R. M., Radojević, M., Wilson, S. J. (Sci. Total Environ. **50** [1986] 129/37).

[115] Blais, J. S., Marshall, W. D. (Appl. Organometal. Chem. **1** [1987] 251/60).

[116] Nerin, C., Olavide, S., Cacho, J. (Anal. Chem. **59** [1987] 1918/21).

1.1.1.1.4.7 With Metals

Reaction of Li and $Pb(CH_3)_4$ in liquid ammonia at $-78\,°C$ causes, in a first step, fission of one Pb-C bond, giving the solvated trimethylplumbide ion $[Pb(CH_3)_3]^-$, and methyl radicals; the latter then yield methane and ethane. Excess lithium cleaves a second methyl group producing $Pb(CH_3)_2$ [4]. From sodium and $Pb(CH_3)_4$ in liquid ammonia [4], or in hexamethylphosphoric acid triamide [7], $[Pb(CH_3)_3]^-$ is analogously produced. Such solutions in liquid ammonia, containing $(CH_3)_3PbNa$ and $NaNH_2$, had been obtained in prior work [2] for synthetic purposes. Reaction with excess sodium also renders transient $Pb(CH_3)_2$ [4].

Reaction of potassium with $Pb(CH_3)_4$ in liquid ammonia at $-78\,°C$ similarly produces at first $[Pb(CH_3)_3]^-$, but further attack by potassium ultimately gives PbNH with evolution of hydrogen, methane, and ethane [4]. Experiments to obtain $Mg(CH_3)_2$ from $Pb(CH_3)_4$ and magnesium remained unsuccesful [6].

Methyl exchange reactions between gaseous $Pb(CH_3)_4$ and ^{210}Bi deposited on a glass wall [5], and between $Pb(CH_3)_4$ in ether solution and ^{210}Bi or ^{212}Pb deposited on metal surfaces have been observed [1, 5].

$Pb(CH_3)_4$ is assumed to be slowly reduced by zinc in the presence of water to give elemental Pb, CH_4, and ZnO [9]. $Pb(CH_3)_4$ poisons platinum-black as a hydrogen catalyst by being adsorbed on the surface. The adsorption is reversible [3]. Poisoning was also observed with a $Pt-Al_2O_3$ catalyst for hydrogenation and H/D exchange of C_6H_6 [8].

References:

[1] Leigh-Smith, A., Richardson, H. O. W. (Nature **135** [1935] 828/9).
[2] Bindschadler, E. (Iowa State Coll. J. Sci. **16** [1941] 33/6).
[3] Maxted, E. B., Moon, K. L. (J. Chem. Soc. **1949** 2171/4).

[4] Holliday, A. K., Pass, G. (J. Chem. Soc. **1958** 3485/8).
[5] Duncan, J. F., Thomas, F. G. (J. Inorg. Nucl. Chem. **29** [1967] 869/90).
[6] Ashby, E. C., Arnott, R. C. (J. Organometal. Chem. **14** [1968] 1/11).
[7] Psarras, T., Sandy, C. A., E. I. du Pont de Nemours and Co., Wilmington (Ger. 1914503 [1969/71]).
[8] Maurel, R., Barbier, J. (J. Chim. Phys. **73** [1976] 995/9).
[9] Hitchen, M. H., Holliday, A. K., Puddephatt, R. J. (J. Organometal. Chem. **172** [1979] 427/44).

1.1.1.1.4.8   With Metal Compounds

A vigorous reaction occurs between $Pb(CH_3)_4$ and $Al(BH_4)_3$ on warming from $-190°C$ to room temperature. On cautious warming, first to $-60°C$, then after removing unreacted $Al(BH_4)_3$ to $-20°C$, a volatile fraction could be separated, which was assumed to contain $(CH_3)_3Pb(BH_4)$ or $(CH_3)_2Pb(BH_4)_2$ and/or $(CH_3)_2Pb(BH_3CH_3)_2$, and which easily decomposes to give elemental lead, methylated boranes, H_2, and $CH_3Al(BH_4)_2$ [26]. $Pb(CH_3)_4$ was reported to react neither with $Li[AlH_4]$ [21] nor with B_2H_6 [26].

$Pb(CH_3)_4$ and $SnCl_4$ react in hexane at ambient temperature to give $(CH_3)_3PbCl$ and $(CH_3)_2PbCl_2$ in a 2:3 ratio; whereas on refluxing for 3 h, the only product is $PbCl_2$ [66]. Methylation of $SnCl_4$ by $Pb(CH_3)_4$ in toluene under reflux is reported to give a 90% yield of $Sn(CH_3)_4$ besides methylchlorostannanes [74]. The same yield was observed when a 1:2 molar mixture of $SnCl_4$ and $Pb(CH_3)_4$ in toluene was stirred for 2 h at room temperature and then heated for 5 h to 100°C. A similar reaction of $SnCl_4$ and $Pb(CH_3)_4$ with a 1:1 mole ratio gives $(CH_3)_3SnCl$ in a yield of about 100% [81]. Also other methylchlorostannanes or methylbromostannanes of the type $(CH_3)_nSnX_{4-n}$ (X = Cl, Br; n = 1, 2, 3) can analogously be prepared by methylation of SnX_4 with $Pb(CH_3)_4$ in toluene on heating to 100°C or under reflux if the proper mole ratio is applied [74,81]. Reaction of $Pb(CH_3)_4$ with $(NH_4)_2PbCl_6$ in C_6H_6 gives excellent yields of $(CH_3)_3PbCl$, which can easily be separated from the other reaction products $PbCl_2$, NH_4Cl, and CH_3Cl [66, 67, 76]. With a large excess of $Pb(CH_3)_4$ only $(CH_3)_3PbCl$ and no $(CH_3)_2PbCl_2$ is obtained [66]. Other hexachloroplumbate(IV) compounds, M_2PbCl_6 (M = K, Rb, Cs), react analogously, but more slowly [76]. Reaction with K_2PbCl_6 in C_6H_6 gave a 91% yield of $(CH_3)_3PbCl$ after 25 h. THF can also be used as solvent for this reaction [66, 67]. Experiments using radioactive tags suggest that lead, deposited presumably as oxide on a metallic surface can exchange with lead in $Pb(CH_3)_4$ in ether solution at room temperature [3].

$Pb(CH_3)_4$ methylates $SbCl_3$ [97, 113] and $SbBr_3$ [113] on refluxing in C_6H_6 under N_2 in 8 to 10 (X = Cl) and 12 h (X = Br) to give CH_3SbCl_2 (yield: 80% [97], 63% [113]) and CH_3SbBr_2 (yield: 55% [113]), respectively, along with the corresponding PbX_2 and CH_3X.

When $Pb(CH_3)_4$ was conducted with a stream of H_2 over ^{224}Ra, methyl exchange with recoil products was observed [69].

$Pb(CH_3)_4$ methylates $TiCl_4$ at low temperatures to give CH_3TiCl_3 and $(CH_3)_3PbCl$ [29, 110]. $Pb(CH_3)_4$ reacts readily via a radical mechanism with $[Fe(phen)_3]^{3+}$ and $[Fe(bipy)_3]^{3+}$ (phen = 1,10-phenanthroline, Formula I; bipy = 2,2'-bipyridine, Formula II) in acetonitrile at 22°C to afford the reduced Fe^{II}-complex, with one ligand being methylated in 4-position. Repeated reaction of the re-oxidized complex $[Fe(4-CH_3-phen)(phen)_2]^{3+}$ with $Pb(CH_3)_4$ leads to additional methylation in the 7-position to give the $[Fe(4,7-(CH_3)_2phen)(phen)_2]^{2+}$ cation. When $[Fe(4,7-(CH_3)_2phen)_3]^{3+}$ is subjected to this reaction, methylation occurs in the 2-position. By contrast, $phenH^+$ is methylated indiscriminately by $Pb(CH_3)_4$ at both the 2- and 4-positions [111]. Electron transfer from $Pb(CH_3)_4$ to Fe^{III} complexes with bipy, phen, and substituted phen ligands proceeds via an outer-sphere mechanism, whereas that to iridate(VI) and tetracyanoethene via inner-sphere processes [103, 104, 105, 107]; see also [106]. The second-order rate constants (in $mol^{-1} \cdot s^{-1}$) for the electron transfer from $Pb(CH_3)_4$ to Fe^{III} complexes $[Fe(^2D\text{-}^2D)_3]^{3+}$ in acetonitrile at 25°C are given for various $^2D\text{-}^2D$: 5.06 for 4,7-diphenyl-phen; 9.09 for bipy; 25.5 for phen; 166 for 5-Cl-phen; and 1.49×10^3 for 5-NO_2-phen [103]; see also [112]. The ionization potential I_D for $Pb(CH_3)_4$ and other alkylmetals are linearly correlated with the rates of electron transfer using phen, 5-Cl-phen, and bipy-complexes of Fe^{III} as oxidants [103, 105]; see also [112]. $Pb(CH_3)_4$ is oxidized by $[IrCl_6]^{2-}$ in acetic acid to give Ir^{III}, CH_3Cl, and $(CH_3)_3PbOOCCH_3$ [82, 83, 86, 87]. The second-order rate constant for the electron transfer from $Pb(CH_3)_4$ to $[IrCl_6]^{2-}$ is given as 0.02 $mol^{-1} \cdot s^{-1}$ at 25°C [103]. From kinetic measurements and comparison with rates of reaction of $(CH_3)_{4-n}Pb(C_2H_5)_n$ (n = 0 to 4) with $[IrCl_6]^{2-}$ in acetic acid and acetonitrile, alkyl transfer by one-electron processes was inferred [82, 83, 86, 87]. The rate of oxidative cleavage of $Pb(CH_3)_4$ and other tetraalkyllead compounds by $[IrCl_6]^{2-}$ correlates linearly with the rates of insertion of tetracyanoethene and with the ionization potentials of the appropriate tetraalkyllead compound [88]; see also [104]. For a linear free energy relationship for electron transfer and a correlation with the Brønsted coefficient α, see [104, 105, 107]; see also [114].

I

II

WO_3 is turned grey-green by traces of $Pb(CH_3)_4$ contained in a stream of hydrogen. Heating in air does not remove the color. The reaction was proposed for qualitative analysis of $Pb(CH_3)_4$ [2]. $Pb(CH_3)_4$ was reported not to react with permanganate [4], but it should be considered that a 0.025% solution of $KMnO_4$ heated to 45 to 60°C has been employed for decontamination of $Pb(C_2H_5)_4$ in transport containers [37].

$Pb(CH_3)_4$ reacts with $Cu(NO_3)_2$ or $Cu(NO_3)_2 \cdot 3 H_2O$ in alcohol in the temperature range between -70 to 30°C to give $(CH_3)_3PbNO_3$ and unstable $CuCH_3$ [18, 19, 22, 24, 30, 31], the latter being formed from intermediate Cu^+ or $CuOCH_3$ and $Pb(CH_3)_4$ [31, 38]. Since $CuCH_3$ reacts with methanol, $CuOCH_3$ is regenerated; finally, these reactions can be interpreted as a cleavage of CH_3OH by $Pb(CH_3)_4$ [31, 38, 53]. $Pb(CH_3)_4$ reacts analogously with $CuSO_4 \cdot 5 H_2O$ [31]. Reaction with $CuCl_2$ [31, 77] and $[CuCl_3]^-$ [87] in alcohol or acetic acid gives CH_3Cl, $(CH_3)_3PbCl$, and $CuCl$ and $[CuCl_2]^-$, respectively. The mechanism of the reaction of $Pb(CH_3)_4$ with $[CuCl_3]^-$ is distinctly different from that with $[IrCl_6]^{2-}$ [87], and it also was established that no methyllead radicals were involved in the reaction [77]. $Pb(CH_3)_4$ reacts faster with $CuCl_2$ than $Pb(C_2H_5)_4$ [77]. For a review of reactions of $Pb(CH_3)_4$ with Cu^{II} salts, see [84]. The reaction of $Pb(CH_3)_4$ and $Cu(OOCCH_3)_2$ was proposed for preparation of $(CH_3)_3PbOOCCH_3$ [54]. Impregnation of silica gel and other adsorbents with $CuCl_2$ is favorable for removing $Pb(CH_3)_4$ from gasoline with such reagents. $CuCl_2$ is more effective in this respect than $FeCl_3$ [89]; see also Section 1.1.1.1.5. CuCl does not undergo reaction with excess $Pb(CH_3)_4$, due to its insolubility [31], but $CuOOCCH_3$ catalyzes acetolysis of $Pb(CH_3)_4$ [78]. Neither CuBr, $CuOOCCH_3$, nor $AlCl_3$ catalyze the alcoholysis of $Pb(CH_3)_4$ [59].

Reaction of $Pb(CH_3)_4$ with $AgNO_3$ in alcoholic solution at room temperature gives metallic Ag and ethane in a 2:1 mole ratio, and the cation $[Pb(CH_3)_3]^+$ [10]. When the metathesis was performed at $-70\,°C$, C_2H_6, some CH_4, and C_2H_4 were evolved and $(CH_3)_3PbNO_3$ was isolated [18]; see also [19]. A yellow solid, obtainable when conducting the reaction at $-30\,°C$ [1], or better at $-80\,°C$, was supposed to be $AgCH_3$ [10, 15, 16] which decomposes to give Ag and methyl radicals [10, 16]. Methyl radicals could be detected in the reaction of $Pb(CH_3)_4$ with $AgNO_3$ by spin trapping [70]. If an increasing excess of $AgNO_3$ was applied in the reaction at low temperature the yellow precipitate approached the formula $AgCH_3 \cdot AgNO_3$ [15]. With an excess of $Pb(CH_3)_4$, a precipitate with the composition $2\,AgCH_3 \cdot AgNO_3$ was obtained [23, 27]; see also [46]. For the analogous reaction of $Pb(CD_3)_4$ with $AgNO_3$, see [32]. Elemental Ag was separated on addition of a 10- to 100-fold excess of Ag^+ to leaded gasoline, and the reaction is used for analytical purposes [55]; see also [117].

$Pb(CH_3)_4$ and $HgCl_2$ react very quickly to give $(CH_3)_3PbCl$. The reaction is faster than the analogous reaction of $Sn(CH_3)_4$ with $HgCl_2$. The second-order rate constant of the reaction of $Pb(CH_3)_4$ and CH_3HgCl giving $(CH_3)_3PbCl$ and $Hg(CH_3)_2$ is $k = 2.8(\pm0.1) \times 10^{-2}\,mol^{-2} \cdot s^{-1}$ [92]. $Pb(CH_3)_4$ and $Hg_2(NO_3)_2$ react in methanol at room temperature to give $(CH_3)_3PbNO_3$, $(CH_3)_2Pb(NO_3)_2$, elemental mercury, and $Hg(CH_3)_2$ [39]. For the analytical application of this reaction, see [46, 60]. Ligand exchange between $Pb(CH_3)_4$ and excess $Hg(CF_3)_2$ in toluene in a sealed tube at $70\,°C$ produces $(CH_3)_3PbCF_3$ and CH_3HgCF_3 [98].

$Pb(CH_3)_4$ and $AuCl_3$ react in methanol at room and at low temperature to give elemental gold and — depending upon the molar ratio — $(CH_3)_3PbCl$, $(CH_3)_2PbCl_2$, and $PbCl_2$ with evolution of CH_4 and C_2H_6 [28, 33]. $Au(CH_3)_3$ was assumed as an intermediate in this reaction. $Pb(CH_3)_4$ and $HAuCl_4$ react in a similar way [40]. No methyluranium compounds were observed in room temperature reactions of $Pb(CH_3)_4$ with UCl_4 or UO_2Cl_2 [25]. Strong quenching of the emission of excited $[UO_2^*]^{2+}$ in acetone was found on addition of $Pb(CH_3)_4$ and was related to electron transfer to give U^{IV}, presumably $[Pb(CH_3)_4^*]^+$ and, ultimately, organic products derived from the latter [115].

$Pt(CH_3)_2 \cdot C_8H_8N_2$ ($C_8H_8N_2 = $ bipy, Formula II) is methylated by $Pb(CH_3)_4$ to give $Pt(CH_3)_4 \cdot C_8H_8N_2$ after separation of elemental lead [99]. $Pb(CH_3)_4$ does not react with solutions of calcium hydridoiron carbonyl [20].

The redistribution of $Pb(CH_3)_4$ and $(C_2H_5)_3PbCl$ at $76\,°C$ in hexane gives in 5 h a mixture of all five possible tetraalkyllead compounds $(CH_3)_{4-n}Pb(C_2H_5)_n$ ($n = 0$ to 4) and all four possible trialkyllead chlorides $(CH_3)_{3-n}(C_2H_5)_nPbCl$ ($n = 0$ to 3). The tetraalkyllead compounds can be separated by fractional distillation [6, 11]. The redistribution reaction of $Pb(CH_3)_4$ and $(C_2H_5)_3PbBr$ is complicated by decomposition of the latter compound [11]. The reaction of $(C_2H_5)_2PbBr_2$ with $Pb(CH_3)_4$ in hexane at 60 to $80\,°C$ was complete in 5 to 7 h without addition of a catalyst [6]. The halides themselves act as catalysts, and the presence of an additional catalyst does not substantially affect the rate of the reaction of the product composition [11]. Methyl exchange between $Pb(CH_3)_4$ and $(CH_3)_3PbCl$ is fast. From kinetic measurements using $Pb(CD_3)_4$, a rate constant of $k = 4.2 \times 10^{-1}\,mol^{-1} \cdot s^{-1}$ at $30\,°C$ in about 0.015 molar solutions was obtained. The rate constant for the analogous exchange with $(CH_3)_3SnCl$ is $k = 3.4 \times 10^{-2}\,mol^{-1} \cdot s^{-1}$ [93]. Rates of reaction of $Pb(CH_3)_4$ with $(CH_3)_3SnCl$ or $(CH_3)_2(C_4H_9\text{-}t)SnCl$, leading to an equilibrium with $(CH_3)_3PbCl$ and $Sn(CH_3)_4$ or $(CH_3)_3SnC_4H_9\text{-}t$ were determined [94]. Reaction of $Pb(CH_3)_4$ and $(CH_3)_3SnCH_2Sn(CH_3)_2Cl$ leads to an equilibrium which is in favor of the products $(CH_3)_3PbCl$ and $((CH_3)_3Sn)_2CH_2$ [118]. $Pb(CH_3)_4$ and $(CH_3)_2(C_4H_9\text{-}t)PbCl$ react to give $(CH_3)_3PbCl$ and $(CH_3)_3PbC_4H_9\text{-}t$. The equilibrium is slightly in favor of the products; the statistically corrected rate constant is $k = 0.13 \pm 0.02\,mol^{-1} \cdot s^{-1}$ [93]. The conproportionation of $Pb(CH_3)_4$ and $(CH_3)_2PbX_2$ is reversible and is much faster than the redistribution of $(CH_3)_3PbX$ ($X = Cl$, Br, I, $OOCCH_3$) [79, 85, 100]. Conproportionation of $Pb(CH_3)_4$ and $(CH_3)_2Pb(M(CO)_5)_2$ ($M = Mn$, Re) in different polar solvents yields $(CH_3)_3Pb(M(CO)_5)$. According to ^1H NMR kinetic studies, the conproportionation is faster than the disproportionation of

$(CH_3)_3Pb(M(CO)_5)$, both reactions being of second-order. The reactions of compounds with M = Mn are faster than those with M = Re [101, 108]. $Pb(CH_3)_4$ and $Pb(OOCCH_3)_4$ react also in the presence of a catalyst, like Hg^{2+}, even at $-60°C$ in a toluene-acetic acid solvent mixture by oxidative cleavage of methyl groups from $Pb(CH_3)_4$. No indication of a redistribution reaction was obtained [62, 63].

Mixtures of $Pb(CH_3)_4$ and other tetraalkyllead compounds show no redistribution reaction in the absence of a catalyst, even after prolonged heating [6, 12]. However, in the presence of a catalyst like $AlCl_3$ and depending upon the temperature, more or less fast exchange of alkyl groups occurs between $Pb(CH_3)_4$ and $Pb(C_2H_5)_4$ [6, 7, 12, 50, 102], $Pb(C_3H_7)_4$ [12], or $Pb(C_4H_9)_4$ [102], between $Pb(CH_3)_4$ and other tetraalkyllead compounds, or between $Pb(CH_3)_4$ and $Pb(C_6H_5)_4$ in a solvent or without a solvent [6, 12, 64]. Redistribution was observed to be complete in a few hours between $Pb(CH_3)_4$ and various tetraalkyllead compounds or $Pb(C_6H_5)_4$ in the presence of a catalyst in refluxing hexane [6, 12, 64], but in 2 minutes between $Pb(CH_3)_4$ and $Pb(C_2H_5)_4$ at 160°C. Above that temperature or at 160°C on more prolonged heating, decomposition occurred [50]. The redistribution of $Pb(CH_3)_4$ and tetrapropenyllead, $Pb(C_3H_5)_4$ (isomer mixture), in the presence of $AlCl_3$ yields mixtures of isomers of composition $(CH_3)_3PbC_3H_5$, $(CH_3)_2Pb(C_3H_5)_2$, and $CH_3Pb(C_3H_5)_3$ [109]. For the redistribution reactions of $Pb(CH_3)_4$ and tetraorganolead compounds, various catalysts other than $AlCl_3$ of differing activity, like BF_3, $AlBr_3$, $ZnCl_2$, $HgCl_2$, $FeCl_3$, $BiCl_3$, $AsCl_3$, PCl_3, $PtCl_4$, or organometal halides have been employed [6, 12, 90]. Irreversible interaction with the catalysts $AlCl_3$ or $(CH_3)_{3-n}AlCl_n$ (n = 1, 2) gives alkyllead chlorides [8, 90]. The redistribution of $Pb(CH_3)_4$ and $Pb(C_2H_5)_4$ is also catalyzed by complexes of CH_3AlCl_2 with $(CH_3)_3PbCl$, $(CH_3)_2PbCl_2$, and $PbCl_2$, and proceeds with almost quantitative yield at room temperature in 10 to 20 min. The redistribution does not occur in the presence of $Al(CH_3)_3$ [90].

The redistribution reaction is quantitative in each case [6]. The compositions of mixtures obtained from redistribution of $Pb(CH_3)_4$ and $Pb(C_2H_5)_4$ [7, 8, 12], of $Pb(CH_3)_4$ and $Pb(C_3H_7)_4$ [12], and of equal portions of $Pb(CH_3)_4$, $Pb(C_2H_5)_4$, and $Pb(C_3H_7)_4$ [12], which had been analyzed by fractional distillation, agree with those predicted for random equilibrium mixtures of all possible products [6]. This was further ascertained by comparing the experimentally determined amounts of each species at equilibrium with the values calculated on the basis of ideal random distribution [64]; see also [50]. The reactions are characterized by equilibrium constants, which are independent of temperature through the temperature range employed [13]. Weighted-average equilibrium constants have been evaluated for the reaction of $Pb(CH_3)_4$ and $Pb(C_2H_5)_4$ with a computer program; the constant for the equilibrium involving the conproportionation of $Pb(CH_3)_4$ and $(CH_3)_2Pb(C_2H_5)_2$ is $K = [Pb(CH_3)_4] \cdot [(CH_3)_2Pb(C_2H_5)_2] \cdot [(CH_3)_3PbC_2H_5]^{-2} = 0.356 \pm 0.020$. Equilibrium constants of the other steps of the redistribution reaction are also given [64]. The controlling factor in determining the equilibrium state for the redistribution reactions is the entropy value [13].

The redistribution of $Pb(CH_3)_4$ and $Pb(C_2H_5)_4$ was intensively studied with the aim of industrially producing mixtures of tetraalkyllead compounds $(CH_3)_{4-n}Pb(C_2H_5)_n$ (n = 0 to 4) which are used as antiknock agents (see Section 1.1.1.1.8). In many cases the reaction was done in the presence of other antiknock additives, like CH_2Cl_2 and CH_2Br_2, and various solvents have been used [34, 47, 48, 56, 61]. The reaction was performed with $BF_3 \cdot (C_2H_5)_2O$ as a catalyst in toluene at room temperature; after a few minutes, the catalyst was removed by washing with 1% NaOH solution [34, 35, 51]. Also, $AlCl_3$ as catalyst and hexane as solvent were employed, and the reaction mixture heated to 85°C for 2 h [5, 9, 17, 43, 44]. Similarly, the redistribution of $Pb(CH_3)_4$ and $Pb(C_2H_5)_4$ was performed in the presence of a silica-magnesia catalyst [61], of activated clay, silica, or alumina, or a zeolite [41]. With these catalysts, the redistribution of $Pb(CH_3)_4$ and tetravinyllead is feasible at room temperature [45]. Other catalysts, proposed for redistribution reactions of $Pb(CH_3)_4$ and tetraalkyllead compounds at room or higher temperatures, are BF_3 [5], $(CH_3)_3Al_2Cl_3$ [51], $FeCl_3$ [5, 17], $HgCl_2$ or $ZrCl_4$ [17], SiF_4 [56], $(CH_3)_3PbBF_4$ in aqueous solution [71, 72], and the complexes of CH_3AlCl_2 with $(CH_3)_3PbCl$, $(CH_3)_2PbCl_2$, or $PbCl_2$ [80]. $(CH_3)_3Al_2Cl_3$ was proposed as a catalyst for the preparation of a redistributed antiknock mixture by reaction of $Pb(CH_3)_4$ and tetravinyllead [48]. The redistribution of $Pb(CH_3)_4$ and $Pb(C_2H_5)_4$ in gasoline is accelerated by trimethyl phosphate. This reaction is hindered by adding $(CH_3C_5H_4)Mn(CO)_3$ [36]. For special technical devices to perform the redistribution reaction, see e.g. [51].

Redistribution of mixtures of $Pb(CH_3)_4$ and $Pb(C_2H_5)_4$ is performed also in the presence of a trialkyllead salt of a cation exchange sulfonate resin at 60°C in 3 h to give an equilibrium mixture of the appropriate tetraalkyllead compounds [49]. Also various metal salts of a strongly acidic cation exchange resin can be employed as catalysts for the redistribution of $Pb(CH_3)_4$ and $Pb(C_2H_5)_4$ at 50°C [52, 57, 58].

Slow exchange of methyl groups and rapid exchange of Cl ligands were observed in toluene solution between $Pb(CH_3)_4$ and CH_3AlCl_2 or $CH_3AlCl_2 \cdot KCl$ [73]. However, $Pb(CH_3)_4$ was later reported not to react with the latter complex [75]. Between $Pb(CH_3)_4$ and $(CH_3)_2AlCl$ at 25°C, slow methyl exchange but no exchange of Cl was observed. Equimolar amounts of benzonitrile stop the exchange with CH_3AlCl_2 and with $(CH_3)_2AlCl$ [73]. $Pb(CH_3)_4$ and CH_3AlCl_2 react to give $(CH_3)_3PbCl$ and $(CH_3)_2AlCl$ [75, 90], and equilibrium exists after 10 minutes at 135°C [75]. A similar reaction was assumed to occur between $Pb(CH_3)_4$ and $(CH_3AlCl)_2S$ [75]. Reaction of $Pb(CH_3)_4$ with $((C_2H_5)_2Al)_2O$ yields $(CH_3)_2(C_2H_5)_2Al_2O$ and $Pb(C_2H_5)_4$ as main products [95]. Alkyl exchange between $Pb(CH_3)_4$ and $(C_2H_5AlCl)_2O \cdot KCl$ gives all possible tetraalkyllead compounds $(CH_3)_{4-n}Pb(C_2H_5)_n$ [91].

Methyl group exchange in mixtures of $Pb(CH_3)_4$ and $Al(CH_3)_3$ at 25 and at 100°C was inferred from NMR spectra [73]. Rapid exothermic alkyl exchange was reported to occur between $Pb(CH_3)_4$ and $Al(C_2H_5)_3$ [42, 65], but according to later work the exchange is slow at 25°C [73]. By employing an excess of $Al(C_2H_5)_3$ (1:4 mole ratio) in the absence of a solvent, only $Pb(C_2H_5)_4$ was formed, and all methyl groups were transferred to aluminium [65]. Heating $Pb(CH_3)_4$ and $Na[Al(C_2H_5)_4]$ in toluene at 60°C gives $Na[Al(CH_3)_4]$ and $Pb(C_2H_5)_4$ [68]. Reaction of $Pb(CH_3)_4$ with $Al(C_2H_5)_3$, $(C_2H_5)_2AlCl$, or $C_2H_5AlCl_2$ under inert gas at 25 to 130°C gives in 0.1 to 140 h a mixture of $(CH_3)_{4-n}Pb(C_2H_5)_n$ (n = 1 to 4). Thus, the reaction with $Al(C_2H_5)_3$ in p-xylene was performed by mixing the components below 0°C under argon, heating at 60°C for 1 h, cooling to 0°C, hydrolyzing with 0.04 N HNO_3, followed by distillation to give the above mixture [96].

$Pb(CH_3)_4$ and $Hg(C_2H_5)_2$ undergo redistribution catalyzed by $AlCl_3$ [6, 14], and yield a random equilibrium mixture which is also obtained from $Pb(C_2H_5)_4$ and $Hg(CH_3)_2$; mercury shows greater relative affinity than lead for methyl with respect to ethyl [14]. $Pb(CH_3)_4$ and $(CH_3)_3PbOOCCH_3$ undergo rapid ligand exchange [116].

References:

[1] Krause, E. (Diss. Univ. Berlin, 1917).
[2] Schultze, G., Müller, E. (Z. Physik. Chem. B **6** [1930] 267/71).
[3] Leigh-Smith, A., Richardson, H. O. W. (Nature **135** [1935] 828/9).
[4] Hein, F. (Angew. Chem. **51** [1938] 503/8).
[5] Calingaert, G., Beatty, H. A., Ethyl Gasoline Corp. (Aust. 106497 [1938/39]; C. **1939** II 300).

[6] Calingaert, G., Beatty, H. A. (J. Am. Chem. Soc. **61** [1939] 2748/54).

[7] Calingaert, G., Beatty, H. A., Neal, H. R. (J. Am. Chem. Soc. **61** [1939] 2755/8).

[8] Calingaert, G., Soroos, H. (J. Am. Chem. Soc. **61** [1939] 2758/60).

[9] Ethyl Gasoline Corp. (Fr. 841534 [1939]; C.A. **1940** 4393).

[10] Semerano, G., Riccoboni, L. (Ric. Sci. **11** [1940] 269/70).

[11] Calingaert, G., Soroos, H., Shapiro, H. (J. Am. Chem. Soc. **62** [1940] 1104/7).

[12] Calingaert, G., Beatty, H. A., Soroos, H. (J. Am. Chem. Soc. **62** [1940] 1099/104).

[13] Stearn, A. E. (J. Am. Chem. Soc. **62** [1940] 1630).

[14] Calingaert, G., Soroos, H., Thomson, G. W. (J. Am. Chem. Soc. **62** [1940] 1542/5).

[15] Semerano, G., Riccoboni, L. (Ber. Deut. Chem. Ges. **74** [1941] 1089/99).

[16] Semerano, G., Riccoboni, L. (Z. Physik. Chem. A **189** [1941] 203/18).

[17] Calingaert, G., Beatty, H. A., Ethyl Gasoline Corp. (U.S. 2270108 [1942]; C.A. **1942** 3190).

[18] Gilman, H., Woods, L. A. (J. Am. Chem. Soc. **65** [1943] 435/7).

[19] Bawn, C. E. H., Whitby, F. J. (Discussions Faraday Soc. No. 2 [1947] 228/36).

[20] Hein, F., Heuser, E. (Z. Anorg. Allgem. Chem. **254** [1947] 138/50).

[21] Wartik, T., Schlesinger, H. I. (J. Am. Chem. Soc. **75** [1953] 835/9).

[22] Costa, G., Camus, A. M., Pauluzzi, E. (Gazz. Chim. Ital. **86** [1956] 997/1013).

[23] Costa, G., Camus, A. (Gazz. Chim. Ital. **86** [1956] 77/86).

[24] Costa, G., de Alti, G. (Gazz. Chim. Ital. **87** [1957] 1273/80).

[25] Gilman, H. (A-28 [1941]; N.S.A. **10** [1956] No. 3508 in: Comyns, A. E., Atomic Energy Research Estab. CM-258 [1957] 1/7).

[26] Holliday, A. K., Jeffers, W. (J. Inorg. Nucl. Chem. **6** [1958] 134/7).

[27] Costa, G., de Alti, G., Sillani, N. (Univ. Studi Trieste Ist. Chim. Pubbl. **1958** No. 26, pp. 1/9).

[28] Tagliavini, G., Schiavon, G., Belluco, U. (Gazz. Chim. Ital. **88** [1958] 746/54).

[29] Bawn, C. E. H., Gladstone, J. (Proc. Chem. Soc. **1959** 227/8).

[30] Costa, G., de Alti, G. (Atti Accad. Nazl. Lincei Rend. Classe Sci. Fis. Mat. Nat. [8] **28** [1960] 845/50).

[31] Bawn, C. E. H., Whitby, F. J. (J. Chem. Soc. **1960** 3926/31).

[32] Costa, G., de Alti, G. (Atti Accad. Nazl. Lincei Rend. Classe Sci. Fis. Mat. Nat. [8] **28** [1960] 627/31).

[33] Belluco, U., Riccoboni, L., Tagliavini, G. (Ric. Sci. **30** [1960] 1255).

[34] Arimoto, F. S., E. I. du Pont de Nemours & Co. (Ger. 1168430 [1961/64]; C.A. **61** [1964] 1893).

[35] Arimoto, F. S., E. I. du Pont de Nemours & Co. (Fr. 1328932 [1961/63]; C.A. **60** [1964] 550/1).

[36] Wood Jr., J. M., Ethyl Corp. (U.S. 3197414 [1961/65]; C.A. **63** [1965] 9730).

[37] Taube, P. R. (Gig. Tr. Prof. Zabol. **6** No. 8 [1962] 53/5; C.A. **57** [1962] 15453).

[38] Costa, G., de Alti, G., Stefani, L., Boscarato, G. (Ann. Chim. [Rome] **52** [1962] 289/304).

[39] Tagliavini, G., Belluco, U. (Ric. Sci. II A [2] **32** [1962] 76/81).

[40] Riccoboni, L., Belluco, U., Tagliavini, G. (Ric. Sci. II A [2] **32** [1962] 323/32).

[41] Closson, R. D., Ethyl Corp. (Fr. 1362696 [1962/64]; C.A. **62** [1965] 4052).

[42] Marlett, E. M., Kobetz, P., Pinkerton, R. C. (Abstr. Papers 142nd Meeting. Am. Chem. Soc., Atlantic City, N. J., 1962, p. 25N, in: Shapiro, H., Frey, F. W., The Organic Compounds of Lead, Wiley, New York, 1968, p. 48, Ref. 471).

[43] Ethyl Corp. (Neth. Appl. 64-12633 [1963/65]; C.A. **63** [1965] 11223).

[44] Ethyl Corp. (Brit. Amended 1088415 [1963/69]; C.A. **72** [1970] No. 81186).

[45] Closson, R. D., Ethyl Corp. (U.S. 3231511 [1963/66]; C.A. **64** [1966] 11251/2).

[46] Pilloni, G., Plazzogna, G. (Ric. Sci. Rend. A [2] **4** [1964] 27/33).

[47] Ethyl Corp. (Neth. Appl. 64-03049 [1964/65]; C.A. **64** [1966] 6694).

[48] Ethyl Corp. (Neth. Appl. 65-05907 [1964/65]; C.A. **64** [1966] 14005).

[49] Imura, S., Tamai, Y., Echiru Kagaku Kogyo K. K. (U.S. 3400142 [1964/68]; C.A. **70** [1969] No. 20226).

[50] Pollard, F. H., Nickless, G., Uden, P. C. (J. Chromatog. **19** [1965] 28/56).

[51] Wall Jr., H. H., Ethyl Corp. (U.S. 3158636 [1963/64]; C.A. **62** [1965] 3870).

[52] Imura, S., Yamanaka, M., Toyo Ethyl Co., Ltd. (U.S. 3442924 [1965/69]; C.A. **71** [1969] No. 39186).

[53] Pedinelli, M., Magri, R., Randi, M. (Chimica [Milan] **48** [1966] 144).

[54] Mayerle, E. A., Craig, R. L., Nalco Chem. Co. (U.S. 3450734 [1966/69]; C.A. **71** [1969] No. 50224).

[55] Leisey, F. A., Standard Oil Co., Indiana (U.S. 3462244 [1966/69]; C.A. **71** [1969] No. 126973).

[56] Johnson, E. E., Walker, A. O., Nalco Chem. Co. (U.S. 3441582 [1966/69]; C.A. **71** [1969] No. 39181).

[57] Imura, S., Yamanaka, Y., Toyo Ethyl Co., Ltd. (Japan. 70-29489 [1966/70]; C.A. **74** [1971] No. 13277).

[58] Imura, S., Yamanaka, Y., Toyo Ethyl Co., Ltd. (Japan. 70-29490 [1966/70]; C.A. **74** [1971] No. 13276).

[59] Willemsens, L. C. (International Lead Zinc Research Organization, Inc., Project LC-18, Progr. Rept. No. 32, 1967, p. 21).

[60] Pilloni, G. (Farmaco Ed. Prat. **22** [1967] 666/76).

[61] Williams Jr., G. H., du Pont de Nemours, E. I., and Co. (U.S. 3527780 [1967/70]; C.A. **73** [1970] No. 109903).

[62] Willemsens, L. C. (International Lead Zinc Research Organization, Inc., Project LC-18, Progr. Rept. No. 36, 1968, p. 14).

[63] Willemsens, L. C., van der Kerk, G. J. M. (J. Organometal. Chem. **13** [1968] 357/61).

[64] Moedritzer, K. (Advan. Organometal. Chem. **6** [1968] 171/271).

[65] Frey, F. W. (unpublished data in: Shapiro, H., Frey, F. W., The Organic Compounds of Lead, Wiley, New York, 1968, p. 48, Ref. 243).

[66] Davidsohn, W. E., Pant, B. C. (International Lead Zinc Research Organization, New York, Project LC-28, Quarterly Rept. No. 31, 1969, pp. 1/5).

[67] Pant, B. C., International Lead Zinc Research Organization, Inc. (Ger. Offen. 2043906 [1969/71]; C.A. **75** [1971] No. 36351).

[68] Kobetz, P., Lindsay, K. L., Cook, S. E., Ethyl Corp. (U.S. 3691221 [1970/72]; C.A. **78** [1973] No. 16301).

[69] Hoffmann, P., Bächmann, K., Klenk, H., Lieser, K. H. (Inorg. Nucl. Chem. Letters **7** [1971] 577/82).

[70] Janzen, E. G. (Accounts Chem. Res. **4** [1971] 31/40).

[71] Shepard Jr., J. C., Nalco Chemical Co. (U.S. 3725447 [1971/73]; C.A. **78** [1973] No. 159870).

[72] Shepard Jr., J. C., Nalco Chemical Co. (U.S. 3804872 [1971/74]; C.A. **81** [1974] No. 13650).

[73] Boleslawski, M., Pasynkiewicz, S., Jaworski, K. (J. Organometal. Chem. **30** [1971] 199/209).

[74] Nederlandse Centrale Organisatie voor Toegepast-Natuurwetenschappelijk Onderzoek (Neth. Appl. 71-00202 [1971/72]; C.A. **78** [1973] No. 4361).

[75] Boleslawski, M., Pasynkiewicz, S., Kunicki, A. (Przemysl Chem. **51** [1972] 446/9).

[76] Pant, B. C., Davidsohn, W. E (J. Organometal. Chem. **39** [1972] 295/9).

[77] Clinton, N. A., Kochi, J. K. (J. Organometal. Chem. **56** [1973] 243/54).

[78] Clinton, N. A., Gardner, H. C., Kochi, J. K. (J. Organometal. Chem. **56** [1973] 227/42).

[79] Gmehling, J. (Diss. Univ. Dortmund, 1973).

[80] Pasynkiewicz, S., Boleslawski, M., Jaworski, K., Politechnika Warszawska (Pol. 89645 [1973/76]; C.A. **87** [1977] No. 135946).

[81] Cosan Chemical Corp. (Japan. Kokai 74-126628 [1973/74]; C.A. **85** [1976] No. 5885).

[82] Gardner, H. C., Kochi, J. K. (J. Am. Chem. Soc. **96** [1974] 1982/4).

[83] Kochi, J. K. (Accounts Chem. Res. **7** [1974] 351/60).

[84] Jukes, A. E. (Advan. Organometal. Chem. **12** [1974] 215/322).

[85] Huber, F., Gmehling, J., Pohl, U. (Proc. 16th Intern. Conf. Coord. Chem., Dublin 1974, Abstr. 3.26, pp. 1/3).

[86] Gardner, H. C. (Diss. Univ. Indiana, Bloomington, 1975; Diss. Abstr. Intern. B **36** [1976] 3959).

[87] Gardner, H. C., Kochi, J. K. (J. Am. Chem. Soc. **97** [1975] 1855/65).

[88] Gardner, H. C., Kochi, J. K. (J. Am. Chem. Soc. **97** [1975] 5026/7).

[89] Zimmerman, A. A., Musser, G. S., Kraus, B. J., Godici, P. E., Siegel, J. R. (SAE [Soc. Automot. Eng.] Tech. Papers No. 750695 [1975] 1/14).

[90] Boleslawski, M., Pasynkiewicz, S., Jaworski, K. (J. Organometal. Chem. **92** [1975] 175/80).

[91] Jaworski, K., Wilkanowicz, L., Kunicki, A. (J. Organometal. Chem. **102** [1975] 431/6).

[92] Arnold, D. P., Wells, P. R. (J. Organometal. Chem. **113** [1976] 311/9).

[93] Arnold, D. P., Wells, P. R. (J. Organometal. Chem. **111** [1976] 285/96).

[94] Arnold, D. P., Wells, P. R. (J. Organometal. Chem. **108** [1976] 345/52).

[95] Boleslawski, M., Pasynkiewicz, S., Kunicki, A., Serwatowski, J. (J. Organometal. Chem. **116** [1976] 285/9).

[96] Pasynkiewicz, S., Jaworski, K., Przybylowicz, J. W., Kunicki, A., Wilkanowicz, L., Politechnika Warszawska (Pol. 107936 [1976/80]; C.A. **95** [1981] No. 25271).

[97] Rheingold, A. L., Choudhury, P., El-Shazly, M. F. (Syn. React. Inorg. Metal-Org. Chem. **8** [1978] 453/65).

[98] Eujen, R., Lagow, R. J. (J. Chem. Soc. Dalton Trans. **1978** 541/4).

[99] Jawad, J. K., Puddephatt, R. J. (Inorg. Chim. Acta **31** [1978] L391/L392).

[100] Huber, F., Schmidt, U., Kirchmann, H. (ACS Symp. Ser. No. 82 [1978] 65/81).

[101] Ködel, W. (Diss. Univ. Dortmund, 1978).

[102] Mitchell, T. N., Gmehling, J., Huber, F. (J. Chem. Soc. Dalton Trans. **1978** 960/4).

[103] Wong, C. L., Kochi, J. K. (J. Am. Chem. Soc. **101** [1979] 5593/603).

[104] Fukuzumi, S., Wong, C. L., Kochi, J. K. (J. Am. Chem. Soc. **102** [1980] 2928/39).

[105] Kochi, J. K. (Pure Appl. Chem. **52** [1980] 571/605).

[106] Klingler, R. J., Kochi, J. K. (J. Am. Chem. Soc. **102** [1980] 4790/8).

[107] Klingler, R. J., Kochi, J. K. (J. Am. Chem. Soc. **103** [1981] 5839/48).

[108] Ködel, W., Huber, F., Haupt, H.-J. (Inorg. Chim. Acta **49** [1981] 209/12).

[109] Mitchell, T. N., Marsmann, H. C. (Org. Magn. Resonance **15** [1981] 263/7).

[110] Dawoodi, Z., Green, M. L. H., Mtetwa, V. S. B., Prout, K. (J. Chem. Soc. Chem. Commun. **1982** 1410/1).

[111] Rollick, K. L., Kochi, J. K. (J. Org. Chem. **47** [1982] 435/44).

[112] Fukuzumi, S., Kochi, J. K. (Tetrahedron **38** [1982] 1035/49).

[113] Wieber, M., Wirth, D., Fetzer, I. (Z. Anorg. Allgem. Chem. **505** [1983] 134/7).

[114] Klingler, R. J., Fukuzumi, S., Kochi, J. K. (ACS Symp. Ser. No. 211 [1983] 117/56).
[115] Ambroz, H. B., Butter, K. R., Kemp, T. J. (Faraday Discussions Chem. Soc. No. 78 [1984] 107/19).
[116] Kubicki, M. M., Kergoat, R., Guerchais, J.-E., L'Haridon, P. (J. Chem. Soc. Dalton Trans. **1984** 1791/3).
[117] Hozman, R. (Sb. Praci Vyzk. Chem. Vyuziti Uhli Dehtu Ropy **17** [1984] 55/69).
[118] Hawker, D. W., Wells, P. R. (Organometallics **4** [1985] 821/5).

1.1.1.1.4.9 Electrochemical Behavior

Tetraalkyllead compounds have been described to be electrochemically inactive [8] and to be reduced only at very high negative potentials in aqueous media, which subsequently excludes direct electrochemical determination [4, 11]. However, in acetonitrile solution well-defined electrochemical oxidation of $Pb(CH_3)_4$ is observed at platinum electrodes [1, 2, 6].

The oxidation potential of $Pb(CH_3)_4$ in a 2.0×10^{-3} molar solution in acetonitrile containing 0.25 F $Li[BF_4]$ at 25°C at a platinum electrode and at a current density of 1.0 mA·cm^{-2} vs. Ag/AgCl reference was determined by thin-layer chronopotentiometry to be 2.13 V [1, 2]. The potential extrapolated to zero current density is 1.90 V [2]. From the data a one-electron process was deduced [2] which is the rate-determining step in the anodic oxidation of $Pb(CH_3)_4$ [6]. This process was found to be irreversible by current-reversal chronopotentiometry [1, 2], suggesting that $[Pb(CH_3)_4]^+$ is unstable and decomposes into $[Pb(CH_3)_3]^+$ and the radical $CH_3^·$ [2]; see also [5, 6, 9]. The one-electron oxidation potentials of $(CH_3)_{4-n}Pb(C_2H_5)_n$ (n = 0 to 4) show a reasonable linear correlation with the rates of reaction with the oxidant $[IrCl_6]^{2-}$; a striking relationship was also found with the vertical ionization potentials determined by He(I) photoelectron spectroscopy [1, 2]. The anodic peak potential measured with a platinum electrode at 20 mV/s in acetonitrile containing $[N(C_2H_5)_4]ClO_4$ at 25°C by cyclic voltammetry (relative to saturated NaCl standard calomel electrode) is 1.80 V. The cyclic voltammogram shows an anodic wave with a well-defined current maximum but no cathodic wave on the reverse scan and also suggests totally irreversible electron transfer. The anodic peak potential is directly related to the activation free energy for heterogeneous electron transfer, and it was concluded that the electrochemical oxidation of $Pb(CH_3)_4$ proceeds by an outer-sphere electron transfer mechanism. The cyclic voltammetric anodic peak potentials of tetraalkylmetals of Si, Ge, Sn, and Pb, i.a. $Pb(CH_3)_4$, in acetonitrile are linearly related to the ionization potential. A close relation between the activated complexes for heterogeneous and homogeneous electron transfer follows from a direct comparison of the rate constants for electrochemical oxidation with those of the oxidation by 1,10-phenanthroline and 2,2′-bipyridine complexes of Fe^{III} [6].

Attempts to measure the reversible oxidation potential E° in acetonitrile by cyclic voltammetry or ac polarography were unsuccessful even at temperatures as low as −35°C [5, 7]. However, by utilizing the Marcus equation, a value E° = 1.46 V vs. SHE (including the work term) was obtained [7].

By means of differential pulse polarography of a 5×10^{-4} molar solution of $Pb(CH_3)_4$ in CH_2Cl_2 (0.2 M $[N(C_4H_9)_4]ClO_4$) at 20°C with a pulse amplitude of 50 mV and a drop time of 0.5 s, a peak potential of +0.41 V vs. Ag/AgCl was measured [10, 11]. The reaction is characterized by a one-electron oxidation process with the two predominant electrode reactions (1) and (2) [10]:

$$Pb(CH_3)_4 + Hg \rightarrow Pb(CH_3)_3^· + [HgCH_3]^+ + e^- \tag{1}$$

$$2\,Pb(CH_3)_3^· \rightarrow Pb_2(CH_3)_6 \tag{2}$$

For a possible application for trace analysis, see [11]. The determination of trace amounts of $Pb(CH_3)_4$ and $Pb(C_2H_5)_4$ in gasoline after appropriate separation using a sensing electrode of a galvanic cell is patented [3].

References:

[1] Gardner, H. C., Kochi, J. K. (J. Am. Chem. Soc. **96** [1974] 1982/4).
[2] Gardner, H. C., Kochi, J. K. (J. Am. Chem. Soc. **97** [1975] 1855/65).
[3] Olson, D. C., Shell Oil Co. (U.S. 4153517 [1978/79]; C.A. **91** [1979] No. 94196).
[4] Fouzder, N. B., Fleet, B. (in: Smith, W. F., Polarography of Molecules of Biological Significance, Academic Press, London 1979, pp. 261/93).
[5] Wong, C. L., Kochi, J. K. (J. Am. Chem. Soc. **101** [1979] 5593/603).
[6] Klingler, R. J., Kochi, J. K. (J. Am. Chem. Soc. **102** [1980] 4790/8).
[7] Fukuzumi, S., Wong, C. L., Kochi, J. K. (J. Am. Chem. Soc. **102** [1980] 2928/39).
[8] Colombini, M. P., Corbini, G., Fuoco, R., Papoff, P. (Ann. Chim. [Rome] **71** [1981] 609/29).
[9] Fukuzumi, S., Kochi, J. K. (Tetrahedron **38** [1982] 1035/49).
[10] Bond, A. M., McLachlan, N. M. (J. Electroanal. Chem. Interfacial Electrochem. **194** [1985] 37/48).

[11] Bond, A. M., McLachlan, N. M. (J. Electroanal. Chem. Interfacial Electrochem. **182** [1985] 367/82).

1.1.1.1.5 Solutions

$Pb(CH_3)_4$ was reported to be practically insoluble in pure water [1, 2, 5]. Detailed measurements indicated a solubility in distilled water of about 20 mg Pb/L [29]. In sea water the steady state concentration is on the order of 9.0 [29, 30] to 22 mg Pb/L [29]; see also [31]. A similar value of 12.30 mg Pb/L was obtained from measurements in the dark [32]. The rate of dissolution of $Pb(CH_3)_4$ in sea water is low. The concentration of $Pb(CH_3)_4$ taken up into flowing water is lower than the saturation value. For a water flow rate of 15 m/min a concentration of 0.020 mg Pb/L was found, and $2.1 \; g \cdot m^{-2} \cdot h^{-1}$ was given as rate of uptake [29]. The water solubility of $Pb(CH_3)_4$ is about 1 to 2 orders of magnitude higher than that of $Pb(C_2H_5)_4$ [29 to 32, 41]. In natural waters, $Pb(CH_3)_4$ like other tetraalkyllead compounds would be expected to concentrate in the lipid tissues of aquatic organisms [33]. Solubility might also be feigned by adsorption on particulate matter. Concentrations of about 12 to 15×10^{-5} mol/L in water may be obtained by regularly shaking a suspension of $Pb(CH_3)_4$ in water [37]. Solubility studies are complicated by breakdown of $Pb(CH_3)_4$ in the aqueous solution phase. Breakdown takes place at rates which give half lives measurable in days [29]. The dispersion and ground level concentration profiles of $Pb(CH_3)_4$ continuously released from a source on the sea bed were calculated in [34].

$Pb(CH_3)_4$ is totally adsorbed from aqueous solutions onto silica gel, the adsorbed species being transformed relatively rapidly into $[Pb(CH_3)_3]^+$ [37]. It is also adsorbed from aqueous samples onto glass walls of bottles [37, 42]. Therefore, extractions from aqueous samples should be carried out inside the glass sampling bottles [42]. Solutions of inorganic lead and mercury salts release tetraalkyllead compounds from silica and sediment surfaces [37].

Hexane is used for extraction of $Pb(CH_3)_4$ and other tetraalkyllead compounds from fish, vegetation, sediment, and water samples [26, 35]. Hexane was found to give more satisfactory

158

recoveries in comparison with cyclohexane, methyl isobutyl ketone, butyl acetate [26, 35], toluene [35], or chloroform [26]. Benzene and octanol gave similar satisfactory results [26]. For extraction of $Pb(CH_3)_4$ by benzene and recovery rates from aqueous system, see [10].

Solubility data in acetic acid and CH_3COOH-H_2O mixtures at 15 and 25°C are reported. A 67.5% solution of acetic acid in water was employed as a selective solvent to separate $Pb(CH_3)_4$ from $Pb(C_2H_5)_4$, the latter being practically insoluble in this solvent system [5].

$Pb(CH_3)_4$ is soluble in absolute ethanol [1, 2], but not in 96% ethanol [3]. It is soluble in ether [1, 2], hydrocarbons, benzene, toluene, and other usual organic solvents [3], but insoluble in liquid ammonia at −78°C [4]. Solutions of $Pb(CH_3)_4$ in C_7H_{14}-n (components are about equally volatile) are used as standards in atomic absorption spectrometry. Such solutions are stable for more than 6 months in contrast to $Pb(C_2H_5)_4$ solutions [27]. For a phase equilibrium study of the systems $Pb(CH_3)_4$-toluene and $Pb(CH_3)_4$-benzene, see [38]. Estimated values of free energies of transfer of $Pb(CH_3)_4$ from methanol to water, alcohols, and several other solvents are given in [12, 21]. A solution of 17 to 90% methanol in 0.1 molar acetate buffer is used as a mobile phase for the separation of $(CH_3)_{4-n}Pb(C_2H_5)_n$ by HPLC [39, 40]. $Pb(CH_3)_4$ and acetonitrile form an azeotrope [6]. $Pb(CH_3)_4$ is quantitatively extracted from dust samples into cold ammoniacal methanol [24]. The lipophilicity of $Pb(CH_3)_4$ is lower than that of $Pb(C_2H_5)_4$ [41].

Solvation of $Pb(CH_3)_4$ in solution increases along the series of solvents cyclohexane < 1,2-dimethoxyethane ≅ dioxane ≅ hexamethylphosphoric acid triamide < pyridine < tetrahydrothiophene < triethylamine < tetrahydrofuran < triethylphosphane < N,N,N′,N′-tetramethylethylenediamine ≅ acetone < dimethylformamide < dimethyl sulfoxide as derived from the increase of the NMR coupling constants $^2J(^1H,^{207}Pb)$. Coordination of only one solvent molecule and trigonal bipyramidal geometry of the complexes was supposed [8]. For studies of the dispersion interaction of $Pb(CH_3)_4$ and various solvents, see [9, 28]. For a correlation of the ionization potential and the solvation energy of $Pb(CH_3)_4$ and other tetraorganometal compounds in acetonitrile, see [36].

$Pb(CH_3)_4$ vapor is absorbed by tap grease [11], and on most organic surfaces such as rubber stoppers and tubing [7].

$Pb(CH_3)_4$ is removed from solutions in hydrocarbons, particularly in gasoline, by treatment with silica, zeolites, and bauxite containing hydrogen chloride [25], with activated carbon or silica gel or alumina, impregnated with $CuCl_2$ or $FeCl_3$ [17, 22, 23] or an amine, like butylamine [23]. Another method comprises contacting the solution with a cyanide-treated chloro-methylated polystyrene or aromatic amine resin, promoted with $SnCl_4$ or $SbCl_5$ [13, 14], or with a cross-linked polystyrene resin modified by chemically bonded sulfur [15, 16], or with strongly acidic ion-exchange resins, such as sulfonated styrene-divinylbenzene copolymers [18]. Another procedure starts with contacting the solution of $Pb(CH_3)_4$ in a hydrocarbon first with $SiCl_4$, $CuCl_2$, $CuBr_2$, I_2, or I_2-HCl, dissolved in a solvent, like isopropanol, and then with activated carbon [20]. $Pb(CH_3)_4$ in gasoline can also be decomposed on a catalyst containing metal oxides, like MoO_3 and CoO, or metal sulfides on alumina with a gas containing hydrogen, low molecular weight hydrocarbons, and hydrogen sulfide at 330 to 375°C [19]; see also Sections 1.1.1.1.4.5, 1.1.1.1.4.6, and 1.1.1.1.4.8.

References:

[1] Cahours, A. (Ann. Chim. [Paris] [3] **62** [1861] 257/350).
[2] Cahours, A. (Liebigs Ann. Chem. **122** [1862] 48/71).
[3] Krause, E. (Diss. Univ. Berlin 1917).

[4] Holliday, A. K., Pass, G. (J. Chem. Soc. **1958** 3485/8).

[5] Pedinelli, M., Randi, M. (Chim. Ind. [Milan] **46** [1964] 172).

[6] Hannan, J. F., E. I. du Pont de Nemours & Co. (U.S. 3362889 [1966/68]; C.A. **68** [1968] No. 51783).

[7] Snyder, L. J. (Anal. Chem. **39** [1967] 591/5).

[8] Petrosyan, V. S., Voyakin, A. S., Reutov, O. A. (Zh. Org. Khim. **6** [1970] 889/93; J. Org. Chem. [USSR] **6** [1970] 895/8).

[9] Laszlo, P., Speert, A. (J. Magn. Resonance **1** [1969] 291/7).

[10] Hayakawa, K. (Nippon Eiseigaku Zasshi **26** [1971] 377/85).

[11] Hoare, D. E., Walsh, A. D., Li, T.-M. (13th Symp. Intern. Combust. Proc., Salt Lake City, Utah, 1970 [1971], pp. 461/9).

[12] Abraham, M. H. (J. Chem. Soc. Perkin Trans. II **1972** 1343/57).

[13] Whitehurst, D. D., Butter, S. A., Rodewald, P. G., Mobil Oil Corp. (U.S. 3944501 [1972/76]; C.A. **85** [1976] No. 8071).

[14] Whitehurst, D. D., Butter, S. A., Rodewald Jr., P. G., Mobil Oil Corp. (U.S. 3791968 [1972/74]; C.A. **81** [1974] No. 80070).

[15] Whitehurst, D. D., Mobil Oil Corp. (U.S. 3785968 [1972/74]; C.A. **80** [1974] No. 147539).

[16] Whitehurst, D. D., Mobil Oil Corp. (U.S. 3875125 [1972/75]; C.A. **83** [1975] No. 100667).

[17] Zimmerman, A. A., Exxon Research and Engineering Co. (Ger. Offen. 2447588 [1973/75]; C.A. **83** [1975] No. 100658).

[18] Obländer, K., Abthoff, J., Langer, H. J., Daimler-Benz A.-G. (Ger. Offen. 2361025 [1973/75]; C.A. **84** [1976] No. 20081).

[19] Kadlec, V., Svajgl, O. (Czech. 165223 [1973/76]; C.A. **87** [1977] No. 70790).

[20] Zimmerman, A. A., Exxon Research and Engineering Co. (U.S. 3893912 [1974/75]; C.A. **83** [1975] No. 150200).

[21] Abraham, M. H., Grellier, P. L. (J. Chem. Soc. Perkin Trans. II **1975** 1856/63).

[22] Zimmerman, A. A., Musser, G. S., Kraus, B. J., Godici, P. E., Siegel, J. R. (SAE [Tech. Pap.] No. 750695 [1975] 1/14).

[23] Zimmerman, A. A., Musser, G. S., Exxon Research and Engineering Co. (U.S. 3998725 [1975/76]; C.A. **86** [1977] No. 75728).

[24] Harrison, R. M. (J. Environ. Sci. Health A **11** [1976] 417/23).

[25] Audeh, C. A., Mobil Oil Corp. (Ger. Offen. 2756222 [1976/78]; C.A. **89** [1978] No. 165966).

[26] Chau, Y. K., Wong, P. T. S., Bengert, G. A., Kramar, O. (Anal. Chem. **51** [1979] 186/8).

[27] Rohbock, E., Müller, J. (Mikrochim. Acta **1979** I 423/34).

[28] Bahbuch, M., Grishin, Yu. K., Ustynyuk, Yu. A., Zemlyanski, N. N. (Vestn. Mosk. Univ. Ser. II Khim. **20** No. 4 [1979] 366/8; Moscow Univ. Chem. Bull. **34** No. 4 [1979] 69/72).

[29] Grove, J. R. (Lead Marine Environ. Proc. Intern. Experts Discuss., Rovinj, Yugoslav., 1977 [1980], pp. 45/52).

[30] Noden, F. G. (Lead Marine Environ. Proc. Intern. Experts Discuss., Rovinj, Yugoslav., 1977 [1980], pp. 83/91).

[31] Harrison, G. F. (Lead Marine Environ. Proc. Intern. Experts Discuss., Rovinj, Yugoslav., 1977 [1980], pp. 305/17).

[32] Charlou, J. L. (Rapport d'activité CNEXO-COB, contrat CNEXO/ENSCR-79/5943 [1980] 1/30).

[33] Chau, Y. K., Wong, P. T. S., Kramar, O., Bengert, G. A., Cruz, R. B., Kinrade, J. O., Lye, J., Van Loon, J. C. (Bull. Environ. Contam. Toxicol. **24** [1980] 265/9).

[34] Cleaver, J. W. (Lead Marine Environ. Proc. Intern. Experts Discuss., Rovinj, Yugoslav., 1977 [1980], pp. 325/43).

160

[35] Cruz, R. B., Lorouso, C., George, S., Thomassen, Y., Kinrade, J. D., Butler, L. R. P., Lye, J., Van Loon, J. C. (Spectrochim. Acta B **35** [1980] 775/83).
[36] Fukuzumi, S., Kochi, J. K. (J. Phys. Chem. **84** [1980] 2246/54).
[37] Jarvie, A. W. P., Markall, R. N., Potter, H. R. (Environ. Res. **25** [1981] 241/9).
[38] Zorin, A. D., Novotorov, Yu. N., Feshchenko, I. A. (Poluch. Anal. Chist. Veshchestv **1982** 33/4; C.A. **99** [1983] No. 141899).
[39] Blaszkewicz, M., Neidhart, B. (Intern. J. Environ. Anal. Chem. **14** [1983] 11/21).
[40] Blaszkewicz, M., Baumhoer, G., Neidhart, B. (Z. Anal. Chem. **317** [1984] 221/5).

[41] Ferreira da Silva, D., Diehl, H. (Xenobiotica **15** [1985] 789/97).
[42] Harrison, R. M., Radojević, M. (Environ. Technol. Letters **6** [1985] 129/36).

1.1.1.1.6 Physiological Properties. Toxicity

$Pb(CH_3)_4$ is highly poisonous. Special care is necessary in handling the compound since its vapor pressure is appreciably high, and its faint smell is almost imperceptible after a short time [1].

$Pb(CH_3)_4$ is absorbed through the skin, the mucosa of the alimentary tract, and the alveoli. Being lipid soluble it is concentrated in the brain, body fat, and liver. Manifestations of poisoning are dominated by involvement of the central nervous system [22]; see also [41, 53, 66, 95, 98, 99]. Symptoms of $Pb(CH_3)_4$ as well as of $Pb(C_2H_5)_4$ intoxication include insomnia, excessive dreaming, toxic psychosis, headaches, hyperactivity, ataxia, emotional instability, erratic behavior, delusions, convulsions, and mania [66, 100]. However, there is also a case reported with a patient showing no such symptoms [50]. Hematological abnormalities are rarely observed [66].

$Pb(CH_3)_4$ itself does not have toxic properties, the toxic effects being due to trimethyllead compounds formed by dealkylation in the liver [58, 59]. Absorption of $Pb(CH_3)_4$ is associated with an increase in urinary δ-aminolevulinic acid levels [45]. Metabolism and toxicokinetics of $Pb(CH_3)_4$ and other alkyllead compounds are reviewed in [101].

$Pb(CH_3)_4$ is less toxic to man than $Pb(C_2H_5)_4$ [50], and this is related to the greater stability of $Pb(CH_3)_4$ compared with $Pb(C_2H_5)_4$ and the slower dealkylation rate of $Pb(CH_3)_4$ [58, 59]. In an early report it was concluded that $Pb(CH_3)_4$ is more toxic than $Pb(C_2H_5)_4$ when chemically pure, owing to its greater volatility [5]. A lesser degree of percutaneous absorption of $Pb(CH_3)_4$, as compared with $Pb(C_2H_5)_4$, was reported [14]. For a comparison of biological action and of pathological findings in intoxications caused by $Pb(CH_3)_4$ and $Pb(C_2H_5)_4$, see [23, 41]. Pathological and histological findings in seven autopsy cases were studied in [42]. Lesser response to respiratory exposure to $Pb(CH_3)_4$ vapor as compared to $Pb(C_2H_5)_4$ is correlated with a greater loss of $Pb(CH_3)_4$ through the expired air [10]. Accordingly, in an inhalation experiment with volunteers, 51% of ^{203}Pb-labelled $Pb(CH_3)_4$ was initially deposited in the lungs, of which about 40% was exhaled in 48 h. Uptake and distribution is governed by gas/liquid phase transfers. Variation of concentrations with time in blood, urine, and organs were estimated [73]. $Pb(CH_3)_4$ is assumed to undergo the same type of degradation as $Pb(C_2H_5)_4$ (see "Organolead Compounds" Vol. 2, Section 1.1.1.2.6, to be published), but at different rates, so that the composition of the urine varies with both the compound absorbed and the lapse of time following absorption [53].

For studies associated with the use of Pb(CH$_3$)$_4$ as antiknock compound in fuels and the replacement of Pb(C$_2$H$_5$)$_4$ by Pb(CH$_3$)$_4$, see [4 to 7, 10, 11, 15, 16, 17, 24, 25, 37, 38, 41, 42, 50]; see also [18, 27, 28, 29, 46, 53]. Toxic effect, toxicokinetics, and environmental health aspects are reviewed in [74].

From studies of persons handling gasoline containing Pb(CH$_3$)$_4$, it was concluded that exposure to lead is negligible from a hygienic standpoint [16, 17]. Also, during gasoline tank truck loading operations, no potential hazardous exposures to Pb(CH$_3$)$_4$ will occur at concentrations of leaded gasoline vapors low enough to avoid hydrocarbon intoxications [67]. However, an epidemiologic study of 29 road gasoline station workers presented evidence of abnormal lead absorptions in five cases [54]. It also was reported that Pb(CH$_3$)$_4$ has been detected in blood samples from three tank cleaners (0.010 to 0.027 µg Pb/mL) and from six gasoline pump servicemen (0.005 to 0.006 µg Pb/mL) [102]. Human exposure to a high level of Pb(CH$_3$)$_4$ by an accident is described in [50]. Workers exposed to Pb(CH$_3$)$_4$ and Pb(C$_2$H$_5$)$_4$ in gasoline had higher average lead urine levels and lower lead blood levels than workers exposed to inorganic lead compounds. There was no direct correlation between blood and urine levels [94]. The alkyllead amount in urine runs parallel to working conditions of workmen handling leaded gasoline. In urine of a worker, one month after poisoning by leaded gasoline, small concentrations of trimethyllead species, but no tetraalkyllead were detected [37]. Respiratory exposure of workmen to the vapor of Pb(CH$_3$)$_4$ yields a lesser response in the form of their urinary lead excretion than that which follows exposure to Pb(C$_2$H$_5$)$_4$ [10]. In the manufacture of tetraalkyllead compounds, lead in urine is used as a health guide, and levels above 150 µg Pb/L are generally a trigger to restrict exposure to lead [91]. Measurement of urinary δ-aminolevulinic acid levels was not considered to be an adequate replacement for urinary lead determination in medical monitoring of workmen exposed to Pb(CH$_3$)$_4$ and Pb(C$_2$H$_5$)$_4$ [45]. It also was stated that blood lead estimations and the urinary excretion of δ-aminolevulinic acid is of little value in determining the severity of absorption of Pb(CH$_3$)$_4$ [50]. The determination of aminolevulinic acid dehydrase is considered to represent the most sensitive test to detect both abnormal lead absorption and possible initial lesions due to lead poisoning [54]. For correlation of Pb(CH$_3$)$_4$ concentrations in air of a manufacturing area, routine medical examinations, and urinary lead excretion levels, see [51]. Atmospheric Pb(CH$_3$)$_4$ concentrations and urinary excretion rates were correlated in [34, 51]. For monitoring methods, see [30, 47, 60, 68, 79]; see also Section 1.1.1.1.9. The use of a personal sampler for Pb(CH$_3$)$_4$ monitoring is described in [51, 68]. A mobile treatment system for handling spilled Pb(CH$_3$)$_4$ is described in [55]. Preventive measures in the occupational setting are reviewed in [103]. A method for analysis of Pb(CH$_3$)$_4$ and other alkyllead compounds in street dust was developed. From measured values of between 0.4 and 7.4 ppm and at least two orders of magnitude less than the inorganic lead content, it was concluded that if there is a health hazard associated with lead in street dust, it is due to the inorganic lead rather than to the organolead compounds present [56]. Exposures and intoxications by Pb(CH$_3$)$_4$ and Pb(C$_2$H$_5$)$_4$ are reviewed in [100].

Threshold limit values (TLV) or maximal allowable concentrations (MAC) for Pb(CH$_3$)$_4$ in the air have been established in many countries. For data, see [80]; examples: U.S.A. 0.070, Japan 0.075, Great Britain 0.150 mg/m^3, Federal Republic of Germany (MAK values) 0.075 mg/m^3, 0.01 mL/m^3 (ppm) [104]; see also [34, 39, 51, 69]. Governmental regulations are reviewed in [105].

First aid in case of contact, inhalation, or ingestion of Pb(CH$_3$)$_4$ is described in [110]. No specific therapeutic measures for treatment of patients intoxicated by Pb(CH$_3$)$_4$ have been reported. Chelation therapy with calcium disodium versenate is described to be justified in patients with a high level of absorption of $Pb(CH_3)_4$ [50]; see also [36]. In mice β-mercapto-guanidine offers no protection against acute $Pb(CH_3)_4$ poisoning. Aggravation of intoxication by this drug is related to direct action of the drug on $Pb(CH_3)_4$, and dealkylation by an oxidation-reduction mechanism is assumed [8]. Action of enzyme inhibitors on intoxication by $Pb(CH_3)_4$ in mice is examined. Protection by "Marsilid" (1-isonicotyl-2-isopropyl-hydrazine) is explained by inhibiting demethylation in liver microsomes, thus preventing the formation of trimethyllead ions [12]. For a correlation between the average quasi-valence number, as a theoretical criterion, and chemical carcinogenicity, see [61]. No studies to assess the cancer threat in humans exposed to $Pb(CH_3)_4$ have been reported, and also animal data are insufficient to classify $Pb(CH_3)_4$ as a carcinogen [43, 75]; see also [58]. For biological data relevant to the evaluation of carcinogenic risk by $Pb(CH_3)_4$ and $Pb(C_2H_5)_4$ to man, see [43]. Toxicity data and dangerous properties are compiled in [75].

$Pb(CH_3)_4$ was found to be less toxic in rats than $Pb(C_2H_5)_4$ [2, 9, 15, 19, 23], other organolead compounds, and lead salts [2]. Signs of poisoning in rats like hyperirritability and tremors are reported to be virtually identical to those produced by $Pb(C_2H_5)_4$ [9, 18, 19], though also an opposite neurological trend, like lethargy and somnolence, has been observed [23]. Up to $Pb(CH_3)_4$ doses of 40 mg/kg given intravenously to rats and a rabbit produced no toxicity symptoms, but poisoning developed in rats upon feeding doses ranging from 125 to 500 mg/kg [13]. The oral LD_{50} value for rats was given as 105 [9], 108 [23], 109 [69], and 109.3 [13] mg/kg. A slight effect, followed by recovery, was seen after an oral dose of 62.5 mg/kg [13]. For toxic effects in rats on administering different concentrations of $Pb(CH_3)_4$, see [19, 40]. Signs of toxicity in rats range from hyperexcitability and weight loss at lower levels to tremors, aggressiveness, spasticity, and paralysis at higher dose levels. A dose of 160 mg/kg was lethal [40]. Administering $Pb(CH_3)_4$ at doses of 22 mg/kg body weight to rats at weakly intervals during gestation and early postnatal life raised the total lead concentration in the brain to about 1 μg/g. Postnatal body growth was stimulated more than brain growth, resulting in a higher body : brain weight ratio. Brain myelination, dendritic growth, granule cell production, retinal receptor development, and other light-histological parameters of neuronal developments did not show any significant effect on treatment of the animals with $Pb(CH_3)_4$. The body:brain weight ratio is the most sensitive measure of the $Pb(CH_3)_4$ toxicity and of detecting the effect of a minimal toxic dose [106, 107]. Growth of 1-day-old chickens is more sensitive to $Pb(CH_3)_4$ than is that of rats [106]. Rats, cats, and mice are possibly poor models for organic lead toxicity in humans [107].

On inhalation of air containing mean $Pb(CH_3)_4$ concentrations of up to 8.87 mg/L during 1 h produced mild symptoms of poisoning in rats with no mortality [13]. $Pb(CH_3)_4$ and $Pb(C_2H_5)_4$ have proven to be more toxic for dogs than for rats in inhalation tests. $Pb(C_2H_5)_4$ was somewhat more toxic in these tests for rats than $Pb(CH_3)_4$, whereas $Pb(CH_3)_4$ was considerably more toxic for dogs than was $Pb(C_2H_5)_4$. An inhalation set-up for tetraalkyllead compounds is also described [18]. Remarkable differences in toxicity to $Pb(CH_3)_4$ and $Pb(C_2H_5)_4$ and in conversion to the appropriate trialkyllead compounds exist also between rat and mouse, and differences in accumulation in different organs of rats and mice were observed [37]. The observed interspecies variation in the relative toxicity of $Pb(CH_3)_4$ and $Pb(C_2H_5)_4$ is related to differing activities of dealkylating enzymes in the liver in different animals [81]. Exchange of ethyl and methyl groups on lead in liver tissue of rats exposed to $Pb(C_2H_5)_4$ was not observed [3]. Slow conversion of $Pb(CH_3)_4$ to trimethyllead compounds did occur in all organs of the rat, and the conversion rate was suggested to be a factor in tetraalkyllead toxicity [13]; see also [19]. For changes in weight of organs and water content of tissues, distribution of lead and lesions in organs and tissues, and for the pathology and histopathology of intoxicated animals, see [18, 23]. Administering multiple doses of $Pb(CH_3)_4$, equivalent cumulatively to a single dose, proved to be more injurious [23]. $Pb(CH_3)_4$ was found to be essentially nonteratogenic in Sprague-Dawley rats. Marked fetal toxicity effects were observed only in severely intoxicated maternal rats [40].

The enzymic dealkylation of $Pb(CH_3)_4$ in the rat-liver microsomal fraction is very rapid with specific activities of 40 to 50 $nmol \cdot min^{-1} \cdot mg^{-1}$ protein and an apparent K_m value of $(1.28 \pm 0.18) \times 10^{-5}$ M. $Pb(CH_3)_4$ binds to cytochrome P-450; the cytochrome P-450-dependent metabolism of $Pb(CH_3)_4$ and $Pb(C_2H_5)_4$ by rat-liver microsomal mono-oxygenase shows considerable differences. Substrate binding and kinetic data are also reported. The metabolism and the related toxicity of $Pb(CH_3)_4$ and $Pb(C_2H_5)_4$ in vivo contrasts markedly with in vitro findings. Whereas $Pb(CH_3)_4$ is metabolized 20 times faster than $Pb(C_2H_5)_4$ in the rat-liver microsomal fraction, the trialkyllead cation concentrations in rat liver are 10 times higher 4 h after administration of $Pb(C_2H_5)_4$ as compared with $Pb(CH_3)_4$. It was inferred that the transformation and the related toxicity in vivo are mainly determined by the volatility and solubility of both tetraalkyllead compounds, conditioning their availability for metabolism in the liver, and not their responsiveness to the mono-oxygenase system [111]; see also [13].

$Pb(CH_3)_4$ administered to mice by intraperitoneal injection is converted into trimethyllead species. The LD_{50} value in mice is given as 14.3 mg/kg [37]. In the muscle of mice, intoxicated subacutely or chronically with $Pb(CH_3)_4$, morphological anomalies concerning mostly the inner part of the blood vessels, sarcoplasmic reticulum, and mitochondria are observed [82]. The effect of microparticles from pyrolysis of $Pb(CH_3)_4$ and other antiknock compounds on lungs of mice was investigated [62]. For studies of intoxication in mice by $Pb(CH_3)_4$, see also [8]. The toxicity of $Pb(CH_3)_4$ solutions to mice and other animals in combination with other compounds, mainly fuel additives, was studied [20, 25, 26].

$Pb(CH_3)_4$ is much less readily absorbed through the skin of animals than is $Pb(C_2H_5)_4$ [10]. Rabbits given subcutaneous doses of 15 mg $Pb(CH_3)_4$/kg body weight for 5 d/week showed no change in the blood picture, but succumbed with neurotoxic symptoms after 12 to 18 doses. For effects of lower doses, see original [21]. $Pb(CH_3)_4$ given subcutaneously caused very rapid increase of lead concentrations in blood of rabbits, but it decreased very rapidly after discontinuing administration. Excretion of lead in urine increased rapidly, in feces only slowly. Excretion of coproporphyrin in urine of rabbits was lower when $Pb(CH_3)_4$ was given subcutaneously than when mixed tetraalkyllead compounds were administered [31]. After subcutaneous injection of $Pb(CH_3)_4$ in rabbits, the lead concentration in whole blood increased much more rapidly than after administration of $Pb(C_2H_5)_4$. The lead concentration was higher in plasma than in blood corpuscles [33]. For differences of affinity of $Pb(CH_3)_4$ and other tetraalkyllead compounds to tissues of rabbits, see [31, 32]. The LD_{Min} values for mature rabbits through the skin are given as 1.7 and 2.0 mL/kg, and through the digestive tract as 0.013 and 0.012 mL/kg in winter and summer, respectively. The toxicity of $Pb(C_2H_5)_4$ through the skin of rabbits is higher than of $Pb(CH_3)_4$, through the digestive tract that of $Pb(CH_3)_4$ is higher than of $Pb(C_2H_5)_4$ [41]. Rabbits excreted $[Pb(CH_3)_3]^+$ more rapidly than $[Pb(C_2H_5)_3]^+$ after injection of $Pb(CH_3)_4$ and $Pb(C_2H_5)_4$, respectively, [37].

In rhesus monkeys (*Macaca mulatta*), $Pb(C_2H_5)_4$ has a greater toxic potential than $Pb(CH_3)_4$ [70, 76]. Intravenous $Pb(CH_3)_4$ doses of 6 $mg \cdot kg^{-1} \cdot d^{-1}$ caused disturbances of patellar reflexes, peripheral nerve damage, and degeneration of skeletal muscle. Transient clinical abnormalities in peripheral nerve reflexes were detected when 1.2 to 2.4 $mg \cdot kg^{-1} \cdot d^{-1}$ were given. All reflexes were normal after a recovery period [70]. Six months at a dose level equivalent to 6 $\mu g \cdot kg^{-1} \cdot d^{-1}$ of Pb did not induce clinical manifestations of toxicity [76]. No evidence was found that sublethal exposure to $Pb(CH_3)_4$ causes permanent damage to any tissue [70, 76]. The rate of clearance of tissue lead is slower for $Pb(CH_3)_4$ than for $Pb(C_2H_5)_4$ [70]. For lead levels in tissue, blood, urine, and feces, see [70, 76].

Toxicity of $Pb(CH_3)_4$ and other organo-group 14 element compounds against Daphnia magna [35, 83, 97], Scenedesmus quadricauda, and Chlorella vulgaris at 18 to 20°C and pH 7.3 to 7.6 is compared in [35]. $Pb(CH_3)_4$ concentration of 0.02 mg/L causes 50% decay of Daphnia magna in 15 d [35]. Reproduction of Daphnia magna is inhibited at 4×10^{-2} mg $Pb(CH_3)_4$/L [44]. Photosynthesis and cell growth of Scenedesmus quadricauda, Chlorella pyrenoidosa, and Ankistrodesmus falcatus decreased by 85, 83, 49, and 32, 74, 32%, respectively, on bubbling gas containing less than 0.5 mg Pb (as $Pb(CH_3)_4$) into the culture [52, 63, 64, 71, 77]. Cells exposed to $Pb(CH_3)_4$ tend to clump together and striking changes in cell fine-structure, degeneration of the cytoplasma, and deposition of lead ions within concretion bodies are observed [63]. $Pb(CH_3)_4$ is more toxic to several algae than other organic and inorganic lead compounds [64]. Later work reports that $Pb(CH_3)_4$ (on the basis of total lead content) is less toxic than commercial solutions of $Pb(C_2H_5)_4$ to algae species, bacteria, nauplii, and fish species; both show higher toxicities by some orders of magnitude than inorganic lead compounds. All fish larvae of Morone Labrax at 20°C in the dark died almost immediately at concentrations of 2250 ppb Pb of commercial $Pb(CH_3)_4$, whereas complete survival was observed with 45 ppb Pb [72]. $Pb(CH_3)_4$ is especially toxic to juvenile rainbow trout (*Salmo gairdneri*); it is rapidly accumulated from water, the highest concentration being in the lipid layer of the intestine, followed by gills, skin, air bladder, liver, and fillet [77, 92]. At a $Pb(CH_3)_4$ concentration of 3.5 µg/L in water the accumulation factor for $Pb(CH_3)_4$ in fish was about 100 × and 700 × in the first and seventh day, respectively, on a whole fish basis. Depuration is initially quite rapid. The concentrations of $Pb(CH_3)_4$ detected in the fishery products thus represent a balance between uptake and depuration of $Pb(CH_3)_4$ in fish. A dosing system for exposing fish to $Pb(CH_3)_4$ was developed [92].

The acute toxicity of commercial $Pb(CH_3)_4$ in sea water to mixed coastal marine bacteria was determined. No effect at 20°C on photosynthesis of the alga Dunaliella tertiolecta at concentrations lower than 450 ppb Pb was observed, while 4500 ppb Pb completely inhibited photosynthetic activity after 16 h contact [72]. Contamination of the Great Lakes by lead compounds, i. a. $Pb(CH_3)_4$ and other alkyllead compounds, and its potential effects on aquatic biota is reviewed in [108].

In connection with the sinking of the cargo ship CAVTAT in 1974 carrying steel drums containing antiknock compounds based on $Pb(CH_3)_4$ and $Pb(C_2H_5)_4$, specific investigations on the toxicity of these compounds to marine animals, on bioaccumulation in marine fauna and on the dispersal, solubility, rate of solution, decomposition, and analysis of these compounds in sea water were undertaken or have been reviewed [78, 81, 84 to 89, 93]. Another incident involving the loss of antiknock cargo of a vessel in 1966 was discussed [81]. For safety controls for diving during salvage of $Pb(CH_3)_4$ using a saturation diving technique, see [90]. For the calculation of concentration profiles giving safe levels downstream from a $Pb(CH_3)_4$ source on the sea bed, see [88]. $Pb(CH_3)_4$ having only a short lifetime in sea water is unlikely to be accumulated over long periods by marine life. For data obtained for short-term exposure, see [86]. The following LC_{50} (96 h) values in mg Pb/L were derived from tests in which $Pb(CH_3)_4$ was used as antiknock formulation containing dichloroethane and dibromoethane: Common Mussel (*Mytilus edulis*) 0.27, Brown Shrimp (*Crangon crangon*) 0.11, Plaice (*Pleuronectes platessa*) 0.05. For the alga Phaeodactylum tricornutum, a EC_{50} (6 h) value (reduction of photosynthetic activity) of 1.3 mg Pb/L was found. For Plaice, a threshold value of 0.04 mg Pb/L was predicted, and a safe level of 0.004 mg Pb/L was estimated. A tentative safe level for the avoidance of acutely toxic effects in the marine environment of 0.001 mg Pb/L was proposed [86]. 1800 ppb Pb from commercial $Pb(CH_3)_4$ at 20°C in the dark caused an almost immediate 100% mortality of nauplii Artemia salina, while no effect was observed after 6 h and 48 h at 900 and 180 ppb Pb, respectively [72]. Following the spill of $Pb(CH_3)_4$ and $Pb(C_2H_5)_4$ into the sea due to sinking of the ship CAVTAT, Mytilus galloprovincialis accumulated more alkyllead derivatives than Halocynthia papillosa. In Phallusia mamillata, 1.06 to 1.24 ppm alkyllead derivatives were found [89]. Fish were found to be able to detect and to avoid $Pb(CH_3)_4$ introduced into water at 0.5 to 18 ppm [65]. $Pb(CH_3)_4$ may be taken up rapidly by marine organisms, probably by diffusion-controlled processes [87].

Application of $Pb(CH_3)_4$ in a concentration of 0.01% to primary roots of garden lettuce (*Lactuca sativa L.*) causes destruction of the Golgi apparatus and its step by step degradation [57]. The number of dividing meristematic cells increased during the first 4 h after application of $Pb(CH_3)_4$ to root cross sections, and then sharply decreased the mitotic index. Also chromosomal aberrations, decreased nuclear differentiation, and disturbances of the dividing figures during karyokinesis were observed, as well as damage in the nucleus, mitochondria, Golgi apparatus, endoplasmic reticulum, and differentiation of proplastides. Formations similar to translosomes arose in the cells at the same time [48]. The differentiation of proplastides under the influence of different concentrations and periods of action of $Pb(CH_3)_4$ was studied in detail [49]. Plant toxicity of a 1:1 mixture of $Pb(CH_3)_4$ and $Pb(C_2H_5)_4$ and uptake by spring wheat (*Triticum aestivum c. v. Kolibri*) is much higher as compared with inorganic lead compounds. In pot experiments the tetraalkyllead compounds were converted into water-soluble products, and as a consequence a relatively large lead enrichment in the vegetative and generative plant parts followed. At 10 ppm Pb in the soil, a lower yield occurred; and at 110 ppm Pb, growth practically stopped [96]. For a review of toxic effects in plant organisms, see [109].

References:

[1] Krause, E. (Diss. Univ. Berlin 1917).
[2] Buck, J. S., Kumro, D. M. (J. Pharmacol. **38** [1930] 161/72; C.A. **1930** 2806).
[3] Stevens, C. D., Feldhake, C. J., Kehoe, R. A. (J. Pharmacol. Exptl. Therap. **117** [1956] 420/4).
[4] Richardson, W. L., Barusch, M. R., Kautsky, G. J., Steinke, R. E. (Am. Chem. Soc. Div. Petrol. Chem. Prepr. **5** No. 3 [1960] 37/48).
[5] Talenti, M., Palla, A. (Nuovi Ann. Ig. Microbiol. **12** [1961] 358/70; C.A. **60** [1964] 10447).
[6] Richardson, W. L., Barusch, M. R., Kautsky, G. J., Steinke, R. E. (J. Chem. Eng. Data **6** [1961] 305/9).
[7] Richardson, W. L., Barusch, M. R., Kautsky, G. J., Steinke, R. E. (Ind. Eng. Chem. **53** [1961] 305).
[8] Salvi, G., Gherardi, M. (Folia Med. [Naples] **44** [1961] 983/6).
[9] Cremer, J. E. (Ann. Occup. Hyg. **3** [1961] 226/30).
[10] de Treville, R. T. P., Wheeler, H. W., Sterling, T. (Arch. Environ. Health **5** [1962] 532/6).

[11] Anonymous (An Investigation of the Potential Hazards of Exposure to Lead Associated with the Handling and Use of Gasoline Containing an Antiknock Fluid in Which Tetramethyl Lead is Substituted for Tetraethyl Lead, The Kettering Laboratory, Univ. of Cincinnati 1962).
[12] Gherardi, M., Salvi, G. (Folia Med. [Naples] **44** [1961] 987/97; C.A. **56** [1962] 14880).
[13] Cremer, J. E., Callaway, S. (Brit. J. Ind. Med. **18** [1961] 277/82; C.A. **56** [1962] 5069).
[14] Machle, W. F. (in: Davis, R. K., Horton, A. W., Laxson, E. E., Stemmer, K. L., Arch. Environ. Health **6** [1963] 473/9).
[15] Magistretti, M., Zurlo, N., Scollo, F., Pacillo, D. (Med. Lav. **54** [1963] 486/95; C.A. **60** [1964] 9817).
[16] Kehoe, R. A., Cholak, J., Spence, J. A., Hancock, W. (Arch. Environ. Health **6** [1963] 239/54).

[17] Kehoe, R. A., Cholak, J., McIlhinney, J. G., Lofquist, G. A., Sterling, T. D. (Arch. Environ. Health 6 [1963] 255/72).

[18] Davis, R. K., Horton, A. W., Larson, E. E., Stemmer, K. L. (Arch. Environ. Health 6 [1963] 473/9).

[19] Springman, F., Bingham, E., Stemmer, K. L. (Arch. Environ. Health 6 [1963] 469/72).

[20] Castellino, N., Rossi, A., Mole, R. (Brit. J. Ind. Med. 20 [1963] 63/5).

[21] Castellino, N., Colicchio, G., Rossi, A. (Folia Med. [Naples] 46 [1963] 980/6; C.A. 61 [1964] 3599).

[22] Gerarde, H. W. (Ann. Rev. Pharmacol. 4 [1964] 223/46).

[23] Schepers, G. W. H. (Arch. Environ. Health 8 [1964] 277/95).

[24] Kehoe, R. A. (Arch. Environ. Health 8 [1964] 296).

[25] Castellino, N., Colicchio, G., Grieco, B., Piccoli, P., Rossi, A. (Arch. Maladies Profess. Med. Travail Secur. Sociale 25 [1964] 203/18; C.A. 61 [1964] 9950).

[26] Colicchio, G., Rossi, A., Grieco, B. (14th Intern. Congr. Occupational Health, Madrid 1963 [1964], pp. 915/6; C.A. 64 [1966] 20497).

[27] Bolanowska, W. (Med. Pracy 16 [1965] 476/83).

[28] Zuliani, G., Perin, G., Rausa, G. (Med. Lavoro 57 [1966] 771/80).

[29] Oettel, H. (Ullmanns Encykl. Tech. Chem. 3rd Ed. 17 [1966] 68/9).

[30] Moss, R., Browett, E. V. (Autom. Anal. Chem. Technicon Symp., New York–London 1965 [1966], pp. 285/90).

[31] Ohmori, K. (Nippon Eiseigaku Zasshi 22 [1967] 376/82).

[32] Ohmori, K., et al. (Yokohama Igaku 18 [1967] 535/9).

[33] Ohmori, K. (Yokohama Igaku 20 [1969] 210/3).

[34] Linch, A. L., Wiest, E. G., Carter, M. D. (Am. Ind. Hyg. Assoc. J. 31 [1970] 170/9).

[35] Stroganov, N. S., Khobotev, V. G., Kolosova, L. V. (Vopr. Vod. Toksikol. 1970 66/74).

[36] Chisolm, J. J. (Modern Treatment 8 [1971] 593/611).

[37] Hayakawa, K. (Nippon Eiseigaku Zasshi 26 [1972] 526/35).

[38] Schwarzbach, E. (Proc. Intern. Symp. Environ. Health Aspects Lead, Luxembourg 1972 [1973], pp. 1117/20).

[39] Blokker, P. C. (Atmos. Environ. 6 [1972] 1/18).

[40] McClain, R. M., Becker, B. A. (Toxicol. Appl. Pharmacol. 21 [1972] 265/74).

[41] Akatsuka, K. (Sangyo Igaku 15 No. 1 [1973] 3/66).

[42] Mizoi, Y., Tatsuno, Y., Hishida, S., Morigaki, T., Nakanishi, K. (Nippon Hoigaku Zasshi 27 [1973] 371/86).

[43] Anonymous (IARC Monogr. Eval. Carcinog. Risk Chem. Man 2 [1973] 150/60).

[44] Kolosova, L. V., Stroganov, M. S. (Eksperim. Vod. Toksikol. No. 5 [1973] 134/45; C.A. 86 [1977] No. 134537).

[45] Robinson, T. R. (Arch. Environ. Health 28 [1974] 133/8).

[46] Waldron, H. A., Stöfen, D. (Sub-Clinical Lead Poisoning, Academic Press, London 1974).

[47] Blears, D. G., Coventry, R. J. (Inst. Chem. Eng. Symp. Ser. No. 39 [1974] 322/37).

[48] Sekerka, V., Bobak, M. (Acta Fac. Rerum Nat. Univ. Comenianae Physiol. Plant. 9 [1974] 1/12; C.A. 82 [1975] No. 165488).

[49] Herich, R., Bobak, M. (Acta Biol. [Budapest] 25 [1974] 289/98).

[50] Gething, J. (Brit. J. Ind. Med. 32 [1975] 329/33).

[51] Linch, A. L. (Am. Ind. Hyg. Assoc. J. 36 [1975] 214/9).

[52] Wong, P. T. S., Chau, Y. K., Luxon, P. L. (Nature 253 [1975] 263/4).

[53] Kehoe, R. A. (Pharmacol. Ther. A 1 [1976] 161/88).

[54] Rotunno, R., Tarantino, M., Bonsignore, D. (Lav. Um. 28 [1976] 65/72).

[55] Gupta, M. K. (PB-256707 [1976] 1/86; C.A. 86 [1977] No. 110830).

[56] Harrison, R. M. (J. Environ. Sci. Health A **11** [1976] 417/23).

[57] Herich, R., Bobak, M. (Cytologia **41** [1976] 477/9).

[58] U.S. Environmental Protection Agency, Air Quality Criteria for Lead, USA 1977.

[59] World Health Organization, Environmental Health Criteria, Pt. 3, Lead, Geneva, Switzerland, 1977).

[60] Anonymous (PB-275834 [1977] 1/334; C.A. **89** [1978] No. 94329).

[61] Veljkovic, V., Lalovic, D. I. (Experientia **33** [1977] 1228/9).

[62] Bouley, G., Dubreuil, A., Arsac, F., Boudène, C. (Compt. Rend. D **285** [1977] 1553/6).

[63] Silverberg, B. A., Wong, P. T. S., Chau, Y. K. (Arch. Environ. Contam. Toxicol. **5** [1977] 305/13).

[64] Chau, Y. K., Wong, P. T. S. (Lead Marine Environ. Proc. Intern. Experts Discuss., Rovinj, Yugoslav., 1977 [1980], pp. 225/31).

[65] Giaccio, M. (Quad. Merceol. **16** [1977] 55/62; C.A. **87** [1977] No. 128378).

[66] Green, V. A., Wise, G. W., Callenbach, J. C. (in: Oehme, F. W., Toxicity of Heavy Metals in the Environment, Pt. 1, Dekker, New York 1978, pp. 123/41).

[67] McDermott, H. J., Killiany Jr., S. E. (Am. Ind. Hyg. Assoc. J. **39** [1978] 110/7).

[68] Coker, D. T. (Ann. Occup. Hyg. **21** [1978] 33/8).

[69] Quinot, E., Moncelon, B., Millard, M. (Cahiers Notes Doc. No. 93 [1978] 547/65).

[70] Heywood, R., James, R. W., Sortwell, R. J., Prentice, D. E., Barry, P. S. I. (Toxicol. Letters **2** [1978] 187/97).

[71] Chau, Y. K., Wong, P. T. S. (ACS Symp. Ser. No. 82 [1978] 39/53).

[72] Marchetti, R. (Marine Pollut. Bull. **9** [1978] 206/7).

[73] Heard, M. J., Wells, A. C., Newton, D., Chamberlain, A. C. (Manage. Control Heavy Met. Environ. Intern. Conf., London 1979, pp. 103/8).

[74] Grandjean, P., Nielsen, T. (Residue Rev. **72** [1979] 97/148).

[75] Irving Sax, N. (Dangerous Properties of Industrial Materials, 5th Ed., Van Nostrand Reinhold, New York 1979, pp. 21, 771).

[76] Heywood, R., James, R. W., Pulsford, A. H., Sortwell, R. J., Barry, P. S. I. (Toxicol. Letters **4** [1979] 119/25).

[77] Wong, P. T. S., Chau, Y. K. (Manage. Control Heavy Met. Environ. Intern. Conf., London 1979, pp. 131/4).

[78] Tiravanti, G., Boari, G. (Environ. Sci. Technol. **13** [1979] 849/54).

[79] Birnie, S. E., Noden, F. G. (Analyst **105** [1980] 110/8).

[80] Anonymous (Occup. Saf. Health Ser. Intern. Labour Off. **37** [1980]).

[81] Harrison, G. F. (Lead Marine Environ. Proc. Intern. Experts Discuss., Rovinj, Yugoslav., 1977 [1980], pp. 305/17).

[82] Marchetti, C., Veneroni, G., Di Franco, S. (Giorn. Ital. Med. Lav. **2** [1980] 223/6; C.A. **95** [1981] No. 126817).

[83] Kolosova, L. V., Nosov, V. N., Dobrovol'skii, I. P. (Samoochishchenie Bioindik. Zagryaz. Vod Tr. 3rd Vses. Soveshch. Sanit. Gidrobiol., Moscow 1977 [1980], pp. 184/93; C.A. **94** [1981] No. 786).

[84] Grove, J. R. (Lead Marine Environ. Proc. Intern. Experts Discuss., Rovinj, Yugoslav., 1977 [1980], pp. 45/52).

[85] Noden, F. G. (Lead Marine Environ. Proc. Intern. Experts Discuss., Rovinj, Yugoslav., 1977 [1980], pp. 83/91).

[86] Maddock, B. G., Taylor, D. (Lead Marine Environ. Proc. Intern. Experts Discuss., Rovinj, Yugoslav., 1977 [1980], pp. 233/61).

[87] Wood, J. M. (Lead Marine Environ. Proc. Intern. Experts Discuss., Rovinj, Yugoslav., 1977 [1980], pp. 299/303).

[88] Cleaver, J. W. (Lead Marine Environ. Proc. Intern. Experts Discuss., Rovinj, Yugoslav., 1977 [1980], pp. 325/43).

[89] Geraci, S., Montanari, M., Di Cintio, R. (Mem. Biol. Mar. Oceanogr. **10** [1980] Suppl., pp. 195/206; C.A. **96** [1982] No. 175670).

[90] Marroni, A., Gething, J., Zannini, D. (Underwater Physiol. **7** [1980/81] 825/31; C.A. **95** [1981] No. 109640).

[91] Ter Haar, G. (Kirk-Othmer Encycl. Chem. Technol. 3rd Ed. **14** [1981] 196/200).

[92] Wong, P. T. S., Chau, Y. K., Kramar, O., Bengert, G. A. (Water Res. **15** [1981] 621/5).

[93] Brondi, M., Dall'Aglio, M., Ghiara, E., Mignuzzi, C., Tiravanti, G. (Sci. Total Environ. **19** [1981] 21/31).

[94] Cabeza Gonzalez de la Fuente, J. M., Garcia Blanco, M. A. (Quim. Ind. [Madrid] **28** [1982] 391/5; C.A. **97** [1982] No. 168173).

[95] Grandjean, P. (in: Rutter, M., Russell Jones, R., Lead Versus Health, Wiley, Chichester 1983, pp. 179/89).

[96] Diehl, K. H., Rosopulo, A., Kreuzer, W., Judel, G. K. (Z. Pflanzenernähr. Bodenkd. **146** [1983] 551/9).

[97] Kolosova, L. V., Nosov, V. N. (Reakts. Gidrobiontov Zagryaz. **1983** 128/34).

[98] Grandjean, P., Grandjean, E. C. (Biological Effects of Organolead Compounds, CRC, Boca Raton 1984).

[99] Turlakiewicz, Z., Chmielnicka, J. (Med. Pracy **35** [1984] 279/87).

[100] Grandjean, P. (from [98] Chapter 17, pp. 227/41).

[101] Jensen, A. A. (from [98] Chapter 8, pp. 97/115).

[102] Andersson, K., Nilsson, C.-A., Nygren, O. (Scand. J. Work Environ. Health **10** [1984] 51/5).

[103] Gething, J., Oxley, G. R. (from [98], Chapter 18, pp. 243/58).

[104] Anonymous (MAK-Werte 1984, Maximale Arbeitsplatzkonzentrationen und Biologische Arbeitsstofftoleranzwerte, TRgA 900, Bundesanstalt für Arbeitsschutz, Dortmund 1984).

[105] Jensen, A. A., Grandjean, P. (from [98], Chapter 19, pp. 259/66).

[106] Cragg, B., Rees, S. (Exp. Neurol. **86** [1984] 113/21).

[107] Ferris, N. J., Cragg, B. G. (Acta Neuropathol. **63** [1984] 306/12).

[108] Hodson, P. V., Whittle, D. M., Wong, P. T. S., Borgmann, U., Thomas, R. L., Chau, Y. K., Nriagu, J. O., Hallett, D. J. (Advan. Environ. Sci. Technol. **14** [1984] 335/69).

[109] Röderer, G. (from [98], Chapter 7, pp. 63/95).

[110] Keith, L. H., Walters, D. B. (Compendium of Safety Data Sheets for Research and Industrial Chemicals, Pt. 3, VCH, Deerfield Beach, Fl., 1985, pp. 1582/3).

[111] Ferreira da Silva, D., Diehl, H. (Xenobiotica **15** [1985] 789/97).

1.1.1.1.7 Radiochemical Studies

Decay of $^{210}Pb(CH_3)_4$ in the gas and in the liquid phase has been studied under different conditions. $Bi(CH_3)_3$ [1, 5] and also $Po(CH_3)_2$ [1] are assumed as decay products, and evidence of molecular survival, i.e., failure of bond rupture in the primary β- and internal conversion process was obtained [5, 6]; see also [7, 10]. $[Bi(CH_3)_4]^+$ is presumed to be the initial ionic species of the decay process [9, 10]. A very detailed study revealed that at low pressures no $Bi(CH_3)_3$ is obtained, the primary ion $[Bi(CH_3)_4]^+$ disproportionates and gives depositions on the wall; however, at higher pressures, production of $Bi(CH_3)_3$ becomes the dominant process since the primary ion undergoes many collisions before reaching the wall [10]. A volatile organolead compound was produced in a methane atmosphere by ^{212}Pb recoiling from

169

α-decay of ^{216}Po; however, the nature of the product was not established [8]. When Pb(CH$_3$)$_4$ or methyl radicals were conducted in a stream of hydrogen or of another inert gas with ^{224}Ra, ^{212}Pb(CH$_3$)$_4$ was obtained as a result of exchange or reaction with recoil products, respectively [11 to 14]. For decay of ^{212}Pb(CH$_3$)$_4$, see [10].

The preparation of Pb(CH$_3$)$_4$ labelled with radioactive isotopes is summarized in Section 1.1.1.1.1; the application of radioactive decay in labelled Pb(CH$_3$)$_4$ as an indicator method to study physical or chemical behavior such as photolysis [3], ligand exchange [2, 4], or uptake and metabolism [15] is described in the relevant sections.

References:

[1] Mortensen, R. A., Leighton, P. A. (J. Am. Chem. Soc. **56** [1934] 2397/8).
[2] Leigh-Smith, A., Richardson, H. O. W. (Nature **135** [1935] 828/9).
[3] Leighton, P. A., Mortensen, R. A. (J. Am. Chem. Soc. **58** [1936] 448/54).
[4] Calingaert, G., Beatty, H. A. (J. Am. Chem. Soc. **61** [1939] 2748/54).
[5] Edwards, R. R., Coryell, C. D. (AECU-50 (BNL-C-7 App.) [1948] 1/85, 63/73; N.S.A. **2** [1949] No. 570, No. 1234).
[6] Edwards, R. R., Day, J. M., Overman, R. F. (J. Chem. Phys. **21** [1953] 1555/8).
[7] Nefedov, V. D., Andreev, V. I. (Zh. Fiz. Khim. **31** [1957] 563/72).
[8] Kay, J., Rowland, F. S. (J. Am. Chem. Soc. **80** [1958] 3165).
[9] Baulch, D., Duncan, J., Thomas, F. (Chem. Eff. Nucl. Transform. Proc. Symp., Prague 1960 [1961], Vol. 2, pp. 169/81).
[10] Duncan, J. F., Thomas, F. G. (J. Inorg. Nucl. Chem. **29** [1967] 869/90).
[11] Hoffmann, P., Bächmann, K., Klenk, H., Lieser, K. H. (Inorg. Nucl. Chem. Letters **7** [1971] 577/82).
[12] Hoffmann, P., Bächmann, K., Bögl, W., Klenk, H., Lieser, K. H. (Radiochim. Acta **16** [1971] 172/9).
[13] Hoffmann, P., Bächmann, K., Klenk, H., Bögl, W., Lieser, K. H. (Angew. Chem. **83** [1971] 909; Angew. Chem. Intern. Ed. Engl. **10** [1971] 835).
[14] Hoffmann, P., Bächmann, K., Klenk, H., Trautmann, W., Lieser, K. H. (Z. Anal. Chem. **267** [1973] 277/80).
[15] Heard, M. J., Wells, A. C., Newton, D., Chamberlain, A. C. (Manage. Control Heavy Met. Environ. Intern. Conf., London 1979, pp. 103/8).

1.1.1.1.8 Uses

Pb(CH$_3$)$_4$ is used as an antiknock agent and is blended into gasoline with other tetraalkyllead compounds. Alkyl halides like 1,2-dichloroethane and/or 1,2-dibromoethane are added to scavenge lead from the motor as volatile lead(II) halides after combustion [1 to 24]. It also increases the antiknock quality of liquefied petroleum gases so as to be comparable to that of methane, ethane, and natural gas [25]. Pb(CH$_3$)$_4$ offers advantages as an antiknock additive in premium gasolines as a result of its higher volatility, higher thermal stability with respect to Pb(C$_2$H$_5$)$_4$, and its superior performance in high octane gasolines with relatively high aromatic hydrocarbon content [2, 3, 12, 21, 26 to 30]; see also [31]. Pb(CH$_3$)$_4$ also helps to reduce engine rumble [12, 32, 33, 34]. Addition of Pb(CH$_3$)$_4$ to diesel fuels for ignition jet gas engines allows greater power output [35]. Pb(CH$_3$)$_4$ is employed as combustion-control additive in the gaseous fuel, e.g., liquefied petroleum, which is used together with diesel fuel to operate a dual fuel-cycle engine [36].

Commercial grade products intended for antiknock use usually contain scavengers and, frequently, also other additives like antioxidants, ignition regulators, and supplementary antiknock compounds. For use of $Pb(CH_3)_4$ in specific motor fuel compositions, see [26, 27, 28, 37 to 56]. Substitution of haloethane by organophosphorus and organoarsine compounds, e.g., $P(C_6H_5)_3$, as scavengers to be applied with $Pb(CH_3)_4$ is proposed to avoid atmospheric pollution by lead(II) halides [57]. The antiknock rating of gasoline is further increased when $RMn(CO)_5$ ($R = CH_3$, or other organo groups) is added to the fuel already containing $Pb(CH_3)_4$ [58]; see also [20]. Numerous papers deal with comparative studies of the antiknock properties of $Pb(CH_3)_4$ and $Pb(C_2H_5)_4$ as well as $(CH_3)_{4-n}Pb(C_2H_5)_n$ ($n = 1$ to 3) and mixtures of these compounds in relation to fuel composition and to operating conditions of engines [2 to 5, 9, 10, 13 to 18, 21, 27 to 31, 40, 44, 59 to 83]. For a comparison of antiknock properties of $Pb(CH_3)_4$ and organolead sulfur compounds, see [84]. Governmental regulations regarding the use and the allowed concentrations of $Pb(CH_3)_4$ and other tetraalkyllead compounds in gasoline are compiled in [85].

Different effectivenesses of antiknock action of $Pb(CH_3)_4$ and other tetraalkyllead compounds in fuels of various compositions and octane levels is preferentially related to relative stabilities of the tetraalkyllead compounds and to their different volatilities [2, 3, 5, 9, 10, 15, 16, 19, 21, 29, 37, 59, 60, 62, 69, 86 to 89]. The role of physical factors in antiknock action of $Pb(CH_3)_4$, like temperature, time, pressure, diffusion, and surface action of particles in the fuel-air mixture in engines is examined in [87], and the mechanism of antiknock activity of $Pb(CH_3)_4$ is discussed in [2, 10, 16, 21, 29, 69, 87, 90]; see also [59, 88, 91 to 96], and the appropriate literature on the mechanism of antiknock action of $Pb(C_2H_5)_4$ in Section 1.1.1.2. The influence of $Pb(CH_3)_4$ on effects of other fuel additives is examined in [12, 16]. For a theoretical study of the distribution of gaseous lead compounds, like PbO, lead(II) halides, etc., in the exhaust gas from the combustion of isooctane-air mixtures containing antiknock additives, see [97]. $Pb(CH_3)_4$ added to diffusion flames burning a hydrogen-argon mixture affords flame ionization by lead ions, like $[Pb]^+$, $[PbOH]^+$, or $[PbOH_2]^+$. Such ions should not affect the saturation current obtainable in a flame ionization detector used to detect organolead compounds [98].

A series of measures has been proposed to improve handling and applicability of $Pb(CH_3)_4$ as an antiknock agent. To inhibit thermal decomposition of $Pb(CH_3)_4$ at temperatures up to 195°C in addition to 1,2-dichloro- and 1,2-dibromoethane or dibromopropane, various admixtures have been suggested: toluene [99 to 102], a mixture of cyclohexene or substituted cyclohexene and cyclohexane or substituted cyclohexane [103], an alkane, e.g., 2,2,4-trimethylpentane, or isooctane [99, 101, 102, 104], other alkylbenzenes, e.g., xylenes and kerosine [104, 105, 106], alkylol amines [107], or diarylamines [108]. Addition of 10 to 150% by weight of a hydrocarbon, such as toluene or n-octane reduces the sensitivity of $Pb(CH_3)_4$ to shock [28, 109]. Improved resistance to detonation by shock is also effected by admixture of ethylbromide to concentrated $Pb(CH_3)_4$ compositions containing dibromoethane as scavenger [110]. Addition of 1- or 2-bromopropane is proposed to prevent ignition of such compositions before blending with gasolines [111].

Use of $Pb(CH_3)_4$ as an antiknock agent and dibromoethane as scavenger minimizes coke deposition in internal combustion engines as compared to $Pb(C_2H_5)_4$ and ethylbromide [33]. $Pb(CH_3)_4$ does not markedly influence the initial rate of combustion of clouds of coal particles in shocked oxygen [96]. It increases the spontaneous ignition temperature of lubricating oils [112]. Addition of $Pb(CH_3)_4$ to diethyl peroxide does not influence the rate of its slow decomposition, but retards its explosive decomposition. No induction period is observed [113]. $Pb(CH_3)_4$ acts as initiator of the oxidation of ethyne in shock waves [114].

The speed of initial uniform flame movement and the mean flame speed of air-hydrocarbon vapor mixtures in a closed tube is lowered by addition of small amounts of $Pb(CH_3)_4$. This effect is correlated with the prevention of knock in the engine cylinder and is ascribed to raise the theoretical flame propagation temperature [91, 115, 116], which has been estimated for $Pb(CH_3)_4$ to be 1680°C [117, 118]. Presence of $Pb(CH_3)_4$ does not change the ignition temperature of air-hydrocarbon mixtures, but decelerates the heat release rate of the surface reaction and increases the heat flux required for ignition [119]. $Pb(CH_3)_4$ retards the second-stage flame of n-heptane, but addition of small amounts of dimethyl sulfide restores the second-stage flame [120]. Organosulfur compounds generally lower the antiknock effectiveness of $Pb(CH_3)_4$ [93]. For an explanation of the antagonistic effect of sulfur compounds, see [93, 120]; see also [84, 87].

The ignition temperature of heptane decreased only slightly when 0.25% $Pb(CH_3)_4$ was added [121]. Addition of up to 0.5% $Pb(CH_3)_4$ lowers the upper limit of inflammability of hydrogen, but further addition raises the upper limit again [117, 118]. $Pb(CH_3)_4$ also influences the limits of inflammability of diethyl ether and hydrocarbons [122]. The speed of air-hydrogen flames is fairly inhibited by $Pb(CH_3)_4$ [123]. For effects of $Pb(CH_3)_4$ on the relation of spontaneous ignition of various fuels and octane and cetane numbers, see [124]. The condenser-discharge-spark energy required to ignite an air-ethyl ether mixture has to be increased on admixture of $Pb(CH_3)_4$ [125]. $Pb(CH_3)_4$, compared to $Pb(C_2H_5)_4$, is the more effective inhibitor in the oxidation of diethyl ether above about 300°C [69]. For flammable gas detection in the presence of $Pb(CH_3)_4$ by pellistor catalytic sensors, see [126].

Thermal dissociation of $Pb(CH_3)_4$ in a stream of an inert gas, like N_2 or H_2, is employed as a source of methyl radicals [127 to 130]. These radicals or $Pb(CH_3)_4$ are used to produce and to separate methyl derivatives of recoil atoms produced by the decay of ^{224}Ra [130 to 133] or ^{252}Cf [133]. An analogous procedure was proposed for the fast separation of elements 113 to 117 [203]. Generation of methyl radicals from $Pb(CH_3)_4$ in an ionization chamber is described in [134]; see also [135].

Pyrolysis of mixtures of $Pb(CH_3)_4$ and $Te(CH_3)_2$ in an H_2 atmosphere at 475°C is applied for epitaxial growth of films of PbTe [136, 206] and the presence of $Sn(C_2H_5)_4$ of $Pb_{1-x}Sn_xTe$ [206]. In a similar way, $Pb_{1-x}Sn_xTe$ films (x = e.g. 0.2) are made from the three-component mixture $Pb(CH_3)_4$, $Sn(CH_3)_4$, and $Te(CH_3)_2$. PbS or PbSe films from $Pb(CH_3)_4$ and H_2S or H_2Se, respectively, are grown at about 550°C [136]. Lead-aluminium-fluoride glass as preform for optical fibers transmitting in the IR are prepared from $Pb(CH_3)_4$ and $Al(CH_3)_3$ by reaction with F_2 with UV radiation [204], see also [207]. Inner surfaces of hollow bodies, e.g., copper wave guides, are coated with lead by thermal decomposition of $Pb(CH_3)_4$, which is diluted with an inert carrier gas [137]. A lead undercoat as electric conductive support for a photoreceptor is deposited in a glow-discharge reaction chamber from $Pb(CH_3)_4$ [138]. Lead microparticles are produced by pyrolysis of antiknock mixtures containing $Pb(CH_3)_4$ and $Pb(C_2H_5)_4$ at 650°C [139]. Supersaturated lead vapor is prepared by shock heating of $Pb(CH_3)_4$ to 990 to 1180 K, finally leading to spherical lead clusters [140]. Polymer films incorporating up to about 90% lead, which are used as coatings for preheat shields for microsphere targets in inertial-confinement nuclear fusion, are obtained by glow-discharge low-pressure polymerization of cyclo-octatetraene [141] or other hydrocarbons [142] in the presence of $Pb(CH_3)_4$. In the manufacture of lead titanate ceramics, $Pb(CH_3)_4$ added to the raw mixture is reported to prevent vaporization of PbO during sintering [143].

$Pb(CH_3)_4$, obtained by appropriate conversion of lead from natural samples, is used for studies of isotopic assay of lead by mass spectral analysis, e.g., also for the determination of the geological age of minerals [144 to 153]; see also [154, 155, 156].

Fluorescence quenching of anthracene in ethanol, benzene, and paraffin [157, 158], and of 9,10-dicyanoanthracene in cyclohexane is effected by $Pb(CH_3)_4$ [159]. Visible chemiluminescence is emitted when a mixture of $Pb(CH_3)_4$, SF_6, and NO [160] or N_2O [161] is subjected to a CO_2 laser pulse. $Pb(CH_3)_4$ has been employed as a chemiluminescence agent for the measurement of upper-atmosphere wind velocities at night [162]. It is used as precursor in flash photolysis to obtain Pb $6p^2(^3P_2)$ and $(^3P_1)$ metastable states [163, 164]. Excitation of these metastable states is also achieved by homogeneous oxidation of ethyne or cyanogen by oxygen in the presence of traces of $Pb(CH_3)_4$ [165]. By pulsed irradiation of $Pb(CH_3)_4$ in the presence of helium using a high-intensity magnetically pinched Garton-Wheaton source, ground state lead atoms Pb (6^3P_0) are generated [166]. Atomic emission from excited lead atoms is induced by photodissociation of $Pb(CH_3)_4$ or $Pb(C_2H_5)_4$ using the ArF laser [167]. By pulsed electrical discharges in $Pb(CH_3)_4$, lead atoms are produced showing laser transitions [168, 169]. Pb^+ ions are produced following Pb 5d core-level ionization in $Pb(CH_3)_4$ with synchrotron radiation [205].

$Pb(CH_3)_4$ vapor has been used to fill Geiger counters [170 to 174], and was suggested for use in liquid ionization chambers [175]. Short-lived transients are generated by pulse radiolysis of $Pb(CH_3)_4$ dosed with naphthalene and toluene, which can be employed for time-resolved dosimetry of low-energy X-rays [176].

$Pb(CH_3)_4$ can be applied to extract aluminium and alkylaluminium alkoxide impurities from organoaluminium complexes used in the electrolytic preparation of tetraorganolead compounds [177]. $Pb(CH_3)_4$ was claimed to prevent corrosion of graphite by CO_2 used as fluid coolant in nuclear reactors [178], and to stabilize halogenated aryl compounds, employed as dielectric, insulating, or cooling agents [179].

$Pb(CH_3)_4$ is used in catalyst compositions for preparing polymers or copolymers of vinyl chloride [180] and 1,3-butadiene [181], polyesters [182], and polyurethanes [183]. $Pb(CH_3)_4$ is a suitable catalyst for chlorosulfonylation [184], and it was proposed for use in a catalyst composition for oxidation of o-xylene to phthalic anhydride [185], and in combination with $TiCl_4$ as catalyst for olefin polymerization [186]. Polymerization of ethene [187, 188], propene [187, 188, 189], and of butene and bivinyl [187] is accelerated by $Pb(CH_3)_4$ under normal and under high pressure at temperatures of about 280 to 450°C. The reaction is induced by methyl radicals obtained by thermal dissociation of $Pb(CH_3)_4$ [188]. Low-rhenium-loading Re_2O_7-$SiO_2 \cdot Al_2O_3$ catalysts with $Pb(CH_3)_4$ as cocatalyst exhibit high activity in the metathesis of functionalized alkanes [190]. $Pb(CH_3)_4$ is also claimed as cocatalyst in olefin metathesis catalyzed by Re_2O_7/Al_2O_3 [191, 192]. Platinum-alumina catalysts active in exchange reactions of deuterium with benzene and cyclopentane are selectively poisoned by $Pb(CH_3)_4$ [193]. By decomposing $Pb(CH_3)_4$ at 200 to 450°C, lead is deposited on iron catalysts which are used in Fischer-Tropsch synthesis [194, 195, 196]. The catalytic activity of NiO and MnO_2 in decomposition of hydrogen peroxide or of nickel in decomposition of methanol is promoted by treatment of the catalyst with $Pb(CH_3)_4$ in the presence of air [197]. Patents describe the synthesis of diamonds by contacting diamond powder at about 700 to 1400 K with $Pb(CH_3)_4$ at a partial pressure of 10^{-4} to 10^{-2} bar [198, 199]. Reaction of $Pb(CH_3)_4$ with halogen, e.g., fluorine or halogen compounds, is claimed to obtain lead halide glass preforms for low-loss optical fibers [200]. $Pb(CH_3)_4$ can be used as calibration standard in analyses of airborne lead tetraalkyl [201].

Petrolatum with 0.2 to 0.8% $Pb(CH_3)_4$ is proposed for treatment of staphylococcal diseases [202].

The use of $Pb(CH_3)_4$ as a methylating agent for nonmetal and metal compounds, and its use in redistribution reactions to obtain mixed methylethyllead compounds, $(CH_3)_{4-n}Pb(C_2H_5)_n$ (n = 1 to 3) is described in the preceding Sections 1.1.1.1.4.6 and 1.1.1.1.4.8, respectively.

References:

[1] Midgley Jr., T. (U.S. 1592954 [1926]; C.A. **1926** No. 3228).
[2] Richardson, W. L., Barusch, M. R., Kautsky, G. J., Steinke, R. E. (Prepr. Am. Chem. Soc. Div. Petrol. Chem. **5** No. 3 [1960] 37/48).
[3] Perry Jr., R. H., Di Perna, C. J., Heath, D. P. (SAE [Soc. Automot. Eng.] Prepr. A No. 207 [1960] 1/37; C.A. **57** [1962] 8802).
[4] Pastell, D. L., Morris, W. E. (SAE [Soc. Automot. Eng.] Prepr. C No. 207 [1960] 1/20; C.A. **57** [1962] 8801).
[5] Korn, T. M., Moss, G. (SAE [Soc. Automot. Eng.] Prepr. D No. 207 [1960] 1/10; C.A. **57** [1962] 8802).
[6] Anonymous (Petrol. Weekly **10** [1960] 58 in: Shapiro, H., Frey, F. W., The Organic Compounds of Lead, Wiley, New York 1968, p. 401, ref. 17).
[7] Stormont, D. H. (Oil Gas J. **58** No. 18 [1960] 74/5).
[8] Anonymous (Oil Daily **1960** 3 in: Shapiro, H., Frey, F. W., The Organic Compounds of Lead, Wiley, New York 1968, p. 402, ref. 22).
[9] Hesselberg, H. E., Howard, J. R. (SAE [Soc. Automot. Eng.] Trans. **69** [1961] 5/16).
[10] Richardson, W. L., Barusch, M. R., Kautsky, G. J., Steinke, R. E. (J. Chem. Eng. Data **6** [1961] 305/9).

[11] Talenti, M., Palla, A. (Nuovi Ann. Ig. Microbiol. **12** [1961] 358/70; C.A. **60** [1964] 10447).
[12] Perry Jr., R. H., Gerard, P. L., Heath, D. P. (SAE [Soc. Automot. Eng.] Prepr. A No. 438 [1961] 1/8).
[13] Stormont, D. H. (Oil Gas J. **60** No. 13 [1962] 189, 192/3, 195).
[14] Morris, W. E. (SAE [Soc. Automot. Eng.] Prepr. C No. 547 [1962] 1/8; C.A. **60** [1964] 11819).
[15] Goodacre, C. L., Foord, D. (Acta Chim. Acad. Sci. Hung. **36** [1963] 235/53).
[16] Richardson, W. L., Ryason, P. R., Kautsky, G. J., Barusch, M. R. (9th Symp. Combust., Ithaca, N.Y., 1962 [1963], pp. 1023/33).
[17] Gursky, J., Vesely, V. (Freiberger Forschungsh. A No. 340 [1964] 303/21).
[18] Griffiths, S. T., Pigott, W. D. (Erdöl Kohle Erdgas Petrochem. **17** [1964] 997/1002).
[19] Marshall, E. F., Wirth, R. A. (Ann. N.Y. Acad. Sci. **125** [1965] 198/217).
[20] Vesely, V., Gursky, J. (Ropa Uhlie **7** [1965] 215/20; C.A. **63** [1965] 17753).

[21] Barusch, M. R., Macpherson, J. H. (in: McKetta Jr., J. J., Advances in Petroleum Chemistry and Refining, Vol. 10, Interscience, New York 1965, pp. 457/546).
[22] Dabelstein, W. (Erdöl Kohle Erdgas Petrochem. **24** [1971] 37/40).
[23] Malyavinskii, L. V., Robert, Yu. A., Timofeev, S. V., Milov, Yu. N., Turovskii, F. V., Grebenshchikov, V. P. (Tr. Vses. Nauchno Issled. Inst. Pererab. Nefti No. 20 [1977] 119/28; C.A. **89** [1978] No. 165780).
[24] Laveskog, A. (in: Grandjean, P., Grandjean, E. C., Biological Effects of Organolead Compounds, CRC, Boca Raton, Fla., 1984, pp. 5/12).
[25] Felt, A. E., Kerley, R. V. (Hydrocarbon Process. Petrol. Refiner **43** No. 4 [1964] 157/64).
[26] Ethyl Corp. (Brit. 961407 [1960/61/64]; C.A. **61** [1964] 6842).
[27] California Research Corp. (Brit. 853515 [1960]; C.A. **55** [1961] 14899).
[28] California Research Corp. (Brit. 941742 [1960/63]; C.A. **60** [1964] 3932).

174

[29] Richardson, W. L., Barusch, M. R., Kautsky, G. J., Steinke, R. E. (Ind. Eng. Chem. **53** [1961] 305).

[30] Gureev, A. A., Mitrofanov, V. A., Chernyak, B. Ya., Goryachii, Ya. V., Azev, V. S., Kol'chenko, B. E. (Avtomob. Prom. **39** No. 7 [1973] 5/7; C.A. **80** [1974] No. 50072).

[31] Robinson, I. C. H. (in: Hancock, E. G., Technology of Gasoline, Soc. Chem. Ind. Blackwell, Oxford 1985, pp. 57/85).

[32] Perry Jr., R. H., Gerard, P. L., Heath, D. P. (SAE [Soc. Automot. Eng.] J. **70** No. 1 [1962] 48/50).

[33] Shifrin, G. G., Gureev, A. A., Sokolov, V. V., Kitskii, B. P., Lebedev, S. R. (Khim. Tekhnol. Topl. Masel **1983** No. 3, pp. 17/8; C.A. **98** [1983] No. 182199).

[34] Socony Mobil Oil Co., Inc. (Brit. 969913 [1960/64]; C.A. **61** [1964] 14447).

[35] Hesselberg, H. E., Ethyl Corp. (Ger. 1145434 [1959/63]; C.A. **58** [1963] 13689).

[36] Lovell, W. G., Ethyl Corp. (U.S. 3202141 [1960/61/63/65]; C.A. **63** [1965] 12957).

[37] Smyers, W. H., Standard Oil Development Co. (U.S. 2310376 [1936/43]; C. **1948** I 428).

[38] Ethyl Gasoline Corp. (Fr. 825981 [1938]; C.A. **1938** 6443).

[39] Ethyl Gasoline Corp. (Brit. 498509 [1939]; C.A. **1939** 4770).

[40] Barusch, M. R., Richardson, W. L., Kautsky, G. J., Olson, D. R., Chevron Research Co. (U.S. 3316071 [1958/67]; C.A. **67** [1967] No. 13586).

[41] Hinkamp, J. B., Warren, J. A., Ethyl Corp. (U.S. 2855905 [1958]; C.A. **1959** 3681).

[42] Kerley, R. V., Felt, A. E., Ethyl Corp. (U.S. 3038792 [1959/62]; C.A. **57** [1962] 7524).

[43] Ethyl Corp. (Brit. 928275 [1959/63]; C.A. **59** [1963] 11168).

[44] Socony Mobil Oil Co., Inc. (Brit. 967665 [1959/64]; C.A. **61** [1964] 13107).

[45] Austen, D. E. G., Coulman, C. E., Taylor, B. G. S., Lowenstein-Lom, W. G., Kinnard, L. M., Esso Research and Engineering Co. (Ger. 1420935 [1960/70]; C.A. **73** [1970] No. 122170).

[46] Rae, N. S., Esso Research and Engineering Co. (Ger. 1144970 [1960/63]; C.A. **58** [1963] 13686).

[47] E. I. du Pont de Nemours & Co. (Brit. 948642 [1960/64]; C.A. **60** [1964] 14315).

[48] Gockel, J. L., Woodruff, R. L., Shell Internationale Research Maatschappij N. V. (Belg. 632840 [1962/63]; C.A. **61** [1964] 4133).

[49] Ethyl Corp. (Neth. Appl. 65-05907 [1964/65]; C.A. **64** [1966] 14005).

[50] Ethyl Corp. (Neth. Appl. 65-11869 [1964/66]; C.A. **65** [1966] 10401).

[51] Schoen, W. F., Atlantic Richfield Co. (U.S. 3751235 [1964/73]; C.A. **80** [1974] No. 5551).

[52] Svajgl, O., Sklenar, K. (Czech. 186870 [1973/81]; C.A. **95** [1981] No. 153565).

[53] Svajgl, O., Sklenar, K. (Czech. 186871 [1973/81]; C.A. **95** [1981] No. 153564).

[54] Svajgl, O. (Czech. 214507 [1978/84]; C.A. **101** [1984] No. 213792).

[55] Rosenthal, W. (U.S. 4430092 [1982/84]; C.A. **100** [1984] No. 142000).

[56] Pass, F. (in: Winnacker-Küchler, Chemische Technologie, 4th Ed., Vol. 5, Organische Technologie I, Hanser, München 1981, pp. 48/163).

[57] Pagliarini, P. (Fr. Demande 2016257 [1968/70]; C.A. **74** [1971] No. 66307).

[58] Brown, J. E., Ethyl Corp. (U.S. 3160592 [1959/64]; C.A. **62** [1965] 3870).

[59] Sturgis, B. M. (Prepr. Am. Chem. Soc. Div. Petr. Chem. **6** No. 3 [1961] A-51/A-66; C.A. **58** [1963] 12348).

[60] Goodacre, C. L., Foord, D. (Riv. Combust. **16** [1962] 340/9).

[61] Morris, W. E. (SAE [Soc. Automot. Eng.] J. **70** [1962] 96).

[62] Goodacre, C. L., Foord, D., Hedde, M. (Bull. Assoc. Franc. Tech. Petrole No. 152 [1962] 253/76).

[63] Mori, T., Maeda, T., Takatori, T., Yamazaki, K. (Bull. Japan Petrol. Inst. **7** [1965] 7/16; C.A. **63** [1965] 17753).

[64] Sterling Jr., J. D. (in: Ullmanns Encykl. Tech. Chem. 3rd Ed. **17** [1966] 64/8).

[65] Glatte, W., Jaskulla, N., Prietsch, W., Gelius, R., Preussner, K. R. (Chem. Tech. [Leipzig] **19** [1967] 294/9).

[66] Hammerich, T. (Erdöl Kohle Erdgas Petrochem. **20** [1967] 488/99).

[67] Clark, M. G. (Rev. Assoc. Franc. Tech. Petrole No. 183 [1967] 67/76).

[68] Foster, J. M., Goodacre, C. L. (Proc. 7th World Petrol. Congr., Barking, Engl., 1967 [1968], Vol. 8, pp. 29/39; C.A. **71** [1969] No. 72561).

[69] Salooja, K. C. (J. Inst. Petrol. **53** No. 521 [1967] 186/93).

[70] Foster, G. M., Goodacre, C. L. (Kach. Mot. Reaktivn. Topl. Masel Prisadok **1970** 157/61; C.A. **75** [1971] No. 89766).

[71] Boddy, J. H. (Petrol. Rev. **26** No. 309 [1972] 284/6).

[72] Porter, F. D. (Chem. Brit. **10** [1974] 61/2).

[73] Guibet, J. C., Duval, A. (Rev. Inst. Franc. Petrole **30** [1975] 499/542).

[74] Azev, V. S., Gureev, A. A., Protasov, V. V., Kol'chenko, B. E., Malykhin, V. D. (Ekspl. Tekh. Svoistva Primen. Avtomob. Topl. Smaz. Mater. Spetszhidk. No. 9 [1977] 88/93; C.A. **87** [1977] No. 170157).

[75] Azev, V. S., Malykhin, V. D., Lebedev, S. R. (Khim. Tekhnol. Topl. Masel **1977** No. 4, pp. 19/21; C.A. **87** [1977] No. 87441).

[76] Svajgl, O. (Chem. Prumysl **28** [1978] 586/9).

[77] Emel'yanov, V. E., Levinson, G. I., Grebenshchikov, V. P., Golosova, V. F. (Sb. Nauchn. Tr. Vses. Nauchno Issled. Inst. Pererab. Nefti No. 37 [1980] 92/6; C.A. **96** [1982] No. 22120).

[78] Azev, V. S., Kitskii, B. P., Malykhin, V. D., Goryachii, Ya. V., Lebedev, S. R. (Khim. Tekhnol. Topl. Masel **1980** No. 4, pp. 24/6; C.A. **93** [1980] No. 75210).

[79] Azev, V. S., Kitskii, B. P., Lebedev, S. R., Malykhin, V. D., Gorina, F. A., Emel'yanov, V. E. (Khim. Tekhnol. Topl. Masel **1980** No. 11, pp. 37/8; C.A. **94** [1981] No. 124243).

[80] Botte, J. M., Le Breton, D. (Pet. Tech. No. 275 [1980] 33/41; C.A. **94** [1981] No. 159381).

[81] Ivanov, A., Apostolov, I. (God. Vissh. Khim. Tekhnol. Inst. "Prof. d-r As. Zlatarov" gr. Burgas **16** [1981/82] 155/61; C.A. **100** [1984] No. 54125).

[82] Shifrin, G. G., Gureev, A. A., Sokolov, V. V., Kitskii, B. P. (Neftepererab. Neftekhim. [Moscow] **1983** No. 1, pp. 10/1; C.A. **98** [1983] No. 92109).

[83] Palmer, F. H., Smith, A. M. (in: Hancock, E. G., Technology of Gasoline, Soc. Chem. Ind. Blackwell, Oxford 1985, pp. 106/36).

[84] Gelius, R., Müller, R. (Erdöl Kohle Erdgas Petrochem. **23** [1970] 817/8).

[85] Jensen, A. A., Grandjean, P. (in: Grandjean, P., Grandjean, E. C., Biological Effects of Organolead Compounds, CRC, Boca Raton, Fla., 1984, pp. 259/66).

[86] Campbell, J. M., General Motors Corp. (U.S. 2304883 [1936/43]; C.A. **1943** 2914).

[87] Rifkin, E. B. (Proc. Am. Petrol. Inst. III **38** [1958] 60/7).

[88] Downs, D., Griffiths, S. T., Wheeler, R. W. (J. Inst. Petrol. **47** No. 445 [1961] 1/21).

[89] Ryason, P. R. (Combust. Flame **7** [1963] 235/43).

[90] Gelius, R., Franke, W. (Brennstoff-Chem. **47** [1966] 280/5).

[91] Nagai, Y. (Kogyo Kagaku Zasshi **33** [1930] 117/20).

[92] Agnew, W. G. (Combust. Flame **4** [1960] 29/44).

[93] Mieville, R. L., Meguerian, G. H. (Ind. Eng. Chem. Prod. Res. Develop. **6** [1967] 253/7).

[94] Hoare, D. E., Walsh, A. D., Ting-Man Li (13th Symp. Intern. Combust. Proc., Salt Lake City, Utah, 1970 [1971], pp. 461/9).

[95] Homer, J. B., Hurle, I. R. (Proc. Roy. Soc. [London] A **327** [1972] 61/79).

[96] Nettleton, M. A., Stirling, R. (Combust. Flame **22** [1974] 407/14).

[97] Ting-Man Li, Simmons, R. F. (Combust. Flame **56** [1984] 113/22).

176

[98] Ham, N. S., McAllister, T. (Aust. J. Chem. **36** [1983] 1299/304).

[99] Ethyl Corp. (Brit. 949268 [1961/64]; C.A. **60** [1964] 14311).

[100] Cook, S. E., Ethyl Corp. (U.S. 3147294 [1963/64]; C.A. **61** [1964] 13345).

[101] Ethyl Corp. (Neth. Appl. 64-12633 [1963/65]; C.A. **63** [1965] 11223).

[102] Ethyl Corp. (Brit. Amended 1088415 [1963/69]; C.A. **72** [1970] No. 81186).

[103] Cook, S. E., Thomas, W. H., Ethyl Corp. (U.S. 3340284 [1964/67]; C.A. **68** [1968] No. 39818).

[104] Ethyl Corp. (Neth. Appl. 64-03049 [1964/65]; C.A. **64** [1966] 6694).

[105] Ethyl Corp. (Brit. 976972 [1960/64]; C.A. **62** [1965] 6326).

[106] Cook, S. E., Sistrunk, T. O., Ethyl Corp. (U.S. 3221039 [1963/65]; C.A. **64** [1966] 4841).

[107] Calcott, W. S., Parmelee, A. E., E. I. du Pont de Nemours & Co. (U.S. 1835140 [1931]; C.A. **1932** 997).

[108] Calcott, W. S., Parmelee, A. E., E. I. du Pont de Nemours & Co. (U.S. 1843942 [1932]; C.A. **1932** 1945).

[109] Richardson, W. L., Kautsky, G. J., Barusch, M. R., Chevron Research Co. (U.S. 3674826 [1960/72]; C.A. **77** [1972] No. 142105).

[110] Goodacre, C. L., Goodacre, U. M. (Brit. 1078259 [1965/67]; C.A. **67** [1967] No. 110323).

[111] Goodacre, C. L., Goodacre, U. M. (Brit. 1092337 [1965/67]; C.A. **68** [1968] No. 42033).

[112] Leonardi, S. J., Oberright, E. A., Socony Mobil Oil Co., Inc. (Fr. 1426693 [1964/66]; C.A. **65** [1966] 8640).

[113] Moriya, K. (Rev. Phys. Chem. Japan S. Horiba Commemoration Volume 1946, pp. 143/52).

[114] Hand, C. W., Kistiakowsky, G. B. (J. Chem. Phys. **37** [1962] 1239/45).

[115] Nagai, Y. (Proc. Imp. Acad. [Tokyo] **4** [1928] 525/8).

[116] Nagai, Y. (Kogyo Kagaku Zasshi **33** [1930] 296/9).

[117] Tanaka, Y., Nagai, Y. (Proc. Imp. Acad. [Tokyo] **3** [1927] 434/6).

[118] Tanaka, Y., Nagai, Y. (Kogyo Kakagu Zasshi **31** [1928] 20/3).

[119] Ise, H., Yamazaki, K. (Kogyo Kagaku Zasshi **72** [1969] 1443/6).

[120] Ballinger, P. R., Ryason, P. R. (13th Symp. Intern. Combust. Proc., Salt Lake City, Utah, 1970 [1971], pp. 271/7).

[121] Ormandy, W. R., Craven, E. C. (J. Inst. Petrol. Technol. **10** [1924] 335/41).

[122] Nagai, Y. (Proc. Imp. Acad. [Tokyo] **3** [1927] 664/9).

[123] Miller, D. R., Evers, R. L., Skinner, G. B. (Combust. Flame **7** [1963] 137/42).

[124] Goodger, E. M., Valvade, A. P. (Inst. Pet. Tech. Pap. IP 80-002 [1980] 1/17; C.A. **94** [1981] No. 142270).

[125] Nagai, Y. (Proc. Imp. Acad. [Tokyo] **3** [1927] 670/1).

[126] Jones, E. (Intern. Environ. Safety **1981** 52/3).

[127] Paneth, F., Hofeditz, W. (Ber. Deut. Chem. Ges. **62** [1929] 1335/47).

[128] Rice, F. O., Johnston, W. R., Evering, B. L. (J. Am. Chem. Soc. **54** [1932] 3529/43).

[129] Paneth, F. A., Hofeditz, W., Wunsch, A. (J. Chem. Soc. **1935** 372/9).

[130] Hoffmann, P., Bächmann, K., Klenk, H., Lieser, K. H. (Inorg. Nucl. Chem. Letters **7** [1971] 577/82).

[131] Hoffmann, P., Bächmann, K., Bögl, W., Klenk, H., Lieser, K. H. (Radiochim. Acta **16** [1971] 172/9).

[132] Hoffmann, P., Bächmann, K., Klenk, H., Bögl, W., Lieser, K. H. (Angew. Chem. **83** [1971] 909; Angew. Chem. Intern. Ed. Engl. **10** [1971] 835).

[133] Hoffmann, P., Bächmann, K., Klenk, H., Trautmann, W., Lieser, K. H. (Z. Anal. Chem. **267** [1973] 277/80).

[134] Hipple, J. A., Stevenson, D. P. (Phys. Rev. [2] **63** [1943] 121/6).

[135] Eltenton, G. C. (J. Chem. Phys. **15** [1947] 465/74).

[136] Manasevit, H. M., Simpson, W. I. (J. Electrochem. Soc. **122** [1975] 444/50).

[137] Padgett, D. W., Shostak, A. A. (U.S. 3560248 [1966/71]; C.A. **74** [1971] No. 78920).

[138] Ricoh Co., Ltd. (Japan. Kokai Tokkyo Koho 84-58434 [1982/84]; C.A. **101** [1984] No. 201470).

[139] Bouley, G., Dubreuil, A., Arsac, F., Boudène, C. (Compt. Rend. D **285** [1977] 1553/6).

[140] Frurip, D. J., Bauer, S. H. (Shock Tube Shock Wave Res. Proc. 11th Intern. Symp., Seattle, Wash., 1977 [1978], pp. 451/8; C.A. **92** [1980] No. 82663).

[141] Sheats, J. E., Hessel, F., Tsarouhas, L., Podejko, K. G., Porter, T. J., Kool, L. B., Nolen Jr., R. L. (Polym. Mater. Sci. Eng. **49** [1983] 363/7).

[142] Liepins, R., Campbell, M., Clements, J. S., Hammond, J., Fries, R. J. (J. Vac. Sci. Technol. **18** [1981] 1218/26).

[143] Toyota Motor Co., Ltd. (Japan. Kokai Tokkyo Koho 83-36975 [1981/83]; C.A. **99** [1983] No. 57768).

[144] Aston, F. W. (Nature **120** [1927] 224).

[145] Aston, F. W. (Proc. Roy. Soc. [London] A **140** [1933] 535/43).

[146] Dibeler, V. H., Mohler, F. L. (J. Res. Natl. Bur. Std. **47** [1951] 337/42).

[147] Collins, C. B., Farquhar, R. M., Russell, R. D. (Phys. Rev. [2] **88** [1952] 1275/6).

[148] Collins, C. B., Russell, R. D., Farquhar, R. M. (Can. J. Phys. **31** [1953] 402/18).

[149] Collins, C. B., Farquhar, R. M., Russell, R. D. (Bull. Geol. Soc. Am. **65** [1954] 1/22).

[150] Bate, G. L., Miller, D. S., Kulp, J. L. (Anal. Chem. **29** [1957] 84/8).

[151] Richards, J. R. (J. Geophys. Res. **67** [1962] 869/84).

[152] Ulrych, T. J., Russell, R. D. (Geochim. Cosmochim. Acta **28** [1964] 455/69).

[153] Whittles, A. B. L., Slawson, W. F. (Geochim. Cosmochim. Acta **29** [1965] 142/3).

[154] Piggot, C. S. (J. Washington Acad. Sci. **18** [1928] 269/73).

[155] Aston, F. W. (Nature **123** [1929] 313).

[156] Richards, J. R. (Vacuum **16** [1966] 310/1).

[157] Vander Donckt, É., Van Bellinghen, J.-P. (Chem. Phys. Letters **7** [1970] 630/2).

[158] Vander Donckt, É., Van Bellinghen, J.-P. (J. Chim. Phys. **68** [1971] 948/53).

[159] Abdullah, K. A., Kemp, T. J. (J. Photochem. **28** [1985] 61/9).

[160] Bauer, S. H., Bar-Ziv, E. (AD-A025002 [1976] 1/31; C.A. **85** [1976] No. 133418).

[161] Bauer, S. H., Bar-Ziv, E., Haberman, J. A. (IEEE J. Quantum Electron. **14** [1978] 237/45).

[162] Hord, R. A., Tolefson, H. B. (Virginia J. Sci. [2] **16** [1965] 105/19; C.A. **63** [1965] 16086).

[163] Ewing, J. J., Trainor, D. W., Yatsiv, S. (J. Chem. Phys. **61** [1974] 4433/9).

[164] Trainor, D. W., Ewing, J. J. (J. Chem. Phys. **64** [1976] 222/7).

[165] Gabai, A., Rokni, M., Shmulovich, J., Yatsiv, S. (J. Chem. Phys. **67** [1977] 2284/9).

[166] Bell, C. F., Husain, D. (J. Photochem. **29** [1985] 267/83).

[167] Karny, Z., Naaman, R., Zare, R. N. (Chem. Phys. Letters **59** [1978] 33/7).

[168] Chou, M. S., Cool, T. A. (Electron. Transition Lasers Proc. 2nd Summer Colloq., Woods Hole, Mass., 1975 [1976], pp. 125/47).

[169] Mau Song Chou, Cool, T. A. (J. Appl. Phys. **47** [1976] 1055/61).

[170] Keston, A. S. (Rev. Sci. Instr. **14** [1943] 293/5; C.A. **1944** 915).

[171] Glassford, H. A., Macklin, R. L. (PB-95754 [1944] in: Shapiro, H., Frey, F. W., The Organic Compounds of Lead, Wiley, New York 1968, p. 425, ref. 142).

[172] Meaker, C. L., Wu, C. S., Rainwater, L. J. (Phys. Rev. [2] **73** [1948] 1240).

[173] Wu, C. A., Meaker, C. L. (A-3846 [1956] 1/18 in: Shapiro, H., Frey, F. W., The Organic Compounds of Lead, Wiley, New York 1968, p. 425, ref. 360; N.S.A. **10** [1956] No. 5322).

178

[174] Bambynek, W. (Z. Physik. Chem. [Frankfurt] **25** [1960] 403/14).

[175] Schmidt, W. F., Sowada, U. (HMI-B-198 [1975] 62/3).

[176] Hosszu, J. L. (AD-A033409 [1976] 1/12; C.A. **86** [1977] No. 179246).

[177] Giraitis, A. P., Ethyl Corp. (U.S. 2944948 [1960]; C.A. **1960** 20591).

[178] Goenvec, H., Commissariat a l'Energie Atomique (Fr. 1461586 [1965/66]; C.A. **67** [1967] No. 17108).

[179] Clark, F. M., General Electric Co. (U.S. 2468544 [1949]; C.A. **1949** 5887).

[180] Societa Edison S.p.A. — Settore Chimico (Neth. Appl. 65-03797 [1964/65]; C.A. **64** [1966] 8340).

[181] Phillips Petroleum Co. (Brit. 931313 [1961/63]; C.A. **59** [1963] 10328).

[182] Caldwell, J. R., Wellman, J. W., Eastman Kodak Co. (U.S. 2720505 [1955]; C.A. **1956** 2205).

[183] Jourquin, L., Du Prez, E., Société Anon. PRB (Ger. Offen. 2710901 [1976/77]; C.A. **87** [1977] No. 202348).

[184] Herold, P., Asinger, F., Badische Anilin- & Soda-Fabrik [I. G. Farbenindustrie A. G. "In Auflösung"] (Ger. 765790 [1953]; C.A. **1955** 3238).

[185] Egbert, R. B., Gluodenis, T. J., Chemical Process Corp. (U.S. 3455962 [1965/69]; C.A. **71** [1969] No. 70337).

[186] Gaylord, N. G., Mark, H. F. (in: Mark, H. F., Polymer Reviews, Vol. 2, Linear and Stereoregular Addition Polymers: Polymerization with Controlled Propagation, Interscience, New York 1959, p. 98).

[187] Romm, F. S. (J. Gen. Chem. [USSR] **10** [1940] 1784/92; C.A. **1941** 3880).

[188] Beeck, O., Rust, F. F. (J. Chem. Phys. **9** [1941] 480/3).

[189] Kooijman, P. L., Ghijsen, W. L. (Rec. Trav. Chim. **66** [1947] 673/9).

[190] Xu Xiaoding, Mol, J. C. (J. Chem. Soc. Chem. Commun. **1985** 631/3).

[191] Chevalier-Seite, B., Commereuc, D., Chauvin, Y., Institut Francais Petrole (Fr. Demande 2521872 [1982/83]; C.A. **100** [1984] No. 34157).

[192] Warwel, S., Janssen, E., Consortium für Elektrochemische Industrie G.m.b.H. (Ger. Offen. 3229419 [1982/84]; C.A. **101** [1984] No. 6641).

[193] Morales, A., Barbier, J., Maurel, R. (Rev. Port. Quim. **18** [1976] 158/62).

[194] Wilson, T. P., Union Carbide Corp. (U.S. 2824116 [1954/58]; C.A. **1958** 7668).

[195] Wilson, T. P., Union Carbide Corp. (Fr. 1135190 [1955/57]; C. **1959** 1938).

[196] Union Carbide Corp. (Brit. 808956 [1954/59]; C.A. **1959** 8593).

[197] Zhabrova, G. M., Fokina, E. A. (Probl. Kinetiki Kataliza Akad. Nauk SSSR No. 6 [1949] 151/6; C.A. **1955** 14454).

[198] Dshevitskii, B. E., Spitsyn, B. V., Kochkin, D. A., Deryagin, B. V., Institute of Physical Chemistry, Academy of Sciences, USSR (Ger. Offen. 2021792 [1970/71]; C.A. **76** [1972] No. 35723).

[199] Dzevitskii, B. E., Spitsyn, B. V., Kochkin, D. A., Deryagin, B. V., Institute of Physical Chemistry, Academy of Sciences, USSR (Can. 940018 [1970/74]; C.A. **82** [1975] No. 66730).

[200] Roba, G., Centro Studi e Laboratori Telecomunicazioni (CSELT) (Eur. Appl. 135903 [1983/85]; C.A. **103** [1985] No. 26163).

[201] Rohbock, E., Georgii, H.-W., Müller, J. (Atmos. Environ. **14** [1980] 89/98).

[202] Madier, J. (Fr. Demande 2246263 [1973/75]; C.A. **83** [1975] No. 209422).

[203] Bächmann, K., Hoffmann, P. (Radiochim. Acta **15** [1971] 153/63).

[204] Parisi, G., Roba, G., Centro Studi e Laboratori Telecomunicazioni (CSELT) (Eur. Appl. 196665 [1985/86]; C.A. **105** [1986] No. 213219).
[205] Nagaoka, S., Suzuki, S., Koyano, J. (Phys. Rev. Letters **58** [1987] 1524/7).
[206] Manasevit, H. M., Ruth, R. P., Simpson, W. I. (J. Cryst. Growth **77** [1986] 468/74).
[207] Modone, E., Roba, G., Centro Studi e Laboratori Telecomunicazioni (CSELT) (Eur. Appl. 196666 [1985/86]; C.A. **106** [1987] No. 71865).

1.1.1.1.9 In the Environment

$Pb(CH_3)_4$ and other tetraalkyllead compounds have been detected in the atmosphere and in the hydrosphere even far away from anthropogenic sources. Evaporation losses during handling of leaded gasoline, and emissions in exhaust gases of cars operating with leaded gasoline, are accepted as anthropogenic sources, but also natural production of $Pb(CH_3)_4$ in minor amounts has been assumed. For reviews, see [1 to 10, 170, 171]; see also [172]. It was estimated that some 2% of the tetraalkyllead in motor gasoline may reach the atmosphere unchanged [11]. Another estimation assumes that 140 of the 12000 t tetraalkyllead used in the U.K. in 1973 escaped unchanged into the atmosphere by volatilization in manufacture, distribution, and use [12]; see also [173].

Concentrations of $Pb(CH_3)_4$ and other tetraalkyllead compounds in atmospheric samples, which have been measured at different locations with species specific analytical methods, are listed in Table 11. Tetraalkyllead concentrations in indoors air were found to correspond closely with those in outdoors air [24]. More data on concentrations of alkyllead compounds in ambient air have been obtained with analytical methods which measure volatile organolead compounds or total organolead [28 to 55, 173]. However, these data, being based on different sampling and analytical procedures, are difficult to compare; see also [56 to 64]. According to [5], the results given in [28, 29, 32, 33] are probably unreliable. In this context it has to be considered that the majority of the nonfilterable (molecular) lead seems not to be organic in nature [58, 59]; recently also tri- and/or dialkyllead compounds have been determined besides $Pb(CH_3)_4$ and other tetraalkyllead compounds in both urban and rural air in the vapor phase and in atmospheric aerosols [27, 174]; see also [182]. For analytical procedures to determine $Pb(CH_3)_4$ in atmospheric samples, see Section 1.1.1.1.1 on p. 75.

Due to diurnal fluctuations, there is a great variability in the concentrations measured over short averaging periods, and roadside concentrations are affected by traffic density, driving mode, and proximity to the vehicular emissions [5, 34, 42 to 48, 61, 173]. Elevated concentrations may exist in the vicinity of service stations, garages, and tank truck filling plants [13, 22, 23, 24, 29, 33, 34, 36, 57, 64 to 67, 173, 175], and cold choked engines [13, 57, 173]; see also [68]. On a global scale, however, the contribution of gasoline stations to general environmental tetraalkyllead levels was considered to be small [22, 23]. The average alkyllead concentration in urban air is about 100 ng/m^3 [5]. Alkyllead compounds typically represent 1 to 4% of total lead in urban air [57], or 1 to 5% of the particulate lead level in the U.S.A. and Canada, and probably 5 to 15% in other countries of the Western World [5], or about 5 to 13% of inorganic lead [22, 23]. Also ranges of concentrations of tetraalkyllead in air samples from 0 to 55% of the total airborne lead have been given [13, 29, 33, 34, 36, 42]. The mean relative atmospheric content of $Pb(CH_3)_4$ and $Pb(C_2H_5)_4$ in a suburban area in Scandinavia was $6.6 \pm 0.5\%$ in winter and $7 \pm 1\%$ of total lead in summer [21]. $Pb(CH_3)_4$ and other tetraalkyllead vapors were essentially undetectable in urban and rural air of Beijing, China, where gasoline is usually unleaded. This would preclude terrestrial biogenic sources [69, 70]. The significance of $Pb(CH_3)_4$ and other tetraalkyllead compounds as urban air pollutants is studied in [22, 23, 28, 57, 71, 72, 73]. In ambient air adjacent to a heavily traveled highway, the only tetraalkyllead compound detectable in a study was $Pb(CH_3)_4$, with concentrations ranging from 1 to 20% of the total lead concentration [74]; and also at rural sites, only $Pb(CH_3)_4$ was found [27].

Table 11

Concentrations of $Pb(CH_3)_4$ and $Pb(CH_3)_4$ with Other Tetraalkyllead Compounds (TAL) in Atmospheric Samples Measured at Different Locations.

location	site	concentration in ng · m^{-3}		Ref.
		$Pb(CH_3)_4$ range (mean)	$Pb(CH_3)_4$ + TAL range (mean)	
Stockholm, Sweden	urban	–	220 to 950 (480)	[13]
	urban	–	100 to 1760 (637)	[14]
	urban	–	120 to 1300 (508)	[17]
	parking garage	–	560 to 3400 (2207)	[17]
	service station	–	– (700)	[17]
	urban (busy street)	61 to 99 (79)	–	[21]
	urban (quiet street)	14 to 28 (21)	–	[21]
	rural	0.6 to 3.2 (1.9)	–	[21]
Copenhagen, Denmark	urban (busy street)	180 to 195 (195)	(195) (in ng Pb/m^3)	[21]
	suburban	1.2 to 75 (26)	<2 to 94 (28) (in ng Pb/m^3)	[21]
	service station	– (1420)	– (1160) (in ng Pb/m^3)	[21]
Antwerp, Belgium	urban (central street)	30 to 66	49 to 109 (83)	[22, 23]
	residential	2.4 to 11	3.2 to 14 (7)	[22, 23]
	highway crossing	5 to 15	14 to 44 (24)	[22, 23]
	tunnel	8 to 112	12 to 162 (39)	[22, 23]
	car-repair	53 to 154	100 to 290 (205)	[22, 23]
	service stations A/B	3 to 40/17 to 188	17 to 410 (149)	[22, 23, 24]
	rural	0.25 to 3.3	0.3 to 3.9 (2)	[22, 23, 24]
	university			
	laboratory	3.6 to 4.0 (3.8)	10.7 to 12.3	[20]
	office room	1.8 to 5.9 (3.8)	2.2 to 9.6	[20]
	residential area	4.6 to 4.8 (4.7)	8.5 to 8.7	[20]
	shopping area	20.6 to 22.7 (21.7)	30.3 to 32.0	[20]
Vienna, Austria	urban (traffic jam)	≤800	≤1500	[15]

Gmelin Handbook
Pb-Org. 1

location			[ref]
Colchester, U.K.	atmospheric aerosol background urban rooftop	0.292	0.799 (in ng Pb/m^3)	[25]
		1.3 to 14.1 (5.7)	2.9 to 16.1 (8.25; 13.5 including ionic alkyllead compounds) (in ng Pb/m^3)	[180]
Lancaster, U.K.	urban	25.2 to 49.7	50.3 to 89.6 (in ng Pb/m^{-3})	[26, 27]
	rural	0.3 to 1.7	0.2 to 1.3 (in ng Pb/m^{-3})	[27]
	university campus	—	1.8 to 7.2 (in ng Pb/m^{-3})	[26]
Fort Collins, Colorado, U.S.A.	highway	33 to 180 (109)	—	[16]
Baltimore, Maryland, U.S.A.	highway	12 to 64 (37)	26 to 75 (53)	[18]
	tunnel	21 to 66 (44)	57 to 130 (92)	[18]
	laboratory	7	26	[18]
Toronto, Canada	urban	—	— (14)	[19]

182

Pb(CH$_3$)$_4$ concentrations in air have been measured at manufacturing sites [75, 76] and at refineries [28]. The average concentration of lead in air during the manufacture of Pb(CH$_3$)$_4$ was found to be almost 3 times that found during manufacture of Pb(C$_2$H$_5$)$_4$ [75], whereas no difference in the appropriate concentrations at gasoline pumps have been measured [65]; see also [52].

The average composition of the tetraalkyllead compounds present in gasoline used in the area of investigation is reflected in the atmospheric tetraalkyllead pattern [22, 23, 24]. However, quantitative differences are affected by different properties of Pb(CH$_3$)$_4$ and Pb(C$_2$H$_5$)$_4$. About 75% of gaseous lead alkyls in urban air was found to be Pb(CH$_3$)$_4$ and about 25% was Pb(C$_2$H$_5$)$_4$, though both compounds were present in 1:1 ratio in the gasoline used in the area [42]. For comments concerning [42], see [43 to 48]. If only a small percentage of a sample of gasoline containing Pb(CH$_3$)$_4$ and Pb(C$_2$H$_5$)$_4$ evaporates, then this portion consists of very little leaded material. Pb(CH$_3$)$_4$ accumulates in gasoline at a slower rate than Pb(C$_2$H$_5$)$_4$. For details, see original [59]. The headspace gas over gasoline in a closed container has a different relative composition, e.g., 62% Pb(CH$_3$)$_4$, in contrast to 16% Pb(CH$_3$)$_4$ in liquid gasoline [63]. For the concentration of alkyllead compounds in saturated gasoline vapor at different temperatures, see [173].

Both evaporation losses and exhaust fumes have been concluded to be important as sources of gaseous lead alkyls in the atmosphere [22, 23, 173]. Lead alkyl emissions of 0.5 mg Pb per day per car [173] and of 1 to 4 mg lead alkyls per day per car [176] have been estimated. Exhaust emission was suggested to be an insignificant contributor to the presence of alkyllead compounds in the atmosphere [19]. This would agree with the conclusion that evaporation of leaded gasoline must be the major source of these compounds in air [24 to 36]. According to another study, however, vehicle exhaust gases probably represent the largest anthropogenic source of tetraalkyllead compounds in the atmosphere [61].

In the exhaust pipe of various vehicles, 240 to 650 ng/m^3 Pb(CH$_3$)$_4$ have been determined besides smaller amounts of other tetraalkylleads [18]; the percentage relative to particulate lead was 1.9 to 2.2%, not much higher than that found alongside a busy highway [18]. In another experiment, less than 70 ng/m^3 alkyllead compounds and a ratio between total and alkyllead exceeding 38000:1 has been found [19]. Another value of 6 to 30 µg Pb/m^3 for a lead in gasoline concentration of 0.8 g/L is given in [78]. The concentration of lead alkyl compounds in the air of a garage was found to be 10 to 30% that of inorganic lead, indicating that in the garage much of the alkyllead originates from evaporation and not from the car exhaust [173].

A car with a warm engine driven at constant speed produced exhaust concentrations of tetraalkyllead compounds of 5 to 100 µg/m^3 which increased on idling to 50 to 1000 µg/m^3. Under choked driving conditions, particularly after a cold start, the tetraalkyllead concentrations could reach 5000 µg/m^3 [13]; see also [15]. The ratio of Pb(CH$_3$)$_4$ to Pb(C$_2$H$_5$)$_4$ in the exhaust gases of a car, which burns gasoline containing these antiknock agents, was found to increase as the motor gets warmer [3]. However, according to [19], the concentrations given in [3] were considered to be doubtful. For a comparison of lead concentrations in and around automobiles burning gasoline with Pb(CH$_3$)$_4$ and Pb(C$_2$H$_5$)$_4$ as antiknock additive, see [71]; see also [79, 80].

Pb(CH$_3$)$_4$ in comparison with Pb(C$_2$H$_5$)$_4$ is relatively stable in air and has a longer residence time [19, 42 to 48]. It is relatively enriched in the atmosphere with respect to other tetraalkyllead species [22, 23]; see also [27, 174]. Upper limit rates of Pb(CH$_3$)$_4$ decay in the middle of the day, in a moderately polluted irradiated atmosphere, are estimated to be in summer (in winter): 8 to 21 (1 to 1.5) %/h by attack by hydroxyl radicals; 8 (2) by photolysis; 1 to 2 (0.5) by attack by ozone; and <0.1 ($\ll 0.1$) %/h by attack by $O(^3P)$, giving a total decay rate of 16 to 29 (3 to 4) %/h [61, 72]; see also [42 to 48, 81, 182]. For rate constants of the reactions of $Pb(CH_3)_4$ with ozone, $O(^3P)$, and OH, see Section 1.1.1.1.4.4.

$Pb(CH_3)_4$ undergoes photolysis in the atmosphere when exposed to sunlight [82]; its decomposition is slower than that of $Pb(C_2H_5)_4$ [61, 72, 83, 84]. Homogeneous breakdown reactions of $Pb(CH_3)_4$ due to photochemical oxidation in the atmosphere are similar to those of hydrocarbons [61], attack by hydroxyl radicals representing the main reaction pathway [61, 72]; see also [174]. The decay in the atmosphere is sensitive to the solar intensity and shows strong diurnal and seasonal fluctuations [61]. The half-life of $Pb(CH_3)_4$ under typical summertime conditions is estimated to be 1 to 30 h, with 9 h as the most plausible value [84]. For other estimations of the half-life of $Pb(CH_3)_4$ in ambient air, see [72, 85]; see also Section 1.1.1.1.4.2.1 on p. 126. $Pb(CH_3)_4$ and $Pb(C_2H_5)_4$ are expected to be relatively stable at night [72]. From uptake studies, it was concluded that neither direct physical adsorption nor a surface reaction of $Pb(CH_3)_4$ on atmospheric particulates could account for measured concentrations of adsorbed organic lead. No uptake of $Pb(CH_3)_4$ from dry purified air or nitrogen onto particulates at a concentration of $Pb(CH_3)_4$ of 590 to 815 µg Pb/m^3 was detected [72].

$Pb(CH_3)_4$ and other tetraalkyllead compounds are eliminated from the atmosphere via formation of tri- and dialkyllead salts [22, 23]; see also [174, 177]. Such alkyllead species have been speculated to cause european forest damage [86, 183]; however, this proposition was rejected as entirely unproven [87] and further research was strongly recommended [172, 183]. The fate of $Pb(CH_3)_4$ and other tetraalkyllead compounds in the atmosphere is reviewed in [2, 61].

Very low levels of organic lead were found in street dust using a procedure which had been successfully tested with $Pb(CH_3)_4$ added to dust samples [88]. In recent studies using GC/AAS, $Pb(CH_3)_4$ was identified in street dust in concentrations on the order of ng Pb/g [171]. According to other studies, $Pb(CH_3)_4$ was absent in street dust [25]. No $Pb(CH_3)_4$ was detected in motorway run off water [25, 89, 174], and the concentration of $Pb(CH_3)_4$ was assumed to be below the detection limit [25, 174]. However, close to known or predictable sources, tetraalkyllead compounds were found to occur in road drainage grid sediments. In drainage grid sediments on garage forecourts, up to 5.3 µg $Pb(CH_3)_4$ per kg dry sediment weight have been identified [90]. Tetraalkyllead compounds could not be detected in any of the river or rain water samples investigated, the detection limit being 0.2 µg/L [24] and 20 ng Pb/L [181] for $Pb(CH_3)_4$. From the absence of detectable amounts of alkyllead compounds in larger bodies of water, it was inferred that these compounds are adequately trapped in grids. The wash-out of airborne alkyllead by rainfall, was assumed not to contribute sufficiently to enable detection in ground waters within the limit of the analytical methods [90]. Recently, alkyllead compounds, but no $Pb(CH_3)_4$, were identified in rainwater, and it was assumed that rainwater scavenging of the atmosphere is the main source of alkyllead species in the road drainage water. The apparent absence of $Pb(CH_3)_4$ was correlated with the relatively poor analytical detection limit and the low scavenging efficiency due to high volatility and low solubility [89]; see also [174, 181]. No $Pb(CH_3)_4$ (unlike ionic alkyllead compounds) was detected in tap water [181].

The fate of $Pb(CH_3)_4$ before degradation in water is not yet fully understood. Accumulation with the sediment in natural systems would be expected due to its insolubility and high density [90], as well as immediate volatilization through the water column due to its high vapor pressure [73]. Photolytic degradation of $Pb(CH_3)_4$ in sea water is slower than that of $Pb(C_2H_5)_4$ [91]; see also [92]. For the presence of tetraalkyllead in surface microlayer of water, see [178], and for degradation of $Pb(CH_3)_4$ in water, see Sections 1.1.1.1.4.2.1 and 1.1.1.1.4.6. Amounts of $Pb(CH_3)_4$ and other tetraalkyllead compounds in the range of 0.01 to 0.1 ppm [93] and

184

0.26 µg/g [94] have been detected in tissues of fish from various lakes and rivers. Similar observations are reported in [96, 97]; see also [62]. In a survey of 107 fish from Ontario Lakes, 17 were found to contain tetraalkyllead in the low ng/g concentration range and 0.5 to 1.7 ppb were determined [97]. Rainbow trout accumulates $Pb(CH_3)_4$ in its organs after exposure to water containing this compound [98]; see also [94]. For the ability of six Adriatic fish species to detect and avoid $Pb(CH_3)_4$ and $Pb(C_2H_5)_4$, see [99]. The contamination of the Great Lakes by $Pb(CH_3)_4$ and other organolead compounds and its potential effects on aquatic biota is studied and reviewed in [100]; see also [101].

After the loss of the cargo vessel CAVTAT on July 14th, 1974, some 6.4 km south-east of Capo d'Otranto in the Adriatic Sea carrying 325 metric tons of tetraalkyllead compounds, of which about 93% were recovered, a series of investigations have been undertaken to study the impact of $Pb(CH_3)_4$ and $Pb(C_2H_5)_4$ on marine life [102 to 107] and the bioaccumulation of such compounds and their degradation products [103 to 106, 108, 109, 110]; see also [111, 112]. Concentrations of $Pb(CH_3)_4$ and $Pb(C_2H_5)_4$, determined in sediments and in water near the wreck site, are reported in [113]. Analysis of water and biota during a period of about 2 to 3 years following the incident have not revealed serious effects [105, 106, 107], and this was explained by the low rate of solution of the tetraalkyllead compounds, the rapid dispersal due to diffusion, to ocean currents, and by the relatively low level of accumulation by aquatic fauna [105]. The major environmental impact of a release of $Pb(CH_3)_4$ and other tetraalkyllead compounds into the marine environment was inferred to be due more likely to acutely toxic effects than those of bioaccumulation [108].

In the soil, $Pb(CH_3)_4$ and $Pb(C_2H_5)_4$ are converted quickly into water-soluble lead compounds, which showed in studies with spring wheat (*Triticum aestivum c.v. Kolibri*), a high plant toxicity and availability and a relatively large lead enrichment in the vegetative and generative plant parts. A close relation between the level of $Pb(CH_3)_4$ applied and the water-extractable lead was observed. From soils treated with tetraalkyllead, the wheat plants can take up appreciable amounts of lead still in the generative phase and can transport it to the grain [114]. For toxic effects of $Pb(CH_3)_4$ in plant organisms, see also [115].

A first report on the biogenesis of $Pb(CH_3)_4$ was given in 1970, when $Pb(CH_3)_4$ from Hawaiian soils incubated with lead(II) acetate was tentatively identified by mass spectrometry [116, 117]. Further evidence that microorganisms in lake and river sediments can biomethylate inorganic lead(II) compounds, namely $Pb(NO_3)_2$ and $PbCl_2$, to give $Pb(CH_3)_4$, was presented in [118 to 124]. The maximum rate of conversion of $Pb(NO_3)_2$ to $Pb(CH_3)_4$ was 6% at 20°C [122], in other experiments 0.03% [125]; see also [98, 126]. $Pb(CH_3)_4$ was also produced in sediments spiked with and without $(CH_3)_3PbOOCCH_3$ [118, 120, 121, 122, 124, 127]. Pure species of *Pseudomonas*, *Alcaligenes*, *Acinetobacter*, *Flavobacterium*, and *Aeromonas* transformed $(CH_3)_3PbOOCCH_3$, but not inorganic lead compounds, into $Pb(CH_3)_4$ [98, 118, 122], and mixed bacterial cultures from a natural lake also produced $Pb(CH_3)_4$ on addition of low concentrations of lead(II) salts [120, 121]. It soon was realized that $Pb(CH_3)_4$, from solutions of trimethyllead and dimethyllead compounds, is partly a product of chemical redistribution (ca. 80%) and partly of biomethylation (ca. 20%), one source for $Pb(CH_3)_4$ production from biomethylation being Pb^{2+} from redistribution of trimethyl- and dimethyllead compounds, another source being trimethyllead species [120, 121]. These results are confirmed in [123]; see also [128]. Other experiments indicated that the biological reaction accounted for 50 to 76% of the total $Pb(CH_3)_4$ produced from $(CH_3)_3PbOOCCH_3$ over the pH range of 3.5 to 7.5 [129]. Also a percentage of only 15 to 20% $Pb(CH_3)_4$ resulting from chemical disproportionation was given [98, 122, 124]. Photolytic production of $Pb(CH_3)_4$ from $(CH_3)_3PbOOCCH_3$ was excluded since, on UV irradiation, $(CH_3)_3PbOOCCH_3$ is not converted to $Pb(CH_3)_4$ in the absence of microorganisms or in autoclaved systems [118, 124]. No direct relationship exists between lead concentration in the sediment and the amount of $Pb(CH_3)_4$ produced [124].

Other experiments indicated that conversion of lead(II) compounds to $Pb(CH_3)_4$ was inconsistent and time-independent [130]. Lead contained in mine tailings is not mobilized in detectable amounts by biomethylation [125, 130]. A volatile organolead compound, presumably the methyllead derivative, was observed upon aerobic, not upon anaerobic, incubation of marine sediment with lead(II) acetate [131]. Marine sediments inoculated with $(CH_3)_3PbOOCCH_3$ produced $Pb(CH_3)_4$ in experiments employing flow-through conditions both under anaerobic and aerobic conditions. Production under anaerobic conditions was approximately threefold greater [130].

Evidence for the existence of a natural source of tetraalkyllead compounds comes from analysis of air from the open sea, from coastal, and from estuarine areas, since 1 to 30 ng/m^3 tetraalkyllead was observed [61]; further evidence was found from abnormally high alkyllead to total lead ratios, which have been observed in atmospheric samples in Morecambe Bay, U.K. [60]. It was speculated that these tetraalkyllead emissions arise from methylation of lead compounds in coastal and estuarine intertidal areas [61]. The hypothesis of a natural source was supported by additional measurements [132, 174]; see also [46, 133, 134]. However, according to recent results on transformation rates and lifetime of alkyllead species in the atmosphere it is not necessary to invoke the hypothesis of the natural alkylation of lead to explain the enhanced alkyllead to total lead ratios [182]. Also no indication of a large-scale natural source for tetraalkyllead compounds was found in other work [22], and determination of atmospheric content of tetraalkyllead compounds gave no indication for the occurrence of natural methylation processes of lead(II) compounds [23]; see also [48, 69, 70, 73]. The origin of trialkyllead species in rural pigeons (*Columba livia*) was associated in part with natural sources of tetraalkyllead in the environment [135], and also the presence of small amounts of $Pb(CH_3)_4$ in fish (Coho Salmon, Yellow Perch, Sucker, Rock Bass, Sunfish) from various lakes and rivers in Ontario, Canada, was taken as an indication for the possibility of environmental methylation or in vivo methylation of lead in fish [95]; see also [93]. It was considered conceivable that bacteria in fish intestines or in fish tissue could methylate lead compounds [93, 94, 101, 178].

The extent and even the existence of a possible natural methylation process is a controversial subject, and indeed, different workers failed to note increased levels of $Pb(CH_3)_4$ above sediments enriched with $Pb(NO_3)_2$ [136] or found no evidence for either bacterial or fungal methylation of lead(II) compounds [24, 137, 138, 139]. It also was concluded from experiments with labelled compounds and radiotracers [140, 141], and from theoretical considerations [142, 143], that there is no basis for the assumption that biological methylation, either of lead(II) [140 to 143] or of organolead compounds, like $(CH_3)_3PbOOCCH_3$ [140, 141], occurs in the environment, whereas chemical conversion of methyllead(IV) compounds into $Pb(CH_3)_4$ by disproportionation is possible [137, 140, 141]; see also [104, 144]. Results of a study of methylation of $(CH_3)_3PbOOCCH_3$ in a sediment from Lake Minnetonka, Minnesota, U.S.A., were explained entirely by disproportionation processes [145]. It was concluded that the methylation processes are abiotic and that the formation of $Pb(CH_3)_4$ from trimethyllead compounds in natural systems arises mainly from disproportionation rather than biomethylation [146], and a chemical mechanism involving intermediate formation and subsequent decomposition of $((CH_3)_3Pb)_2S$ was assumed [137]. Lastly, a sharp differentiation between biotic and abiotic processes becomes difficult, and it was argued that methylation of lead compounds may occur through an abiotic route mediated by biological processes [147].

The ions $[Pb(CH_3)_3]^+$ and $[Pb(CH_3)_2]^{2+}$ present in the environment as a result of methylation of lead(II) compounds, as decomposition products of spilled antiknock additives [148], or as emissions from auto exhausts may be converted into $Pb(CH_3)_4$ by chemical [149] or microbial methylation [148, 149]. In general, an increase of anthropogenically created methyllead compounds in natural aquatic environments via biomethylation of lead compounds is questionable [150], and it seems unlikely that significant quantities of organic lead will be produced by this route [151]. Regarding the ecological significance of biomethylation of lead(II) compounds, see [9, 70, 126, 151, 152]. For a critical review, see [153]. A biogeochemical cycle for organolead compounds involving $Pb(CH_3)_4$ was proposed in [104, 154, 174]; see also [9, 73, 150, 171]. Biological methylation is reviewed in [6, 155 to 158]; see also [2, 124, 159, 160, 179].

Small amounts of $Pb(CH_3)_4$ can be produced by transformation of elemental lead by CH_3I [161], which is present in the environment [162, 163, 164]. Also lead(II) salts were reported to react in aqueous solution with CH_3I occurring in natural waters to produce $Pb(CH_3)_4$ [138, 165, 166]. However, it was also reported that, in the headspace gas of reactions of CH_3I with lead(II) salts with [162] or without [162, 163] the presence of elemental magnesium, no $Pb(CH_3)_4$ was detected. But the reaction of CH_3I with $PbCl_2$ in aqueous solution produced $Pb(CH_3)_4$ in the presence of magnesium metal [163]. According to [167], $Pb(CH_3)_4$ is not formed by direct reaction of lead(II) acetate with CH_3I, but from a secondary reaction of CH_3I with metallic lead, which was produced by contact of the lead(II) acetate with aluminium foil present in the experiment. It was concluded that methylation of lead(II) acetate does not occur in the environment; see also [168]. From model studies, the possibility was inferred that methyl-cobalamin will methylate Pb^{2+} under certain environmental conditions which enhance its $[CH_3]^-$ donor capability [148, 169].

References:

[1] Brinckman, F. E., Bellama, J. M. (ACS Symp. Ser. No. 82 [1978]).
[2] Grandjean, P., Nielsen, T. (Residue Rev. **72** [1979] 97/148).
[3] Chau, Y. K., Wong, P. T. S. (NBS-SP-618 [1981] 65/80).
[4] Harrison, R. M., Laxen, D. P. H. (Lead Pollution, Causes and Control, Chapman & Hall, London 1981).
[5] De Jonghe, W. R. A., Adams, F. C. (Talanta **29** [1982] 1057/67).
[6] Craig, P. J. (Spec. Publ. Royal Soc. Chem. No. 44 [1983] 277/322).
[7] Berg, S., Jonsson, A. (in: Grandjean, P., Grandjean, E. C., Biological Effects of Organolead Compounds, CRC, Boca Raton, Fla., 1984, pp. 33/42).
[8] Nielsen, T. (in: Grandjean, P., Grandjean, E. C., Biological Effects of Organolead Compounds, CRC, Boca Raton, Fla., 1984, pp. 43/62).
[9] De Jonghe, W. R. A., Adams, F. C. (Advan. Environ. Sci. Technol. **17** [1986] 561/94).
[10] Craig, P. J. (Organometallic Compounds in the Environment, Longmans, Harlow 1986).

[11] Hewitt, C. N., Harrison, R. M. (from [10] pp. 160/97).
[12] Anonymous (Lead in the Environment and Its Significance to Man, HMSO, London 1974).
[13] Laveskog, A. (Proc. 2nd Intern. Clean Air Congr., Washington 1970 [1971], pp. 549/57).
[14] Laveskog, A. (TPM-BIL-64 [1972] from [8]).
[15] Tausch, H. (SGAE-BER-2636 [1976] 1/11).
[16] Corrin, M. L. (in: Edwards, H. W., Environmental Contamination Caused by Lead, Final Report, Colorado State Univ., Fort Collins, Colo., 1977, pp. 179/200 from [5]).
[17] Allvin, B., Berg, S. (SNV-PM-907 [1977] 1/16).
[18] Reamer, D. C., Zoller, W. H., O'Haver, T. C. (Anal. Chem. **50** [1978] 1449/53).
[19] Radziuk, B., Thomassen, Y., Van Loon, J. C., Chau, Y. K. (Anal. Chim. Acta **105** [1979] 255/62).
[20] De Jonghe, W. R. A., Chakraborti, D., Adams, F. C. (Anal. Chem. **52** [1980] 1974/7).

[21] Nielsen, T., Egsgaard, H., Larsen, E., Schroll, G. (Anal. Chim. Acta **124** [1981] 1/13).

[22] De Jonghe, W. R. A., Chakraborti, D., Adams, F. C. (Environ. Sci. Technol. **15** [1981] 1217/22).

[23] De Jonghe, W., Chakraborti, D., Adams, F. (Heavy Metals Environ. 3rd Intern. Conf., Amsterdam 1981, pp. 72/5).

[24] Chakraborti, D., Jiang, S. G., Surkijn, P., De Jonghe, W., Adams, F. (Anal. Proc. [London] **18** [1981] 347/50).

[25] Harrison, R. M., Radojević, M. (Environ. Technol. Letters **6** [1985] 129/36).

[26] Hewitt, C. N., Harrison, R. M. (Anal. Chim. Acta **167** [1985] 277/87).

[27] Harrison, R. M., Radojević, M., Hewitt, C. N. (Sci. Total Environ. **44** [1985] 235/44).

[28] Kehoe, R. A., Cholak, J., McIlhinney, J. G., Sterling, T. D. (Arch. Environ. Health **6** [1963] 255/72).

[29] Cholak, J. (Arch. Environ. Health **8** [1964] 314/24).

[30] Working Group on Lead Contamination (PB-170739 [1965]).

[31] Snyder, L. J. (Anal. Chem. **39** [1967] 591/5).

[32] Purdue, L. J., Enrione, R. E., Thompson, R. J., Bonfield, B. A. (Anal. Chem. **45** [1973] 527/30).

[33] Colwill, D. M., Hickman, A. J. (TRRL-LR-545 [1973]; PB-221663 [1973] 1/8; C.A. **80** [1974] No. 18930).

[34] Harrison, R. M., Perry, R., Slater, D. H. (EUR-5360 [1974/75] 1783/8).

[35] Robinson, J. W., Wolcott, D. K. (Environ. Letters **6** [1974] 321/33).

[36] Harrison, R. M., Perry, R., Slater D. H. (Atmos. Environ. **8** [1974] 1187/94).

[37] Hancock, S., Slater, A. (Analyst [London] **100** [1975] 422/9).

[38] Harrison, R. M., Laxen, D. P. H. (Atmos. Environ. **11** [1977] 201/3).

[39] Rabinowitz, M. B., Wetherill, G. W., Kopple, J. D. (J. Lab. Clin. Med. **90** [1977] 238/48).

[40] Laxen, D. P. H. (Diss. Univ. Lancaster, Lancaster, England, 1978).

[41] Rohbock, E., Müller, J. (Mikrochim. Acta **1979** I 423/34).

[42] Rohbock, E., Georgii, H.-W., Müller, J. (Atmos. Environ. **14** [1980] 89/98).

[43] Robinson, J. W. (Atmos. Environ. **14** [1980] 1207).

[44] Rohbock, E., Georgii, H.-W., Müller, J. (Atmos. Environ. **14** [1980] 1207/8).

[45] De Jonghe, W. R. A., Chakraborti, D., Adams, F. C. (Atmos. Environ. **15** [1981] 421/2).

[46] Rohbock, E., Georgii, H.-W., Müller, J. (Atmos. Environ. **15** [1981] 422).

[47] Harrison, R. M., Laxen, D. P. H. (Atmos. Environ. **15** [1981] 422/3).

[48] Rohbock, E., Georgii, H.-W., Müller, J. (Atmos. Environ. **15** [1981] 423/4).

[49] Birch, J., Harrison, R. M., Laxen, D. P. H. (Sci. Total Environ. **14** [1980] 31/42).

[50] De Jonghe, W. R. A., Adams, F. C. (Atmos. Environ. **14** [1980] 1177/80).

[51] Gibson, M. J., Farmer, J. G. (Environ. Technol. Letters **2** [1981] 521/30).

[52] Jiang, S. G., Chakraborti, D., De Jonghe, W., Adams, F. (Z. Anal. Chem. **305** [1981] 177/80).

[53] Rohbock, E. (Atomspektrom. Spurenanal. Vortr. Kolloq., Konstanz, FRG, 1981 [1982], pp. 267/74).

[54] van Loon, J. C., Balilcki, M. R., Nimjee, M. C., Brzezinska, A., Douglas, D. (Heavy Metals Environ. 4th Intern. Conf., Heidelberg 1983, pp. 78/81).

[55] Dmitriev, M. T., Braude, A. Yu., Bykhovskii, M. Ya., Emel'yanov, B. V., Pautova, L. F., Rotin, V. A. (Gig. Sanit. **1984** No. 9, pp. 55/7; C.A. **102** [1985] No. 31071).

[56] Sawicki, C. R. (EPA-650/2-75-003 [1975]; PB-240620 [1975] 1/19).

[57] Harrison, R. M., Perry, R. (Atmos. Environ. **11** [1977] 847/52).

[58] Robinson, J. W., Kiesel, E. L., Goodbread, J. P., Bliss, R., Marshall, R. (Anal. Chim. Acta **92** [1977] 321/8).

[59] Robinson, J. W., Kiesel, E. L. (J. Environ. Sci. Health A **12** [1977] 411/22).

[60] Harrison, R. M., Laxen, D. P. H. (Nature **275** [1978] 738/40).

[61] Harrison, R. M., Laxen, D. P. H., Birch, J. (Manage. Control Heavy Metals Environ. Intern. Conf., London 1979, pp. 257/61).

[62] Chau, Y. K., Wong, P. T. S. (NATO Conf. Ser. I **6** [1983] 87/103).

[63] Harrison, R. M., Hewitt, C. N. (Intern. J. Environ. Anal. Chem. **21** [1985] 89/104).

[64] Febo, A., Di Palo, V., Possanzini, M. (Sci. Total Environ. **48** [1986] 187/94).

[65] Magistretti, M., Zurlo, N., Scollo, F., Pacillo, D. (Med. Lavoro **54** [1963] 486/95).

[66] Kehoe, R. A., Cholak, J., Spence, J. A., Hancock, W. (Arch. Environ. Health **6** [1963] 239/54).

[67] Zuliani, G., Perin, G., Rausa, G. (Med. Lavoro **57** [1966] 771/80).

[68] Schwarzbach, E. (Proc. Intern. Symp. Environ. Health Aspects Lead, Luxembourg 1972, pp. 1117/20; C.A. **79** [1973] No. 96414).

[69] Jiang, S., Ma, C., Liu, H., Ge, J., Li, M., Adams, F. C., Winchester, J. W. (1st Annual Scientific Meeting of the Society for Environmental Geochemistry and Health, East Carolina University, Greenville, N.C., 1982).

[70] Jiang, S.-G., Ma, C.-G., Liu, H.-C., Ge, J.-R., Li, M., Adams, F. C., Winchester, J. W. (Atmos. Environ. **18** [1984] 2553/6).

[71] Brief, R. S. (Arch. Environ. Health **5** [1962] 527/31).

[72] Harrison, R. M., Laxen, D. P. H. (Environ. Sci. Technol. **12** [1978] 1384/92).

[73] De Jonghe, W., Adams, F. (COST 61a bis, 2nd Discussion Meeting Working Party 1, Detection, Identification Analysis Air Pollutants, Vienna 1982, pp. 17/23).

[74] Corrin, M. L., Natusch, D. F. S. (NSF/RA-770214 [1977] 7/31).

[75] DeTreville, R. T. P., Wheeler, H. W., Sterling, T. (Arch. Environ. Health **5** [1962] 532/6).

[76] Linch, A. L., Wiest, E. G., Carter, M. D. (Am. Ind. Hyg. Assoc. J. **31** [1970] 170/9).

[77] Sturges, W. T., Harrison, R. M. (Atmos. Environ. **20** [1986] 845/50).

[78] Jacobs, E. S. (personal communication from [61]).

[79] Hirschler, D. A., Gilbert, L. F. (Arch. Environ. Health **8** [1964] 297/313).

[80] Blokker, P. C. (Atmos. Environ. **6** [1972] 1/18).

[81] Hewitt, C. N., Harrison, R. M. (Environ. Sci. Technol. **20** [1986] 797/802).

[82] Anonymous (U.S. Environmental Protection Agency, Air Quality Criteria for Lead, 1977).

[83] Charlou, J. L., Caprais, M. P., Blanchard, G., Martin, G. (Environ. Technol. Letters **3** [1982] 415/24).

[84] Nielsen, O. J., Nielsen, T., Pagsberg, P. (Risoe-R-463 [1982] 1/17).

[85] Nielsen, O. J. (Risoe-R-480 [1984] 1/126).

[86] Faulstich, H., Stournaras, C. (Nature **317** [1985] 714/5).

[87] Unsworth, M. H., Harrison, R. M. (Nature **317** [1985] 674).

[88] Harrison, R. M. (J. Environ. Sci. Health A **11** [1976] 417/23).

[89] Harrison, R. M., Radojević, M., Wilson, S. J. (Sci. Total Environ. **50** [1986] 129/37).

[90] Potter, H. R., Jarvie, A. W. P., Markall, R. N. (Water Pollut. Control [Maidstone, Engl.] **76** [1977] 123/8).

[91] Charlou, J. L. (CNEXO-COB-EL-LC-17-79 [1979], Contrat CNEXO-ENSCR No. 78/5678, 1/15).

[92] Charlou, J. L. (CNEXO-COB-EL-Chimie-JLC-199-80 [1980], Contrat CNEXO/ENSCR No. 79/5943, 1/30).

[93] Sirota, G. R., Uthe, J. F. (Anal. Chem. **49** [1977] 823/5).

[94] Chau, Y. K., Wong, P. T. S., Bengert, G. A., Kramar, O. (Anal. Chem. **51** [1979] 186/8).

[95] Chau, Y. K., Wong, P. T. S., Kramar, O., Bengert, G. A., Cruz, R. B., Kinrade, J. O., Lye, J., van Loon, J. C. (Bull. Environ. Contam. Toxicol. **24** [1980] 265/9).

[96] Chau, Y. K., Wong, P. T. S., Bengert, G. A., Dunn, J. L. (Anal. Chem. **56** [1984] 271/4).

[97] Cruz, R. B., Lorouso, C., George, S., Thomassen, Y., Kinrade, J. D., Butler, L. R. P., Lye, J., van Loon, J. C. (Spectrochim. Acta B **35** [1980] 775/83).

[98] Wong, P. T. S., Chau, Y. K. (Manage. Control Heavy Metals Environ. Intern. Conf., London 1979, pp. 131/4).

[99] Giaccio, M. (Quad. Merceol. **16** [1977] 55/62).

[100] Hodson, P. V., Whittle, D. M., Wong, P. T. S., Borgmann, U., Thomas, R. L., Chau, Y. K., Nriagu, J. O., Hallett, D. J. (Advan. Environ. Sci. Technol. **14** [1984] 335/69).

[101] Chau, Y. K., Wong, P. T. S., Bengert, G. A., Dunn, J. L., Glen, B. (J. Great Lakes Res. **11** [1985] 313/9).

[102] Grove, J. R. (Lead Marine Environ. Proc. Intern. Experts Discuss., Rovinj, Yugoslav., 1977 [1980], pp. 45/52).

[103] Noden, F. G. (Lead Marine Environ. Proc. Intern. Experts Discuss., Rovinj, Yugoslav., 1977 [1980], pp. 83/91).

[104] Wood, J. M. (Lead Marine Environ. Proc. Intern. Experts Discuss., Rovinj, Yugoslav., 1977 [1980], pp. 299/303).

[105] Harrison, G. F. (Lead Marine Environ. Proc. Intern. Experts Discuss., Rovinj, Yugoslav., 1977 [1980], pp. 305/17).

[106] Tiravanti, G., Boari, G. (Environ. Sci. Technol. **13** [1979] 849/54).

[107] Brondi, M., Dall'Aglio, M., Ghiara, E., Mignuzzi, C., Tiravanti, G. (Sci. Total Environ. **19** [1981] 21/31).

[108] Maddock, B. G., Taylor, D. (Lead Marine Environ. Proc. Intern. Experts Discuss., Rovinj, Yugoslav., 1977 [1980], pp. 233/61).

[109] Cleaver, J. W. (Lead Marine Environ. Proc. Intern. Experts Discuss., Rovinj, Yugoslav., 1977 [1980], pp. 325/43).

[110] Geraci, S., Montanari, M., Di Cintio, R. (Mem. Biol. Mar. Oceanogr. Suppl. **10** [1980] 195/206).

[111] Tiravanti, G., Rozzi, A., Dall'Aglio, M., Delaney, W., Dadone, A. (Prog. Water Technol. **12** [1980] 49/65).

[112] Colombini, M. P., Corbini, G., Fuoco, R., Papoff, P. (Ann. Chim. [Rome] **71** [1981] 609/29).

[113] Tiravanti, G., Passino, R. (NATO Conf. I **11** [1985] 25/45; C.A. **105** [1986] No. 66030).

[114] Diehl, K. H., Rosopulo, A., Kreuzer, W., Judel, G. K. (Z. Pflanzenernähr. Bodenk. **146** [1983] 551/9).

[115] Röderer, G. (in: Grandjean, P., Grandjean, E. C., Biological Effects of Organolead Compounds, CRC, Boca Raton, Fla., 1984, pp. 63/95).

[116] Eshleman, A., Siegel, S. M. (Hawaii Air Pollut. Semin., Honolulu, Hawaii, 1970; in: Siegel, S. M., Eshleman, A., Umeno, I., Puerner, N., Smith, C. W., Mercury West. Environ. Proc. Workshop, Portland, Oreg., 1971 [1973], pp. 119/34).

[117] Eshleman, A., Siegel, S. M. (Western Soc. Naturalists Ann. Meet., Univ. Hawaii, Honolulu, Hawaii, 1970; in: Siegel, S. M., Eshleman, A., Umeno, I., Puerner, N., Smith, C. W., Mercury West. Environ. Proc. Workshop, Portland, Oreg., 1971 [1973], pp. 119/34).

[118] Wong, P. T. S., Chau, Y. K., Luxon, P. L. (Nature **253** [1975] 263/4).

[119] Schmidt, U., Huber, F. (Nature **259** [1976] 157/8).

[120] Schmidt, U. (Diss. Univ. Dortmund 1977).

[121] Huber, F., Schmidt, U. (Organometal. Coord. Chem. Germanium, Tin, Lead, 2nd Intern. Conf., Nottingham 1977, p. A7).

[122] Chau, Y. K., Wong, P. T. S. (Lead Marine Environ. Proc. Intern. Experts Discuss., Rovinj, Yugoslav., 1977 [1980], pp. 225/31).

[123] Dumas, J.-P., Pazdernik, LeRoy, Belloncik, S., Bouchard, D., Vaillancourt, G. (Water Pollut. Res. Can. **12** [1977] 91/100).

[124] Chau, Y. K., Wong, P. T. S. (ACS Symp. Ser. No. 82 [1978] 39/53).

[125] Thompson, J. A. J., Crerar, J. A. (Marine Pollut. Bull. **11** [1980] 251/3).

[126] Huber, F., Schmidt, U., Kirchmann, H. (ACS Symp. Ser. No. 82 [1978] 65/81).

[127] Chau, Y. K., Snodgrass, W. J., Wong, P. T. S. (Water Res. **11** [1977] 807/9).

[128] Macaskie, L. E., Ainsworth, M. A., Dean, A. C. R. (Environ. Technol. Letters **6** [1985] 237/50).

[129] Baker, M. D., Wong, P. T. S., Chau, Y. K., Mayfield, C. I., Inniss, W. E. (Heavy Met. Environ. 3rd Intern. Conf., Amsterdam 1981, pp. 645/8).

[130] Thompson, J. A. J. (Heavy Met. Environ. 3rd Intern. Conf., Amsterdam 1981, pp. 653/6).

[131] Berdicevsky, I., Shachar, M., Yannai, S. (Arch. Toxicol. Suppl. **6** [1983] 285/91).

[132] Hewitt, C. N., Harrison, R. M., De Mora, S. J. (Marine Chem. **15** [1984] 189/90).

[133] Chester, R., Sharples, E. J., Murphy, K., Saydam, A. C., Sanders, G. S. (Marine Chem. **15** [1984] 191).

[134] Brinckman, F. E., Olson, G. J., Thayer, J. S. (Marine Estuarine Geochem. Proc. Symp., Honolulu 1984 [1985], pp. 227/38).

[135] Johnson, M. S., Pluck, H., Hutton, M., Moore, G. (Arch. Environ. Contam. Toxicol. **11** [1982] 761/7).

[136] Bentz, J., Brandt, M., Pallay, A., Ter Haar, G. (personal communication, in: Ahmad, I., Chau, Y. K., Wong, P. T. S., Carty, A. J., Taylor, L., Nature **287** [1980] 716/7).

[137] Jarvie, A. W. P., Markall, R. N., Potter, H. R. (Nature **255** [1975] 217/8).

[138] Jarvie, A. W. P., Whitmore, A. (Proc. 10th Intern. Conf. Organometal. Chem., Toronto 1981, Abstr. 1D 08, p. 58).

[139] Jarvie, A. W. P., Whitmore, A. P., Markall, R. N., Potter, H. R. (Environ. Pollut. Ser B **6** [1983] 81/94).

[140] Reisinger, K., Stoeppler, M., Nürnberg, H. W. (Nature **291** [1981] 228/30).

[141] Reisinger, K., Stoeppler, M., Nürnberg, H. W. (Heavy Met. Environ. 3rd Intern. Conf., Amsterdam 1981, pp. 649/52).

[142] Wood, J. M. (Science **183** [1974] 1049/52).

[143] Ridley, W. P., Dizikes, L. J., Wood, J. M. (Science **197** [1977] 329/32).

[144] Wood, J. M. (Chem. Eng. News **55** No. 27 [1977] 37).

[145] Craig, P. J. (Environ. Technol. Letters **1** [1980] 17/20).

[146] Jarvie, A. W. P., Whitmore, A. P., Markall, R. N., Potter, H. R. (Environ. Pollut. Ser. B **6** [1983] 69/79).

[147] Chau, Y. K. (Sci. Total Environ. **49** [1986] 305/23).

[148] Rapsomanikis, S., Ciejka, J. J., Weber, J. H. (Inorg. Chim. Acta **89** [1984] 179/83).

[149] Grandjean, P. (in: Rutter, M., Russell Jones, R., Lead Versus Health, Wiley, Chichester 1983, pp. 179/89).

[150] Röderer, G. (Heavy Met. Environ. 3rd Intern. Conf., Amsterdam 1981, pp. 250/3).

[151] Jarvie, A. W. P., Markall, R. N., Potter, H. R. (Environ. Res. **25** [1981] 241/9).

[152] Waldichuk, M. (Marine Pollut. Bull. **16** [1985] 7/11).

[153] Craig, P. J. (in: Hutzinger, O., Handb. Environ. Chem., Vol. 1, Pt. A, Springer, Berlin 1980, pp. 169/227).

[154] Thayer, J. S. (Organometallic Compounds and Living Organisms, Academic, New York 1984, pp. 234/5).

[155] Craig, P. J., Wood, J. M. (Environ. Lead Proc. 2nd Intern. Symp. Environ. Lead Res., Cincinnati 1978 [1981], pp. 333/49).
[156] Thayer, J. S., Brinckman, F. E. (Advan. Organometal. Chem. **20** [1982] 313/56).
[157] Jiang, S. (Huanjing Huaxue **2** [1983] 9/14; C.A. **99** [1983] No. 109895).
[158] Beijer, K., Jernelöv, A. (in: Grandjean, P., Grandjean, E. C., Biological Effects of Organolead Compounds, CRC, Boca Raton, Fla., 1984, pp. 13/20).
[159] Craig, P. J., Rapsomanikis, S. (NBS-SP-618 [1981] 54/64).
[160] Whitmore, A. P. (Diss. Univ. Aston, Birmingham 1981).

[161] Craig, P. J., Rapsomanikis, S. (Environ. Sci. Technol. **19** [1985] 726/30).
[162] Craig, P. J., Rapsomanikis, S. (J. Chem. Soc. Chem. Commun. **1982** 114).
[163] Craig, P. J., Moreton, P. A., Rapsomanikis, S. (Heavy Met. Environ. 4th Intern. Conf., Heidelberg 1983, pp. 788/92).
[164] Thayer, J. S., Olson, G. J., Brinckman, F. E. (Environ. Sci. Technol. **18** [1984] 726/9).
[165] Ahmad, I., Chau, Y. K., Wong, P. T. S., Carty, A. J., Taylor, L. (Nature **287** [1980] 716/7).
[166] Chau, Y. K., Wong, P. T. S., Carty, A. J., Taylor, L. (Proc. 10th Intern. Conf. Organometal. Chem., Toronto 1981, Abstr. 1D 09, p. 59).
[167] Snyder, L. J., Bentz, J. M. (Nature **296** [1982] 228/9).
[168] Jarvie, A. W. P., Whitmore, A. P. (Environ. Technol. Letters **2** [1981] 197/204).
[169] Rhode, S. F., Weber, J. H. (Environ. Technol. Letters **5** [1984] 63/8).
[170] Caforio, A. (Inquinamento **28** No. 11 [1986] 54/7).

[171] Radojević, M., Harrison, R. M. (Sci. Total Environ. **59** [1987] 157/80).
[172] Faulstich, H., Stournaras, C., Endres, K. P. (Experientia **43** [1987] 115/27).
[173] Røyset, O., Thomassen, Y. (Atmos. Environ. **21** [1987] 655/8).
[174] Harrison, R. M., Hewitt, C. N., Radojević, M. (Heavy Met. Environ. 5th Intern. Conf., Athens 1985, Vol. 1, pp. 82/4).
[175] Røyset, O., Thomassen, Y. (Anal. Chim. Acta **188** [1986] 247/55).
[176] State Pollution Control Authority of Norway, Report No. 37, 1984 (from [173]).
[177] Van Cleuvenbergen, R., Chakraborti, D., Van Mol, W., Adams, F. (Heavy Met. Environ. 5th Intern. Conf., Athens 1985, Vol. 1, pp. 153/5).
[178] Chau, Y. K., Wong, P. T. S. (Proc. Intern. Conf. Chem. Environ., Lisbon, Portugal, 1986, pp. 77/82).
[179] Brinckman, F. E. (Environ. Inorg. Chem. Papers U.S. Italy Workshop, San Miniato, Italy, 1983 [1985], pp. 195/238; C.A. **103** [1985] No. 218393).
[180] Hewitt, C. N., Harrison, R. M., Radojević, M. (Anal. Chim. Acta **188** [1986] 229/38).

[181] Radojević, M., Harrison, R. M. (Environ. Technol. Letters **7** [1986] 519/24).
[182] Hewitt, C. N., Harrison, R. M. (Environ. Sci. Technol. **21** [1987] 260/6).
[183] Faulstich, H., Stournaras, C. (Nature **319** [1986] 17).

1.1.1.1.10 Coordination Compounds

Well-defined coordination compounds of $Pb(CH_3)_4$ are not known. However, from the increase of the NMR coupling constants $^2J(^{207}Pb, {}^1H)$ in various solvents, solvation of $Pb(CH_3)_4$ by one solvent molecule and formation of complexes with trigonal bipyramidal geometry were inferred. The ability of the solvent to solvate $Pb(CH_3)_4$ increased along the series: C_6H_{12} < 1,2-dimethoxyethane $\cong$ dioxane $\cong$ hexamethylphosphortriamide < pyridine < tetrahydro-thiophene < triethylamine < THF < triethylphosphine < N,N,N′,N′-tetramethylethylenedi-amine $\cong$ acetone < dimethylformamide < dimethyl sulfoxide [2]. In another paper $^2J(^{207}Pb, {}^1H)$ was considered to be independent of solvent donor strength and this is consistent with $Pb(CH_3)_4$ having no acceptor properties [3]. No correlation of solvating power of the solvents and chemical shift of the methyl protons was observed on changing from one solvent to another [2]; see also [1, 6, 9]. For a correlation of the ionization potential and the solvation energy of $Pb(CH_3)_4$ and other tetraorganometal compounds in acetonitrile, see [7].

From charge-transfer bands in the visible spectral region observed immediately upon mixing $Pb(CH_3)_4$ and tetracyanoethylene (TCNE) in $CHCl_3$, CH_2Cl_2, or 1,2-dichloropropane, formation of a 1:1 electron donor-acceptor complex ($Pb(CH_3)_4^+ \cdot TCNE^-$) was inferred [4, 5, 8]; see also Section 1.1.1.1.4.6, p. 138.

References:

[1] Laszlo, P., Spert, A. (J. Magn. Resonance **1** [1969] 291/7).
[2] Petrosyan, V. S., Voyakin, A. S., Reutov, O. A. (Zh. Org. Khim.**6** [1970] 889/93; J. Org. Chem. [USSR] **6** [1970] 895/8).
[3] Puddephatt, R. J., Thistlethwaite, G. H. (J. Organometal. Chem. **40** [1972] 143/50).
[4] Gardner, H. C., Kochi, J. K. (J. Am. Chem. Soc. **97** [1975] 1855/65).
[5] Gardner, H. C., Kochi, J. K. (J. Am. Chem. Soc. **98** [1976] 2460/9).
[6] Bakhbukh, M., Grishin, Yu. K., Ustynyuk, Yu. A., Zemlyanski, N. N. (Vestn. Mosk. Univ. Ser. II Khim. **20** No. 4 [1979] 366/8; Moscow Univ. Chem. Bull. **34** No. 4 [1979] 69/72).
[7] Fukuzumi, S., Kochi, J. K. (J. Phys. Chem. **84** [1980] 2246/54).
[8] Fukuzumi, S., Kochi, J. K. (Tetrahedron **38** [1982] 1035/49).
[9] Kerschl, S., Sebald, A., Wrackmeyer, B. (Magn. Resonance Chem. **23** [1985] 514/20).

Table of Conversion Factors

Following the notation in Landolt-Börnstein [7], values that have been fixed by convention are indicated by a bold-face last digit. The conversion factor between calorie and Joule that is given here is based on the thermochemical calorie, cal_{thch}, and is defined as 4.1840 J/cal. However, for the conversion of the "Internationale Tafelkalorie", cal_{IT}, into Joule, the factor 4.1868 J/cal is to be used [1, p. 147]. For the conversion factor for the British thermal unit, the Steam Table Btu, BTU_{ST}, is used [1, p. 95].

Force	N	dyn	kp
1 N (Newton)	1	10^5	0.1019716
1 dyn	10^{-5}	1	1.019716×10^{-6}
1 kp	9.80665	9.80665×10^5	1

Pressure	Pa	bar	kp/m^2	at	atm	Torr	lb/in^2
1 Pa (Pascal)=$1N/m^2$	1	10^{-5}	1.019716×10^{-1}	1.019716×10^{-5}	0.986923×10^{-5}	0.750062×10^{-2}	145.0378×10^{-6}
1 bar=10^6 dyn/cm^2	10^5	1	10.19716×10^3	1.019716	0.986923	750.062	14.50378
1 kp/m^2=1mm H_2O	9.80665	0.980665×10^{-4}	10^4	10^{-4}	0.967841×10^{-4}	0.735559×10^{-1}	1.422335×10^{-3}
1 at=1kp/cm^2	0.980665×10^5	0.980665	10^4	1	0.967841	735.559	14.22335
1 atm=760 Torr	1.01325×10^5	1.01325	1.033227×10^4	1.033227	1	760	14.69595
1 Torr=1mm Hg	133.3224	1.333224×10^{-3}	13.59510	1.359510×10^{-3}	1.315789×10^{-3}	1	19.33678×10^{-3}
1 lb/in^2=1psi	6.89476×10^3	68.9476×10^{-3}	703.069	70.3069×10^{-3}	68.0460×10^{-3}	51.7149	1

Work, Energy, Heat	J	kWh	kcal	Btu	MeV
1 J (Joule) = 1 Ws = 1 Nm = 10^7 erg	1	2.778×10^{-7}	2.39006×10^{-4}	9.4781×10^{-4}	6.242×10^{12}
1 kWh	3.6×10^6	1	860.4	3412.14	2.247×10^{19}
1 kcal	4184.0	1.1622×10^{-3}	1	3.96566	2.6117×10^{16}
1 Btu (British thermal unit)	1055.06	2.93071×10^{-4}	0.25164	1	6.5858×10^{15}
1 MeV	1.602×10^{-13}	4.450×10^{-20}	3.8289×10^{-17}	1.51840×10^{-16}	1

1 eV $\cong$ 23.0578 kcal/mol = 96.473 kJ/mol

Power	kW	PS	kp m/s	kcal/s
1 kW = 10^{10} erg/s	1	1.35962	101.972	0.239006
1 PS	0.73550	1	75	0.17579
1 kp m/s	9.80665×10^{-3}	0.01333	1	2.34384×10^{-3}
1 kcal/s	4.1840	5.6886	426.650	1

References:

[1] A. Sacklowski, Die neuen SI-Einheiten, Goldmann, München 1979. (Conversion tables in an appendix.)
[2] International Union of Pure and Applied Chemistry, Manual of Symbols and Terminology for Physicochemical Quantities and Units, Pergamon, London 1979; Pure Appl. Chem. 51 [1979] 1/41.
[3] The International System of Units (SI), National Bureau of Standards Spec. Publ. 330 [1972].
[4] H. Ebert, Physikalisches Taschenbuch, 5th Ed., Vieweg, Wiesbaden 1976.
[5] Kraftwerk Union Information, Technical and Economic Data on Power Engineering, Mülheim/Ruhr 1978.
[6] E. Padelt, H. Laporte, Einheiten und Größenarten der Naturwissenschaften, 3rd Ed., VEB Fachbuchverlag, Leipzig 1976.
[7] Landolt-Börnstein, 6th Ed., Vol. II, Pt. 1, 1971, pp. 1/14.
[8] ISO Standards Handbook 2, Units of Measurement, 2nd Ed., Geneva 1982.

Key to the Gmelin System of Elements and Compounds

System Number	Symbol	Element
1		Noble Gases
2	H	Hydrogen
3	O	Oxygen
4	N	Nitrogen
5	F	Fluorine
6	**Cl**	**Chlorine**
7	Br	Bromine
8	I	Iodine
8a	At	Astatine
9	S	Sulfur
10	Se	Selenium
11	Te	Tellurium
12	Po	Polonium
13	B	Boron
14	C	Carbon
15	Si	Silicon
16	P	Phosphorus
17	As	Arsenic
18	Sb	Antimony
19	Bi	Bismuth
20	Li	Lithium
21	Na	Sodium
22	K	Potassium
23	NH_4	Ammonium
24	Rb	Rubidium
25	Cs	Caesium
25a	Fr	Francium
26	Be	Beryllium
27	Mg	Magnesium
28	Ca	Calcium
29	Sr	Strontium
30	Ba	Barium
31	Ra	Radium
32	**Zn**	**Zinc**
33	Cd	Cadmium
34	Hg	Mercury
35	Al	Aluminium
36	Ga	Gallium

System Number	Symbol	Element
37	In	Indium
38	Tl	Thallium
39	Sc, Y	Rare Earth
	La—Lu	Elements
40	Ac	Actinium
41	Ti	Titanium
42	Zr	Zirconium
43	Hf	Hafnium
44	Th	Thorium
45	Ge	Germanium
46	Sn	Tin
47	Pb	Lead
48	V	Vanadium
49	Nb	Niobium
50	Ta	Tantalum
51	Pa	Protactinium
52	**Cr**	**Chromium**
53	Mo	Molybdenum
54	W	Tungsten
55	U	Uranium
56	Mn	Manganese
57	Ni	Nickel
58	Co	Cobalt
59	Fe	Iron
60	Cu	Copper
61	Ag	Silver
62	Au	Gold
63	Ru	Ruthenium
64	Rh	Rhodium
65	Pd	Palladium
66	Os	Osmium
67	Ir	Iridium
68	Pt	Platinum
69	Tc	Technetium[1]
70	Re	Rhenium
71	Np,Pu . . .	Transuranium Elements

HCl

$CrCl_2$

$ZnCrO_4$

$ZnCl_2$

Material presented under each Gmelin System Number includes all information concerning the element(s) listed for that number plus the compounds with elements of lower System Number.

For example, zinc (System Number 32) as well as all zinc compounds with elements numbered from 1 to 31 are classified under number 32.

[1] A Gmelin volume titled "Masurium" was published with this System Number in 1941.

A Periodic Table of the Elements with the Gmelin System Numbers is given on the Inside Front Cover